Biomass Conversion and Organic Waste Utilization

Biomass Conversion and Organic Waste Utilization

Editors

Shicheng Zhang
Gang Luo
Yan Shi
Andrzej Białowiec
Abdul-Sattar Nizami

Basel • Beijing • Wuhan • Barcelona • Belgrade • Novi Sad • Cluj • Manchester

Editors

Shicheng Zhang
Fudan University
Shanghai
China

Gang Luo
Fudan University
Shanghai
China

Yan Shi
Fudan University
Shanghai
China

Andrzej Białowiec
Wroclaw University of
Environmental and
Life Sciences
Wrocław
Poland

Abdul-Sattar Nizami
Government College
University
Lahore
Pakistan

Editorial Office
MDPI AG
Grosspeteranlage 5
4052 Basel, Switzerland

This is a reprint of articles from the Special Issue published online in the open access journal *Processes* (ISSN 2227-9717) (available at: https://www.mdpi.com/journal/processes/special_issues/biomass_waste_utilization).

For citation purposes, cite each article independently as indicated on the article page online and as indicated below:

Lastname, A.A.; Lastname, B.B. Article Title. *Journal Name* **Year**, *Volume Number*, Page Range.

ISBN 978-3-7258-1893-8 (Hbk)
ISBN 978-3-7258-1894-5 (PDF)
doi.org/10.3390/books978-3-7258-1894-5

© 2024 by the authors. Articles in this book are Open Access and distributed under the Creative Commons Attribution (CC BY) license. The book as a whole is distributed by MDPI under the terms and conditions of the Creative Commons Attribution-NonCommercial-NoDerivs (CC BY-NC-ND) license.

Contents

About the Editors . . . . **vii**

Yan Shi, Gang Luo and Shicheng Zhang
Special Issue on "Biomass Conversion and Organic Waste Utilization"
Reprinted from: *Processes* **2023**, *11*, 3070, doi:10.3390/pr11113070 . . . . **1**

Yi Wen, Dingxiang Chen, Yong Zhang, Huabin Wang and Rui Xu
Cadmium Elimination via Magnetic Biochar Derived from Cow Manure: Parameter Optimization and Mechanism Insights
Reprinted from: *Processes* **2023**, *11*, 2295, doi:10.3390/pr11082295 . . . . **5**

Xinyan Zhang, Shanshan Liu, Qingyu Qin, Guifang Chen and Wenlong Wang
Alkali Etching Hydrochar-Based Adsorbent Preparation Using Chinese Medicine Industry Waste and Its Application in Efficient Removal of Multiple Pollutants
Reprinted from: *Processes* **2023**, *11*, 412, doi:10.3390/pr11020412 . . . . **23**

Yixin Lu, Yujie Liu, Chunlin Li, Haolin Liu, Huan Liu, Yi Tang, et al.
Adsorption Characteristics and Mechanism of Methylene Blue in Water by NaOH-Modified Areca Residue Biochar
Reprinted from: *Processes* **2022**, *10*, 2729, doi:10.3390/pr10122729 . . . . **38**

Huiting Cheng, Yuanjuan Gong, Nan Zhao, Luji Zhang, Dongqing Lv and Dezhi Ren
Simulation and Experimental Validation on the Effect of Twin-Screw Pulping Technology upon Straw Pulping Performance Based on Tavares Mathematical Model
Reprinted from: *Processes* **2022**, *10*, 2336, doi:10.3390/pr10112336 . . . . **61**

Tae-Sung Shin, Seong-Yeun Yoo, In-Kook Kang, Namhyun Kim, Sanggyu Kim, Hun-Bong Lim, et al.
Analysis of Hydrothermal Solid Fuel Characteristics Using Waste Wood and Verification of Scalability through a Pilot Plant
Reprinted from: *Processes* **2022**, *10*, 2315, doi:10.3390/pr10112315 . . . . **82**

Kean Long Lim, Wai Yin Wong, Nowilin James Rubinsin, Soh Kheang Loh and Mook Tzeng Lim
Techno-Economic Analysis of an Integrated Bio-Refinery for the Production of Biofuels and Value-Added Chemicals from Oil Palm Empty Fruit Bunches
Reprinted from: *Processes* **2022**, *10*, 1965, doi:10.3390/pr10101965 . . . . **95**

Xiaoyu Huo, Chao Jia, Shanshan Shi, Tao Teng, Shaojie Zhou, Mingda Hua, et al.
One-Step Synthesis of High-Performance N/S Co-Doped Porous Carbon Material for Environmental Remediation
Reprinted from: *Processes* **2022**, *10*, 1359, doi:10.3390/pr10071359 . . . . **115**

Hengsong Ji, Xiang Li, Mei Zhang, Zhenqiang Li, Yan Zhou and Xiang Ma
Molecular Dynamics Simulation for Structural Evolution of Mixed Ash from Coal and Wheat Straw
Reprinted from: *Processes* **2022**, *10*, 215, doi:10.3390/pr10020215 . . . . **127**

Elena Goldan, Valentin Nedeff, Narcis Barsan, Mihaela Culea, Mirela Panainte-Lehadus, Emilian Mosnegutu, et al.
Assessment of Manure Compost Used as Soil Amendment—A Review
Reprinted from: *Processes* **2023**, *11*, 1167, doi:10.3390/pr11041167 . . . . **139**

Muhammad Waqas, Sarfraz Hashim, Usa Wannasingha Humphries, Shakeel Ahmad, Rabeea Noor, Muhammad Shoaib, et al.
Composting Processes for Agricultural Waste Management: A Comprehensive Review
Reprinted from: *Processes* **2023**, *11*, 731, doi:10.3390/pr11030731 . . . . **155**

About the Editors

Shicheng Zhang

Prof. Shicheng Zhang is a full professor at the department of environmental science and engineering, Fudan University, Shanghai, China, and the Director of Shanghai Technical Service Platform for Pollution Control and Resource Utilization of Organic Wastes. He currently serves as a core group member of the Sustainable Waste Management Program for the Association of Pacific Rim Universities (APRU SWMP), the steering committee member of World Society of Engineering Thermochemistry (WSETC), and a fellow member of Australasia Practical Zero Emissions Society (APZES). His research group is focusing on biomass waste and organic waste utilization by chemical, biological, and biochemical methods, especially hydrothermal conversion of organic waste to high value-added products. Prof. Zhang has published 8 book chapters and more than 220 journal papers, with over 12000 SCI citations (H index 62). He has received several awards for his scholarly achievements, such as the Second Prize of the Natural Science Award for Scientific Research in Colleges and Universities of the Ministry of Education, and so on. He serves as Associate Editor of Environmental Engineering Research and npj Materials Sustainability, Editorial Board Member of Green Chemical Technology, Processes, Biomass, Environmental Sanitation Engineering (in Chinese), and Energy Environmental Protection (in Chinese), and Guest Editor of Green Chemistry and Bioresources Technology. He has been recognized in the World's Top 2% Scientists (2020–2023) and as a Highly Cited Researcher in the field of Cross-Field (2023).

Gang Luo

Prof. Gang Luo is currently is full professor at Fudan University. His current research interests include anaerobic digestion, anaerobic microbiology, wastewater treatments, hydrothermal conversion, and biological syngas conversion. His projects were mainly founded from NSFC, MOST, and Science and Technology Commission of Shanghai Municipality. He has published more than 100 ISI publications in well-known journals including Environmental Science & Technology and Water Research, with total citations of >5000 and an H Index of 39. He has also published four book chapters and obtained two patents. He is serving as the Associate Editor of Cleaner Water and as a Youth Editorial Advisory Board Member of Front Environ Sci Eng.

Yan Shi

Dr. Yan Shi is currently a postdoctoral researcher at Harvard John A. Paulson School of Engineering and Applied Sciences, Harvard University. His recent research focuses on the hydrothermal conversion of organic wastes, mineral synthesis, industrial wastes disposal and recycling, and CO_2 mineralization. He has published more than 10 papers in renowned journals such as Construction and Building Materials, Journal of Hazardous Material, Chemical Engineering Journal, and Waste Management, with a total citation number above 400.

Andrzej Białowiec

Dr hab. Eng. Andrzej Białowiec, WUELS Assoc. Prof., has over 20 years of experience in research on waste management and environmental engineering and the development and implementation of technologies, including the use of hydrophyte systems for the disposal of sewage sludge, leachate, and sewage; the bio-stabilization of waste (anaerobic digestion of waste with recovery biogas, aerobic composting and waste bio-drying); and the conversion of waste into solid and gaseous fuels using torrefaction, pyrolysis, and gasification technologies. He has initiated an innovative approach to waste management—Waste to Carbon. His solutions have been implemented on a full scale 15 times. Has one international patent (EU) and three inventions (Poland). He is the author of 62 expert opinions and technical reports. He has spent about 36 months working or visiting high-ranked worldwide universities, including Iowa State University (9m), Cardiff University (18m), Stanford University (2m), Tsinghua University (1m), University of Beira Interior (3m), Politechnica di Milano (1m). He has completed 43 research and R&D projects, and in 34 he was the PI. He is the author and co-author of 71 scientific articles, 27 chapters in monographs, 7 monographs, and 35 articles published in conference materials, with an H Index of 11 and 433 citations.

Abdul-Sattar Nizami

Dr. Abdul-Sattar Nizami has Master of Science in Engineering from the Chalmers University of Technology, Sweden. He has a Ph.D. in Sustainable Gaseous Biofuel from the School of Civil and Environmental Engineering, University College Cork, Ireland. He worked at the University of Toronto, Canada, as a Postdoctoral Fellow on alternative fuels and life cycle studies in the Department of Chemical Engineering & Applied Chemistry. Later, he served as an Assistant Professor and Head of the Solid Waste Management Unit at the Center of Excellence in Environmental Studies (CEES) of King Abdulaziz University, Jeddah, Saudi Arabia. He is currently working as a Professor (Associate) at the Sustainable Development Study Centre (SDSC), Government College University, Lahore, Pakistan. He has published over 150 papers on renewable energy, alternative fuels, waste-to-energy, catalytic pyrolysis, anaerobic digestion, and resource recovery. He has delivered 46 invited talks to various national and international forums. His work has been cited more than 10 thousand times in the peer- reviewed press, with a total impact factor of over 1100 and an H Index of 53. He is a Senior Editor in Renewable & Sustainable Energy Reviews (Elsevier Impact Factor 16.8), Energy & Environment (Sage Impact Factor 3.15), and Frontiers in Energy Research (IF 4.01). He serves as an Editorial Board Member in Bioresource Technology Reports (Elsevier) and Energy Sources Part B (Taylor & Francis IF 3.21). He is also a Guest Editor of several special issues and a reviewer for many high-impact Journals of Elsevier, ACS, Springer, Wiley, and Taylor and Francis. He is actively involved in community and consultation services for various international organizations, including the European Commission based IF@ULB, National Research Agency (NRA) of France, National Science Centre Poland, World Bank, and UNEP. He is ranked among the Top 2% Scientists worldwide by Stanford University, USA.

Editorial

Special Issue on "Biomass Conversion and Organic Waste Utilization"

Yan Shi [1,*], Gang Luo [1,2] and Shicheng Zhang [1,2,*]

1. Shanghai Key Laboratory of Atmospheric Particle Pollution and Prevention (LAP3), Shanghai Technical Service Platform for Pollution Control and Resource Utilization of Organic Wastes, Department of Environmental Science and Engineering, Fudan University, Shanghai 200438, China; gangl@fudan.edu.cn
2. Shanghai Institute of Pollution Control and Ecological Security, Shanghai 200092, China
* Correspondence: yans@fudan.edu.cn (Y.S.); zhangsc@fudan.edu.cn (S.Z.)

The recycling and utilization of biomass and organic wastes have emerged as effective strategies for saving energy and resources. Biomass and organic wastes are produced in huge amounts, and they can be utilized to produce biomaterials, biofuels, and biochemicals through biological and thermochemical approaches [1–3]. The physical and chemical properties of biomass and organic wastes undergo substantial changes during biorefinery processes, making them highly promising as environmentally friendly materials. Pyrolysis stands out as a typical thermochemical technique for the utilization of biomass and organic wastes, which can produce different valuable products including bio-oil, biochar, and syngas [4–6]. However, a major challenge with this technology is the relatively low quality of bio-oil, necessitating further treatment. Additionally, other technologies, like hydrothermal conversion (HTC) and composting, can be applied to biomass and organic wastes with a high-water content. This Special Issue of *Processes* on "Biomass Conversion and Organic Waste Utilization" collected recent research works related to the disposal and reutilization of biomass wastes that document the application of different techniques and process simulation, as well as the properties of the treatment products.

1. Thermal Conversion and Adsorption Properties

Adsorption is one of the key applications for the thermochemical products of various biomass wastes. Nilavazhagi et al. [7] revealed that sulfuric acid-modified and subsequently thermally treated raw jamun fruit seeds demonstrated a maximum Langmuir adsorption capacity of 266.9 mg/g, following pseudo-first-order kinetics. The thermodynamic separation of Fe(II) ions by adsorbents was not only exothermic, but also a feasible and spontaneous process. Two articles in this Special Issue explored the adsorption properties of different thermochemical products. Wen et al. used cow manure as a raw material for the synthesis of magnetic cow manure biochar. The pseudo-second-order model and the Langmuir model were used to fit the Cd(II) adsorption data, and the adsorption capacity was 612.43 mg·g^{-1}. The adsorption was dominated by chemisorption, with the mechanisms of ion-exchange, electrostatic attraction, pore-filling, co-precipitation, and the formation of complexes. The study of Huo et al. examined the impacts of several inorganic activators (KOH, K_2CO_3, and $ZnCl_2$) on the characteristics (porosity and N/S content) of N/S co-doped porous carbon (NSC) materials. Their study showed that the NSC materials acquired through cooperative activation using potassium salts (KOH or K_2CO_3) exhibited significantly larger surface areas than those activated solely with KSCN (1403 m^2/g). Furthermore, KSCN could synergize with K_2CO_3 to produce samples with excellent specific surface area (2900 m^2/g) or N/S content. The as-prepared NSC materials demonstrated enhanced adsorption capabilities for chloramphenicol (833 mg/g) and Pb^{2+} (303 mg/g).

2. The Conversion of Biomass Wastes to Biofuel

The conversion of biomass wastes to biofuel is also an important strategy to solve the problems caused by environmental pollution and natural resource depletion. Lim et al. [8] investigated the preliminary techno-economic feasibility of expanding an existing pellet production plant into an integrated biorefinery plant to produce xylitol and bioethanol. The feedstock used was empty fruit bunches (EFBs), which were divided into two production streams: one stream for pellet production and the other for xylitol and bioethanol production. The analysis indicated that an EFB splitting ratio of less than 40% for pellet production was economically feasible. In the study of Shin et al. [9], biofuel with a calorific value equivalent to that of coal for power generation was produced in a pilot plant through waste wood HTC. The outcomes show that it was possible to convert waste wood into solid fuel with a calorific value of over 27,000 kJ/kg through the HTC process of the pilot plant. In addition, the analyses of heavy metal and hazardous substances proved that the product can be used as a safe biosolid fuel.

3. Simulation Studies on Biorefinery

In this Special Issue, some simulation studies were also conducted on the disposal and properties of different biomass wastes. Ji et al. carried out molecular dynamics simulations to explore the effect of wheat straw (WS) content on the structural properties of ash slag. It was demonstrated that WS introduced many metal ions to the ash system, enhancing its overall activity. The number of bridging and non-bridging oxygen atoms changed upon straw addition, which affected the stability of the system. The number of low-coordination Si atoms was highest for a WS content of 30%, at which point the density reached the minimum value. In the work of Cheng et al., a discrete element model of straw crushing was created to describe the pulping process of rice straw in a twin-screw pulping machine. This model was based on the Tavares mathematical model, and the number of broken straw particles was analyzed. The simulation results reveal that the highest number of broken straw particles was achieved when the twin-screw spiral casing combination was negative–positive–negative–positive, and the tooth groove angle arrangement of the negative spiral casing was 45°–30°–15°.

4. Composting of Biomass Wastes

Due to the large production volume and complex composition of biomass wastes, various treatment techniques such as thermal conversion processes, anaerobic digestion, and composting have been applied for their effective disposal. Composting is an adaptable and fruitful approach towards biomass waste management and circular economy. Isibika et al. [10] found that co-composting the fruit peels with fish waste increased the biomass conversion efficiency (BCE). The highest BCE on a volatile solid basis (25%) was achieved when 75% fish waste was incorporated. Two review papers of this Special Issue summarized new research work about the composting process and the properties of compost. The review by Waqas et al. examined the management of agricultural waste through various composting processes—conventional composting (vermicomposting, aerobic composting, and anaerobic composting) and emerging composting (two stage composting). This reveals that in conventional composting, vermicomposting humified in 3 to 4 months, which cannot fulfill fertilizer demand. Rapid compost preparation with a fine and higher degree of humification can only be achieved through aerobic composting. Anaerobic composting followed by aerobic composting will minimize the area, labor, and time required for aerobic composting, along with the capital cost and power depletion. Goldan et al. conducted an evaluation of manure and compost as soil amendment. The outcome reveals that the properties of the soil, such as aeration, density, porosity, pH, electrical conductivity, water retention capacity, etc., can be improved by altering the structure and composition of manure. Additionally, the nutrient content of the compost served as a gradual nutrient source for plant growth and development. However, it is crucial to use compost in moderate quantities and to test the soil regularly to avoid excessive application of nutrients, which can be harmful to both plants and the environment.

5. Conclusions

The papers presented in this Special Issue showcase various techniques and strategies for the disposal and reutilization of biomass wastes. Due to the complex composition and volatile properties of biomass wastes, there remains a pressing need for the creation of innovative techniques and processes. We hope that this Special Issue will shed light on some current and emerging research activities related to biomass wastes, with the aim of advancing the development of waste management and the circular economy.

Author Contributions: Investigation, Y.S.; writing—original draft preparation, Y.S.; writing—review and editing, Y.S., G.L. and S.Z. All authors have read and agreed to the published version of the manuscript.

Conflicts of Interest: The authors declare no conflict of interest.

List of Contributions:

1. Wen, Y.; Chen, D.; Zhang, Y.; Wang, H.; Xu, R. Cadmium Elimination via Magnetic. Biochar Derived from Cow Manure: Parameter Optimization and Mechanism Insights. *Processes* **2023**, *11*(8), 2295.
2. Huo, X.; Jia, C.; Shi, S.; Teng, T.; Zhou, S.; Hua, M.; Zhu, X.; Zhang, S.; Xu, Q. One-Step Synthesis of High-Performance N/S Co-Doped Porous Carbon Material for Environmental Remediation. *Processes* **2022**, *10*(7), 1359.
3. Lim, K. L.; Wong, W. Y.; James Rubinsin, N.; Loh, S. K.; Lim, M. T. Techno-Economic Analysis of an Integrated Bio-Refinery for the Production of Biofuels and Value-Added Chemicals from Oil Palm Empty Fruit Bunches. *Processes* **2022**, *10*(10), 1965.
4. Shin, T. S.; Yoo, S. Y.; Kang, I. K.; Kim, N.; Kim, S.; Lim, H. B.; Choe, K.; Lee, J. C.; Yang, H. I. Analysis of Hydrothermal Solid Fuel Characteristics Using Waste Wood and Verification of Scalability through a Pilot Plant. *Processes* **2022**, *10*(11), 2315.
5. Ji, H.; Li, X.; Zhang, M.; Li, Z.; Zhou, Y.; Ma, X. Molecular Dynamics Simulation for Structural Evolution of Mixed Ash from Coal and Wheat Straw. *Processes* **2022**, *10*(2), 215.
6. Cheng, H.; Gong, Y.; Zhao, N.; Zhang, L.; Lv, D.; Ren, D. Simulation and Experimental Validation on the Effect of Twin-Screw Pulping Technology upon Straw Pulping Performance Based on Tavares Mathematical Model. *Processes* **2022**, *10*(11), 2336.
7. Waqas, M.; Hashim, S.; Humphries, U. W.; Ahmad, S.; Noor, R.; Shoaib, M.; Naseem, A.; Hlaing, P. T.; Lin, H. A. Composting Processes for Agricultural Waste Management: A Comprehensive Review. *Processes* **2023**, *11*(3), 731.
8. Goldan, E.; Nedeff, V.; Barsan, N.; Culea, M.; Panainte-Lehadus, M.; Mosnegutu, E.; Tomozei, C.; Chitimus, D.; Irimia, O. Assessment of Manure Compost Used as Soil Amendment-A Review. *Processes* **2023**, *11*(4), 1167.

References

1. Ahorsu, R.; Medina, F.; Constantí, M. Significance and challenges of biomass as a suitable feedstock for bioenergy and biochemical production: A review. *Energies* **2018**, *11*, 3366. [CrossRef]
2. Awasthi, M.K.; Sarsaiya, S.; Patel, A.; Juneja, A.; Singh, R.P.; Yan, B.; Awasthi, S.K.; Jain, A.; Liu, T.; Duan, Y.; et al. Refining biomass residues for sustainable energy and bio-products: An assessment of technology, its importance, and strategic applications in circular bio-economy. *Renew. Sustain. Energy Rev.* **2020**, *127*, 109876. [CrossRef]
3. Kumar, B.; Verma, P. Biomass-based biorefineries: An important architype towards a circular economy. *Fuel* **2021**, *288*, 119622. [CrossRef]
4. Lee, X.J.; Ong, H.C.; Gan, Y.Y.; Chen, W.H.; Mahlia, T.M.I. State of art review on conventional and advanced pyrolysis of macroalgae and microalgae for biochar, bio-oil and bio-syngas production. *Energy Convers. Manag.* **2020**, *210*, 112707. [CrossRef]
5. Sekar, M.; Mathimani, T.; Alagumalai, A.; Chi, N.T.L.; Duc, P.A.; Bhatia, S.K.; Brindhadevi, K.; Pugazhendhi, A. A review on the pyrolysis of algal biomass for biochar and bio-oil–Bottlenecks and scope. *Fuel* **2021**, *283*, 119190. [CrossRef]
6. Kan, T.; Strezov, V.; Evans, T.J. Lignocellulosic biomass pyrolysis: A review of product properties and effects of pyrolysis parameters. *Renew. Sustain. Energy Rev.* **2016**, *57*, 1126–1140. [CrossRef]
7. Nilavazhagi, A.; Felixkala, T. Adsorptive removal of Fe(II) ions from water using carbon derived from thermal/chemical treatment of agricultural waste biomass: Application in groundwater contamination. *Chemosphere* **2021**, *282*, 131060. [CrossRef] [PubMed]

8. Lim, K.L.; Wong, W.Y.; James Rubinsin, N.; Loh, S.K.; Lim, M.T. Techno-Economic Analysis of an Integrated Bio-Refinery for the Production of Biofuels and Value-Added Chemicals from Oil Palm Empty Fruit Bunches. *Processes* **2022**, *10*, 1965. [CrossRef]
9. Shin, T.S.; Yoo, S.Y.; Kang, I.K.; Kim, N.; Kim, S.; Lim, H.B.; Choe, K.; Lee, J.C.; Yang, H.I. Analysis of Hydrothermal Solid Fuel Characteristics Using Waste Wood and Verification of Scalability through a Pilot Plant. *Processes* **2022**, *10*, 2315. [CrossRef]
10. Isibika, A.; Vinnerås, B.; Kibazohi, O.; Zurbrügg, C.; Lalander, C. Co-composting of banana peel and orange peel waste with fish waste to improve conversion by black soldier fly (*Hermetia illucens* (L.), Diptera: Stratiomyidae) larvae. *J. Clean. Prod.* **2021**, *318*, 128570. [CrossRef]

Disclaimer/Publisher's Note: The statements, opinions and data contained in all publications are solely those of the individual author(s) and contributor(s) and not of MDPI and/or the editor(s). MDPI and/or the editor(s) disclaim responsibility for any injury to people or property resulting from any ideas, methods, instructions or products referred to in the content.

Article

Cadmium Elimination via Magnetic Biochar Derived from Cow Manure: Parameter Optimization and Mechanism Insights

Yi Wen [1,2,†], Dingxiang Chen [1,2,†], Yong Zhang [1,2], Huabin Wang [1,2,*] and Rui Xu [1,2,*]

[1] School of Energy and Environment Science, Yunnan Normal University, Kunming 650500, China; wyaquarius@foxmail.com (Y.W.); ynnuchendx@foxmail.com (D.C.); yongzhang7805@126.com (Y.Z.)
[2] Yunnan Key Laboratory of Rural Energy Engineering, Kunming 650500, China
* Correspondence: hbwang@ynnu.edu.cn (H.W.); ecowatch_xr@163.com (R.X.); Tel./Fax: +86-27-65940928 (R.X.)
† These authors contributed equally to this work.

Abstract: Designing an efficient and recyclable adsorbent for cadmium pollution control is an urgent necessity. In this paper, cow manure, an abundant agricultural/animal husbandry byproduct, was employed as the raw material for the synthesis of magnetic cow manure biochar. The optimal preparation conditions were found using the response surface methodology model: 160 °C for the hydrothermal temperature, 600 °C for the pyrolysis temperature, and Fe-loading with 10 wt%. The optimal reaction conditions were also identified via the response surface methodology model: a dosage of 1 g·L^{-1}, a pH of 7, and an initial concentration of 100 mg·L^{-1}. The pseudo-second-order model and the Langmuir model were used to fit the Cd(II) adsorption, and the adsorption capacity was 612.43 mg·g^{-1}. The adsorption was dominated by chemisorption with the mechanisms of ion-exchange, electrostatic attraction, pore-filling, co-precipitation, and the formation of complexations. Compared to the response surface methodology model, the back-propagation artificial neural network model fit the Cd(II) adsorption better as the error values were less. All these results demonstrate the potential application of CM for Cd(II) removal and its optimization through machine-learning processes.

Keywords: magnetic biochar; cow manure; cadmium removal; response surface methodology; artificial neural network

1. Introduction

The Cd-containing effluent is a severe issue that endangers human health and the environment. Cd(II) can accumulate in cereal grains when wastewater flows into farmland [1], and an intake of 0.125 mg of Cd per kg^{-1} of body weight can lead to severe health problems [2], such as renal illness, cancer, and bone damage. Researchers use biological, chemical, and physical adsorption as well as other methods to control Cd(II) contamination. It has been reported that the exogenous application of salicylic acid decreased the adverse effects of Cd(II) on photosynthesis in maize [3]. Studies also showed that an increased exogenous supply of metal ions, e.g., Fe(II), lead to decreased Cd(II) uptake in rice and Arabidopsis [4,5].

As adsorbents, biomaterials have several advantages, including being low-cost and environmentally friendly and having efficient phosphorous-removal abilities [6]. Researchers have prepared biochar derived from different biomass materials for Cd(II) adsorption, such as corn cob, rice husk [7], and tobacco stalk [8]. However, absorption efficiency needs to be improved, as the biochar can only be separated from wastewater and recycled with difficulty [9,10]. As a matter of fact, biochar only achieves adsorption rather than removal. In order to tackle this problem, diverse types of magnetic biochar, which has the capability of adsorbing heavy metals efficiently and can be separated from water using an external magnetic field, were designed. The typical techniques for synthesizing magnetic biochar include liquid reductio, impregnation, and precipitation methods; in addition, one of the most common modification techniques is to load iron onto biochar [6].

Manure production increased dramatically as a result of the recent rapid expansion of concentrated animal feeding operations in many parts of the world [11]. It is imperative to find solutions for the processing and usage of cow manure (CM) produced in these sizable feedlots [12]. The CM biochar formed after pyrolysis has a plentiful pore structure; thus, CM biochar has the potential for heavy metal adsorption [13].

The adsorption effect of CM biochar on heavy metals can be enhanced via modifications, such as via acid or metal oxide. Additionally, during the preparation of magnetic CM biochar, three factors, namely pyrolysis temperature (PT), hydrothermal temperature (HT), and iron content, are key [14]. Previous research has demonstrated that the specific surface area (SSA) and the functional groups on the surfaces of biochar are disparate when preparation conditions are different. Meanwhile, the increase in Cd(II) adsorption capacity can be attributed to the increased amount of SSA and functional groups, especially phenolic hydroxyl [15,16]. Some researchers have explored the influence of a single factor on adsorption capacity, but there are few researchers focusing on mutual interference between the three factors mentioned above. Moreover, the optimum preparation conditions of magnetic CM biochar used for Cd(II) adsorption have not been summarized. Similarly, during adsorption, three factors, namely dosage, pH, and initial concentration, are key [17], and their combined effects on Cd(II) adsorption has also not yet been clarified [18,19]. Therefore, it is necessary to discuss the relationship between multiple parameters through model analysis in order to optimize the preparation and adsorption.

In traditional methods of optimization, only one parameter is varied at a time, which makes experimentation time-consuming and costly. Model fitting based on certain data can simplify the experimental design, thus reducing costs and saving time. The response surface methodology (RSM) adopts a reasonable experimental design method and obtains data through the test. By constructing a multiple quadratic regression equation to fit the function relationship between factors and response value, the optimal process parameters were sought by analyzing the regression equation [20]. The back-propagation artificial neural network (BP–ANN) is a soft computing technique that studies the process through the modification of network weights to produce the required response [21]. BP–ANN is better than the traditional nonlinear model when dealing with the complex relationship between the input and output variables [22].

However, the RSM and BP–ANN models have some disadvantages. When using the RSM model, the preparation and reaction conditions must be within the range of the test values. Moreover, there are nonlinear and uncertain relationships among the factors affecting the adsorption effect, so it can be difficult to predict the strengths of specific multivariate nonlinear function forms. Analogously, the BP–ANN model requires a large number of experiments for training. Currently, there is no method for determining the minimal number of experiments required for BP–ANN training; therefore, this methodology is troublesome for the design of experiments [23]. Although the two models have been widely used in adsorption studies, the prediction results are different. Some researchers found that the RSM model fitted better [24], while others found that BP–ANN fitted better in terms of the adsorption of heavy metals [25].

This study used CM as a raw material and utilized a combined hydrothermal and pyrolysis preparation [26], along with an added iron source ($FeSO_4$ and $FeCl_3$) for magnetic modification, to prepare the magnetic CM biochar, which was inexpensive and recyclable. In this study, RSM and BP–ANN models were utilized for the analysis of the adsorption effects of various types of magnetic CM biochar, and the preparation method of the biochar with the best adsorption effect was thus obtained. RSM and BP–ANN were also used to evaluate the adsorption effects under different reaction conditions, and the optimum reaction conditions of Cd(II) adsorption were thus identified. The adsorption kinetics, thermodynamics, and mechanism of Cd(II) were also investigated. The work can improve the adsorption efficiency, and provide ideas for the high-value utilization of CM, which provides a reference for the adsorption of Cd(II). Meanwhile, the prediction effects of RSM and BP–ANN on the adsorption of Cd(II) by magnetic biochar were compared.

2. Experiments and Models

2.1. Chemicals and Materials

The primary chemicals, e.g., $FeCl_3$, $FeSO_4 \cdot 7H_2O$, NaOH, and $Cd(NO_3)_2$, used in this study were purchased from Sinopharm Co., Ltd. (Shanghai, China) and used without further purification. CM was collected from the Laohadu Demonstration Ranch in Mengzi City, Yunnan Province, China. The CM was oven–dried at 85 °C for 24 h and sieved through 100–mesh.

2.2. Synthesis of Magnetic CM Biochar

As shown in Figure 1, the magnetic CM biochar was synthesized in three parts. Chemical synthesis modification was the first step. The original CM (4, 6, and 8 g), 1.12 g $FeSO_4 \cdot 7H_2O$, 2.19 g $FeCl_3 \cdot 6H_2O$, and 38 mL NaOH (1 mol·L^{-1}) were mixed in the 100 mL deionized water. The three mixtures were severally sonicated for 30 min and stirred for 2 h. Hydrothermal carbonization was the second step. The three mixtures were placed in hydrothermal synthesis reactors separately and heated at different HT for 24 h (HT = 120, 160, and 200 °C). The carbonization was the final step. Three hydrothermal products were carbonized under nitrogen flow in a tube furnace at setting PT for 1 h to obtain three kinds of magnetic modified CM biochar (PT = 400, 600, 800 °C). The Fe_3O_4 or Fe_2O_3 nanoparticles had loaded on the products, and the products were marked as MCBa–b–c, where a represented HT (120, 160, and 200 °C), b represented PT (400, 600, and 800 °C), and c represented iron content (5%, 10%, and 15%), and the magnetic CM biochar that possessed the most ideal characteristics was marked as MCB.

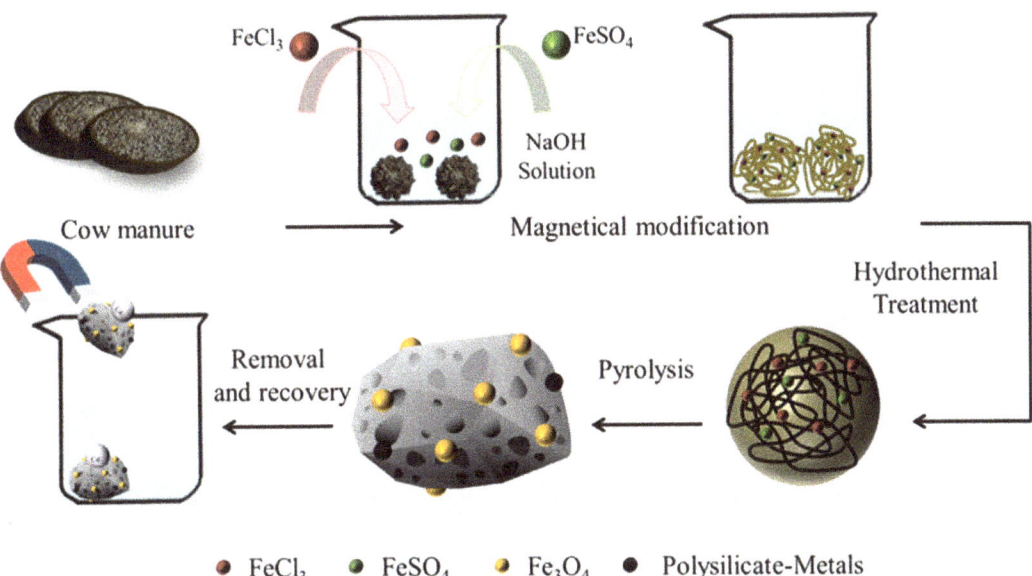

Figure 1. Preparation process for magnetic CM biochar.

2.3. Application of RSM

Box–Behnken design model was used to optimize synthesis parameters with Cd(II) adsorption capacity of magnetic CM biochar as the response and examine the relationship between various synthesis factors or reaction conditions. In the first part, RSM studied the effects of PT (X_1), HT (X_2), and iron content (X_3) on the adsorption capacity (Y) when the initial Cd(II) concentration was 100 mg·L^{-1}. In the second part, RSM studied the effects of dosage (X_1), pH (X_2), and the initial concentration (X_3) on the removal rate of Cd(II) (Y). The experimental findings were examined and fitted to a quadratic equation using Design Expert.8.0.

2.4. Application of BP–ANN

BP–ANN is an information processing system designed to mimic the analytic and processing functions of the human brain, and it uses supervised learning technology and trains by minimizing the square error of the network's output [21]. The BP–ANN model typically consists of an input layer, one or more hidden layers, and an output layer. In this work, one of the three-layer BP–ANN models were established for optimum preparation parameters, and the other for optimal adsorption conditions, respectively, and they were both established by MATLAB R2018b.

To optimize the preparation conditions of MCB, a BP–ANN model was established, whose quantities of neurons in the input layer and output layer were three and one, respectively. The input variables were PT, HT, and iron content, while the output variable was adsorption capacity. In order to explore the optimal reaction conditions of Cd(II) adsorption by MCB, another BP–ANN model was established. The number of neurons in the input layer was five, and the input parameters were dosage, pH, initial concentration, temperature, and time; the output parameter was the removal rate of Cd(II). The selection formula of the optimal number of neuron nodes in both hidden layers is shown in Equations (1)–(3) [27].

$$I < n - 1 \tag{1}$$

$$I < \sqrt{m + n} + \alpha \tag{2}$$

$$I = \log_2 n \tag{3}$$

where n represents the number of neuronal nodes in the input layer, I represents the number of neuronal nodes in the hidden layer, m represents the number of neuronal nodes in the output layer, and α represents a constant between 0 and 10.

In this work, the data were selected as a training data group (Training), a verification data group (Verification), and a test data group (Test) according to the proportion of 75%, 15%, and 15%, respectively. The BP–ANN models were accepted when their correlation coefficients were higher than 0.95.

2.5. Adsorption Experiments

To obtain the MCB, 1 g·L^{-1} of the various magnetic CM biochar was added into Cd(II) solution, respectively (20 mL, 100 mg·L^{-1}). And in the following adsorption experiment, MCB was used as an adsorbent to explore the optimal reaction conditions and adsorption mechanism of Cd(II). To obtain the optimal reaction conditions, different dosages of MCB (0.5, 1, and 1.5 g·L^{-1}) were placed in Cd(II) solutions with different initial concentrations (50, 100, and 150 mg·L^{-1}) and pH (5, 7, and 9), respectively.

For the adsorption kinetics, the MCB (1 g·L^{-1}) was added into the Cd(II) solution (20 mL, 100 mg·L^{-1}). After the set adsorption time, the supernatants were collected to quantify the residual content of Cd(II). For isotherm studies, the MCB (1 g·L^{-1}) was added to a series of Cd(II) solutions (100–1000 mg·L^{-1}). For the effect of pH on adsorption, the MCB (1 g·L^{-1}) was added into different solutions (pH = 1, 3, 5, 7, 9, and 11). The calculation of adsorption capacity was presented in Text S1. All experiments were repeated three times, and the data were averaged with the actual data of the three times.

2.6. Analysis Methods

The elemental composition of biochar was analyzed with an elemental analyzer (EA) (Elementar Vario EL cube, Langenselbold, Germany). The morphology was analyzed with scanning electron microscopy (SEM) at 5 kV (Mira LMS, Tescan, Brno, Czech Republic). The SSA of biochar was detected by N_2 adsorption isotherms at 77 K using a Micropore Analyzer (ASAP 2460, Micrometrics, Norcross, GA, USA). The crystal structure was characterized by X-ray diffractometry (XRD) (D8 Advance Sox-1, Bruker Co., Ltd, Billerica, MA, USA), and all samples were scanned over the region of 10–80°. The electron binding energy and elemental valence were analyzed using X-ray photoelectron spectroscopy (XPS) (K–Alpha, Thermo Scientific, Waltham, MA, USA). The functional groups were qualitatively examined using a Fourier transform infrared spectrometer (FTIR) (Niolet iN10EA, Thermo Scientific). The contents of the elements in the solutions were determined by inductively coupled plasma mass spectrometry (ICP) (Thermo Fisher–X series, Waltham, MA, USA).

3. Results and Discussion

3.1. Preparation Conditions Optimization

3.1.1. RSM Analysis

Experimental factors were summarized in Table S1, and the actual values of adsorption quantity were summarized in Table S2. According to the results obtained by the regression analysis (Table S3), the R squared (R^2) was calculated as 0.9620, indicating the model was significant. The p-value and F-value were 0.0004 (significant) and 1.75 (not significant), respectively, indicating the model was valid [20]. Shown in Equation (4), there was a quadratic polynomial equation of the relationship between the response value and the three experimental parameters:

$$Y = 73.7295 + 0.02475 X_1 + 0.22355 X_2 + 28.9042 X_3 + 8.9262 X_1 X_2 + 0.2112 X_2 X_3 - 8.07 X_1 X_3 - 2.93 X_1^2 - 1.02 X_2^2 - 261.45 X_3^2 \quad (4)$$

As the mutual effects between variables according to Equation (4), the relationships between the response value and the three parameters can be represented by the contour maps. The two-dimensional contour maps and three-dimensional response surfaces obtained by RSM were presented in Figure 2. The adsorption capacity was sensitive to the three parameters. Within a certain range of PT (<600 °C), the increase in adsorption capacity of the various magnetic CM biochar could be seen with the increasing PT. This indicated that the increasing PT promoted the SSA of biochar, which provided potential adsorption sites for Cd(II) [28]. Accordingly, the increasing PT may improve the physical adsorption effect of magnetic CM biochar via pore–filling. Nevertheless, the adsorption capacity was weakened when the PT was higher than 600 °C, which may be because the complexation ability was weakened. The high PT can effectively attenuate the existence of oxygen-contained functional groups [29]. Meanwhile, as shown in Figure 2f, the contour plot tended to a circle, indicating the interaction between PT and iron content might be negligible [30]. According to the analysis of two-dimensional contour maps and three-dimensional response surfaces, the optimal strategy for preparing MCB, the adsorbent for Cd(II), can be derived. Considering the adsorption capacity for Cd(II) and preparation efficiency, the optimum parameters were procured as follows: PT = 600 °C, HT = 160 °C, and iron content = 10 wt%.

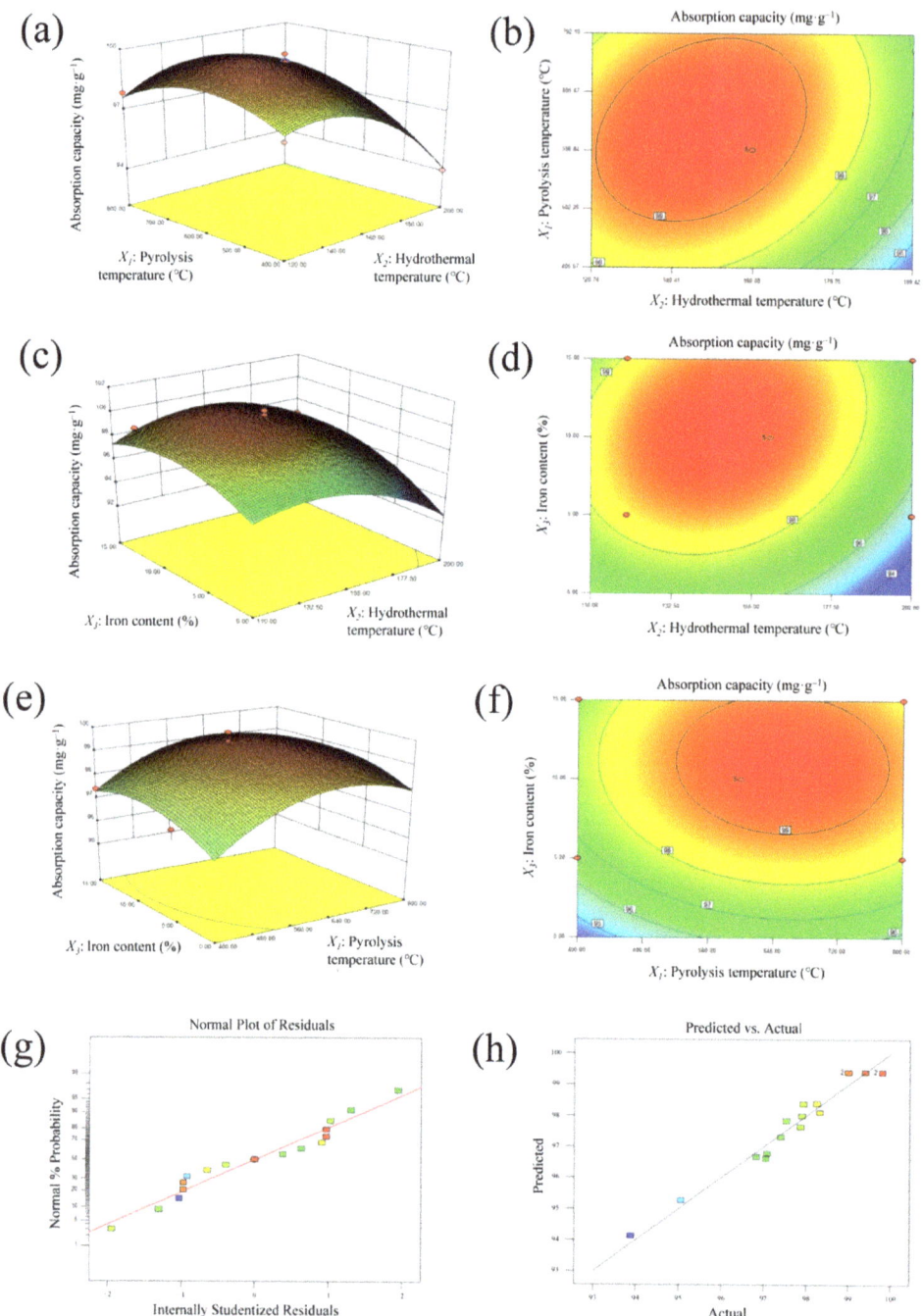

Figure 2. Duel effect on Cd(II) removal rate, Normal Plot of Residuals, and Predicted vs. Actual values: response surfaces (**a**,**c**,**e**), contour plots (**b**,**d**,**f**), Normal Plot of Residuals (**g**), and Predicted vs. Actual values (**h**). Experimental conditions: [initial concentration of Cd(II)] = 100 mg·L^{-1}, [dosage] = 1 g·L^{-1}, [time] = 24 h, [temperature] = 25 °C.

3.1.2. BP–ANN Model

The Levenberg–Marquardt training algorithm was used in the training of the data set, and the Training of the BP–ANN model was done until the error between the experimental and predicted values of responses reached the minimum. The weights and biases were all together known as neural network parameters. The trained model was validated via Validation (experimental data that were not used for training). The development of the BP–ANN model was carried out by dividing the data set into three groups: 70% for Training, 15% for Testing, and 15% for Validation. The weight values of the synaptic joints between the input and hidden layers, and that between the hidden and output layers, were calculated by well versed BP–ANN model for optimization.

The BP–ANN model used in this study (Figure 3a) had three input parameters (PT, HT, and iron content) and one output parameter, and the number of neurons varied from two to ten for the hidden layer. The Training, Validation, and Test showed that the BP–ANN model fitted well ($R^2 > 0.9$); thus, the accepted function had a great correlation coefficient between the target and simulated output values in Training, Validation, and Test (Figure 3d–g). The BP–ANN model tended to be stable at epoch 4, and the model had both a quick contingency speed and good stability.

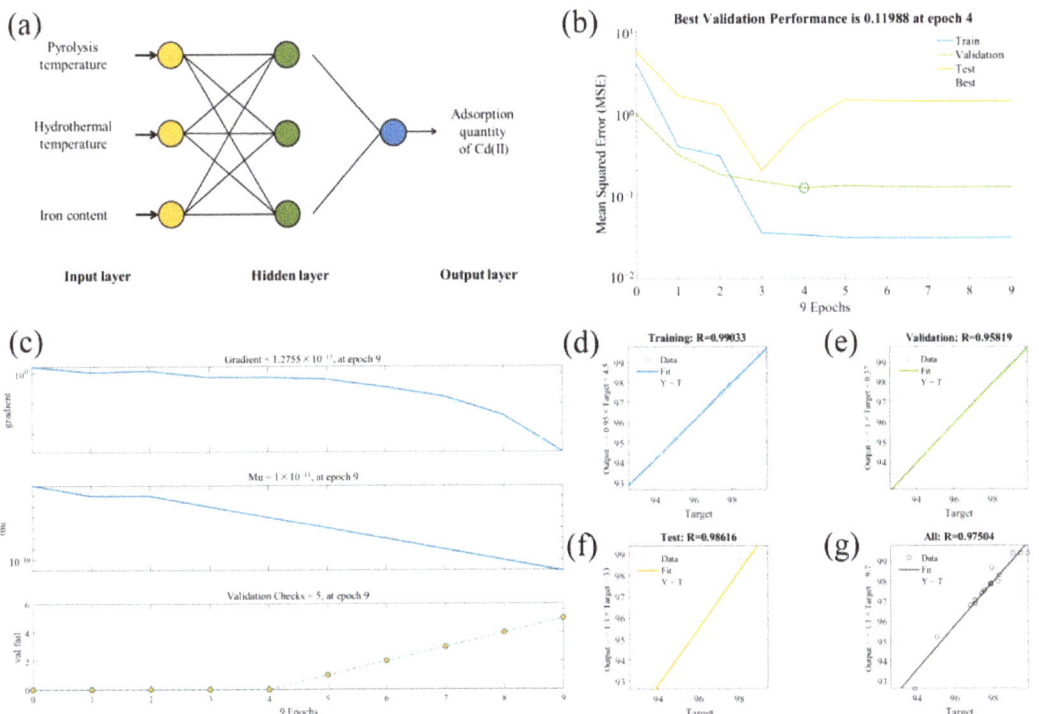

Figure 3. BP–ANN model of experimental data. Architecture of the BP–ANN model (**a**), error convergence curve during the iteration training of BP–ANN model (**b**), comparison between the output of BP–ANN model and the measured value (**c**), correlation between the experimental and simulated rate of removal during BP–ANN Training, Test, and Validation by magnetic CM biochar (**d**–**g**).

3.2. Reaction Conditions Analysis

3.2.1. RSM

Experimental factors were summarized in Table S1, and the actual values of adsorption quantity were summarized in Table S4. Similarly to the previous analysis, the R^2 was 0.9924, and the p-value and Lack of Fit of the model were <0.0001 and 102.1500, respectively, indicating that the model was valid. In this RSM model, the values of Adeq Precision and C.V.% were 37.83 and 0.25%, respectively. Thus, the RSM model could also fit the data well (Table S5). The mathematical relationships between the response value and the three experimental parameters can be quantified by the following formula, Equation (5).

$$Y = 95.64 + 1.5\ X_1 + 1.99\ X_2 - 0.51\ X_3 \quad (5)$$

The two-dimensional contour maps and three-dimensional response surfaces obtained by RSM model were presented in Figure 4. The removal rate of MCB was strongly sensitive to pH and dosage, while the initial concentration had little effect. According to RSM model, the optimal reaction conditions could be worked out. Considering the removal rate of Cd(II), the optimal reaction conditions were procured as follows: Dosage = 1 g·L^{-1}, pH = 7, and initial concentration = 100 mg·L^{-1}.

3.2.2. BP–ANN Model

The BP–ANN model used in this study (Figure 5a) had five input parameters (dosage, pH, initial concentration, temperature, and time), one hidden layer with five neurons, and one output parameter. The BP–ANN model fitted well ($R^2 > 0.9$); thus, the accepted function had a great correlation coefficient between the target and simulated output values in Training, Validation, and Test (Figure 5d–g). The BP–ANN model tended to be stable at epoch 21; hence, the BP–ANN model used in this work had both a quick contingency speed and good stability [31].

3.3. Comparison of the Developed RSM and BP–ANN Models

The performances of the RSM and BP–ANN models were compared using statistical parameters, such as R^2 and root mean square error (RMSE). For exploring the optimum preparation conditions of MCB, it can be seen from Table S6 that the BP–ANN model had a higher R^2 value and lower RMSE value compared to those of the RSM model (0.98 vs. 0.96 and 1.13 vs. 4.17). This indicated that the BP–ANN model had better-predicting ability than the RSM model for the investigation of optimum preparation conditions. For exploring the reaction conditions, both models had a good fitting effect for their R^2 values, both 0.99, while the BP–ANN model had a lower RMSE value compared to the RSM model (1.02 vs. 5.90).

As shown in Figure 6, the experimental values were plotted against the predicted values of the RSM model and BP–ANN model. It can be seen that the BP–ANN model performed better than the RSM model in exploring the optimum preparation conditions and reaction conditions. In terms of data fitting, the BP–ANN model fitted slightly better than the RSM model. In addition, in terms of fitting speed, the BP–ANN model achieved stability with a few cycles. Thus, the BP–ANN model had the dual advantages of faster fitting speed and higher precision. In the aspect of data prediction, the BP–ANN model had a better effect than the RSM model. When the number of BP–ANN data was sufficient, they can often present a better fitting and prediction effect than the RSM model. However, in this paper, the amount of data was 30 groups and 40 groups, respectively, indicating that the BP–ANN model can achieve better fitting and prediction results than the RSM model even with a small amount of data.

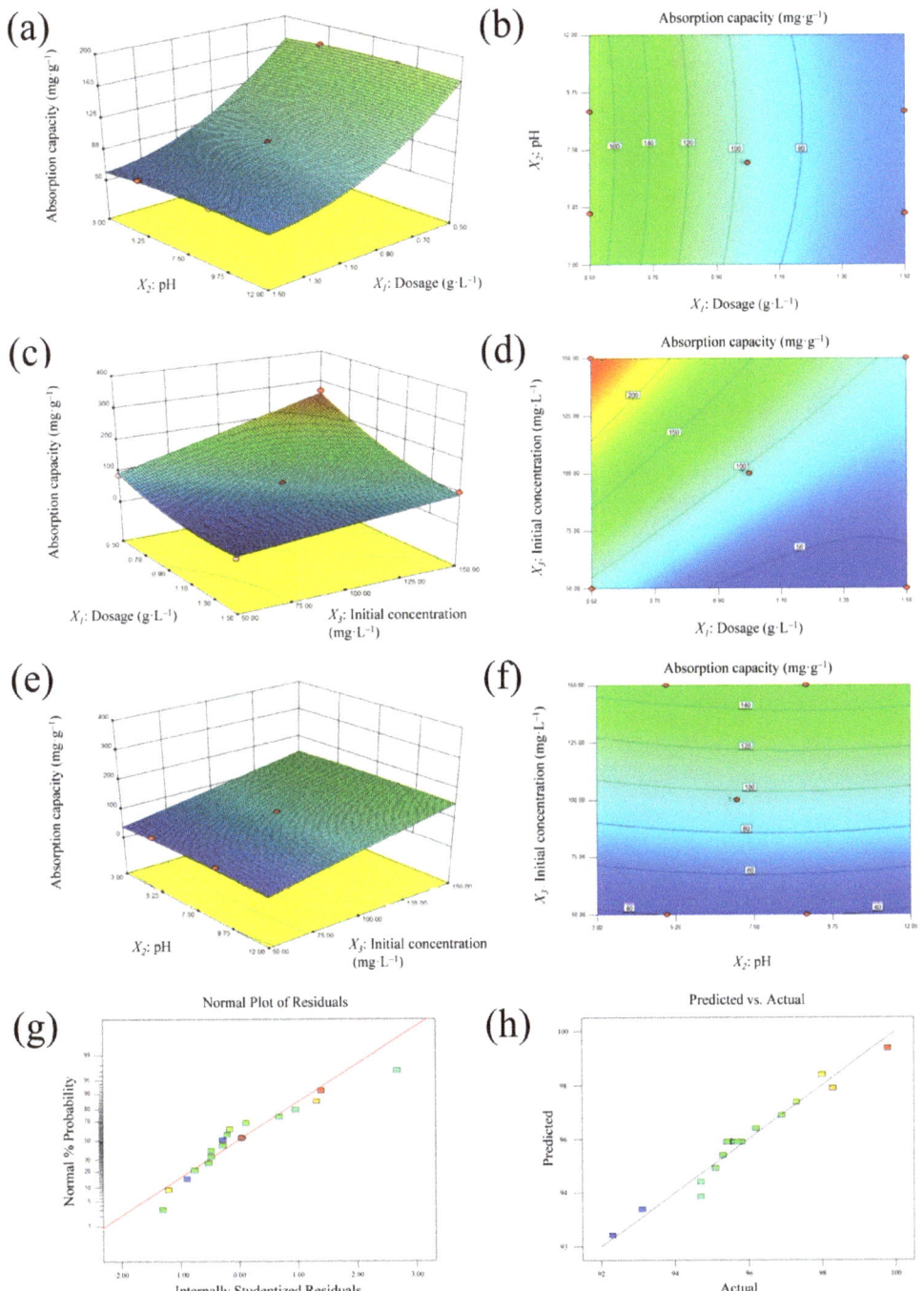

Figure 4. Duel effect on Cd(II) adsorption quantity, Normal Plot of Residuals, and Predicted vs. Actual values: response surfaces (**a,c,e**), contour plots (**b,d,f**), Normal Plot of Residuals (**g**), and Predicted vs. Actual values (**h**). Experimental conditions: [time] = 24 h, [temperature] = 25 °C.

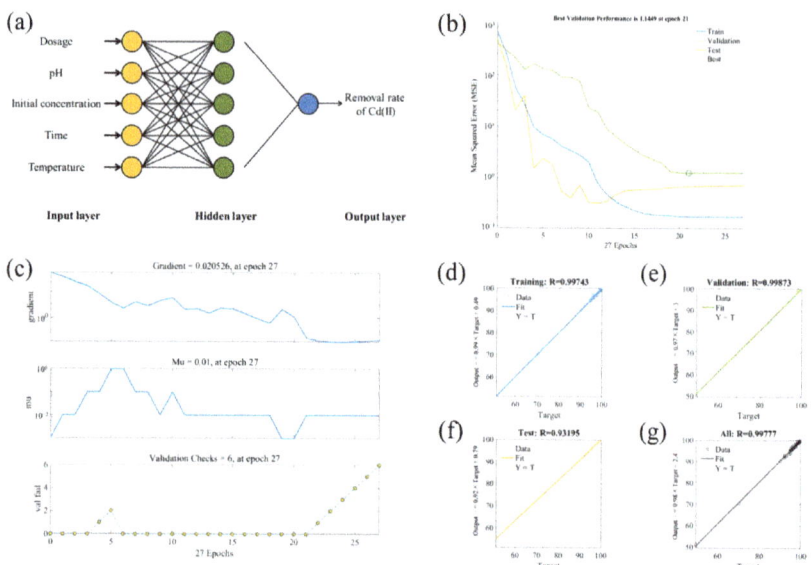

Figure 5. BP–ANN model of experimental data. Architecture of the BP–ANN model (**a**), error convergence curve during the iteration training of BP–ANN model (**b**), comparison between the output of BP–ANN model and the measured value (**c**), correlation between the experimental and simulated rate of removal during BP–ANN Training, Test, and Validation by MCB (**d**–**g**).

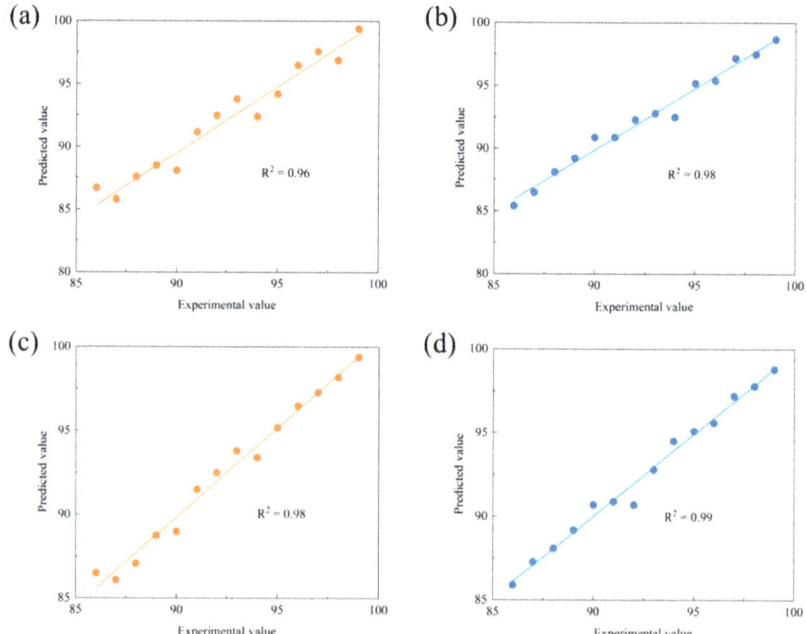

Figure 6. Comparison of the performances of the BP–ANN and RSM models. RSM (**a**) and BP–ANN (**b**) models for optimum preparation conditions, RSM (**c**) and BP–ANN (**d**) models for optimal reaction conditions.

3.4. Characteristics

3.4.1. Morphological Analysis

The basic elements of MCB were analyzed using an elemental analyzer. The carbon content was 51.41%, the oxygen content was 29.09%, the hydrogen content was 1.38%, and the nitrogen content was 0.55%. The H/C ratio was used to characterize the aromaticity of the adsorbent, and the O/C ratio was employed to gauge the level of carbonization [32]. Generally, the H/C ratio of CM biochar is 0.93 normally, while the MCB featured better aromatic and hydrophilic properties (0.93 vs. 0.03), which indicated that the MCB may have a better adsorption effect [33]. Moreover, the O/C ratio of MCB was 0.57, thus, the functional groups with oxygen were prevalent on the surface of MCB, which promoted adsorption.

The SSA of MCB was 37.43 $m^2 \cdot g^{-1}$ and the average pore size of MCB was 11.02 nm. The iron oxides loaded on the surface of MCB made the surface of the biochar rougher, so the SSA of MCB is larger than that of biochar without modification. A larger SSA may provide more opportunities for biochar to come into contact with Cd(II), resulting in enhanced physical adsorption capacity. The average pore size of MCB was small (11.02 nm), which may be because the iron oxides occupied part of the porous structure of MCB. Compared with the iron–modified rice straw biochar (8.28 $m^2 \cdot g^{-1}$, 25.83 nm) [34], the MCB prepared in this work had larger SSA and less pore size, which meant that MCB may have more excellent physical qualities. Moreover, the N_2 adsorption–desorption isotherms and pore size distribution curve were shown in Figure S1.

The SEM and SEM–EDS imagines of MCB were shown in Figure S2. There was porous tubular and carbon skeleton morphology in MCB via the scanning of the electron microscope, which may be formed from the straw fiber. Part of the carbon structure in MCB collapsed after 600 °C pyrolysis, while part of the carbon skeleton was still retained, which may be because the residence time of pyrolysis was only one hour. The surface of the MCB was slightly rough, and the retained carbon skeleton of MCB provided the basis for the physical adsorption including the pore filling of MCB.

The SEM–EDS images showed that the MCB was successfully loaded with Fe; thus, these images confirmed the conclusion in EA, that the content of Fe was high, achieving 78.51%. In addition, SEM–EDS images also showed that Fe existed in some microporous structures. Moreover, the MCB also contained abundant Ca and Mg, which were evenly distributed on the surface of the carbon skeleton of MCB.

3.4.2. XRD and FTIR Analysis

The XRD spectra of MCB was shown in Figure 7a. The analysis of XRD spectra was applied to determine the existence of crystalline minerals and analyze their structure of MCB. The fluctuating peaks of the typical amorphous carbon and graphite carbon structures were about 22° and 44°, respectively, and these peaks were common in biochar characterization, indicating that MCB belonged to biochar material [35]. But the amorphous carbon structure of MCB disappeared for the disappearance of these characteristic peaks, which indicated that the modification substantially changed the carbon structure in MCB [10]. This conjecture was mutually corroborated the SSA and SEM–EDS imagines mentioned above, and the modification affected the physical and chemical properties of MCB, then affected its adsorption of Cd(II). The Fe_3O_4 (PDF#89–2355) and Fe_2O_3 (PDF#87–1166) were loaded on MCB (Figure 7a), indicating the iron was loaded onto the MCB and the MCB was magnetic. The Fe^0 had strong reducibility and the peak at $2\theta = 45°$ indicated Fe^0 was loaded on MCB, which indicating the MCB was reductive slightly, and REDOX reaction may occur during the adsorption. Once the metal was loaded on MCB, the loaded metal would grow as the prominent active site, moreover, the doped heteroatom could work as active sites or induce other Lewis base sites on MCB to promote the catalysis process [36]; thus, modification may increase the number of active sites of MCB available and enhance the ability to adsorb Cd(II).

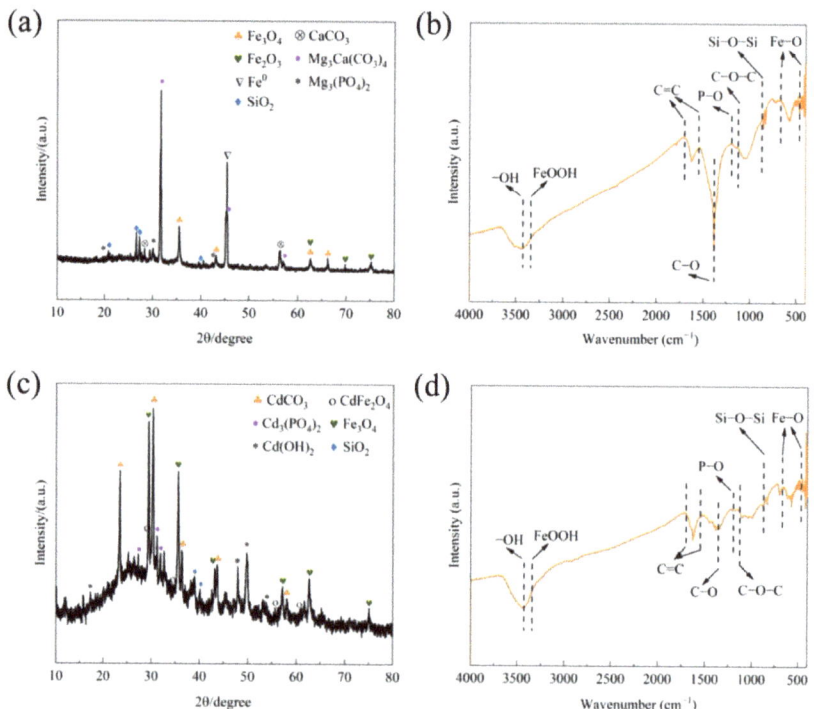

Figure 7. XRD and FTIR spectra of MCB (a,b) and Cd(II)–loaded MCB (c,d).

The MCB contained abundant Ca and Mg, which were evenly distributed on the surface of the carbon skeleton of MCB (Figure S2), and XRD spectra also confirmed the presence of metallic compounds on the surface of MCB (Figure 7a). The peaks indicated the presence of $CaCO_3$ (PDF#41–1475), $Mg_3(PO_4)_2$ (PDF#75–1491), and $Mg_3Ca(CO_3)_4$ (PDF#14–0409). These metal salts would react interfacially with Cd(II) in the solution on the surface of MCB, and the metal elements in the metal salts may be precipitated and undergo replacement reactions with the Cd(II) [37], resulting in substances with less solubility, such as $CdCO_3$ and $Cd_3(PO_4)_2$. Namely, Cd(II) may be removed via ion–exchange by MCB. Moreover, the MCB contained SiO_2 (PDF#45–1045), which corresponded to the results of EA.

FTIR spectra of MCB was shown in Figure 7b. The functional groups in MCB included –OH (3400 cm^{-1}), C=O (1735 cm^{-1}), Si–O–Si (873 cm^{-1}), and C–O–C (1099 cm^{-1}). The modification also affected the carbon structure of MCB. After the Fe modification, the peaks of –CH2 (2921 cm^{-1}), C–O, and –CH (946 and 873 cm^{-1}) in MCB were changed, thus, the modification may affect the carbon–containing functional groups. Though stretching vibrations of the aromatic C–C ring in MCB were retained after modification, the aromatic structure weakened due to Fe having invaded the carbon structure [38]. After modification, Fe–O (470–669 cm^{-1}) and FeOOH (3340 cm^{-1}) had been detected in MCB [39,40], which once again confirmed that iron was successfully loaded onto MCB. The vibration peaks at 927 and 786 cm^{-1} were the aromatic and hetero–aromatic structure, respectively, which provided pi electrons. As the pi electrons had a potential advantage to capture heavy metal ions, MCB had the potential to adsorb Cd(II) via π bonds. The vibrational transitions in the 1094–1182 cm^{-1} region were due to P=O and P–O stretching vibrations of the P–O–P and P=O–OH groups, and those in the 900–926 cm^{-1} region were ascribed to the asymmetric stretching of the P–O–P group [41], and this result was consistent with the results of XRD spectra analysis above, that MCB contained phosphate compounds.

3.4.3. Magnetic Properties

The hysteresis loop was used to characterize the response of magnetic materials, and the hysteresis loop of the MCB was displayed in Figure S3. The saturation magnetization value of the MCB was 12.8 emu·g^{-1}, which indicated that MCB can be rapidly magnetically separated from solution in a magnetic field environment.

3.5. Batch Experiments

3.5.1. Adsorption Kinetics and Isotherms

The fitting results were shown in Table S7 and Figure S4. The R^2 of the pseudo-first-order model (PFO), pseudo-second-order model (PSO), Elovich model, and internal diffusion model fitting for the adsorption of Cd(II) on MCB were 0.9848 (Figure S4a), 0.9999 (Figure S4a), 0.7268 (Figure S4c), and 0.9050 (Figure S4b). Thus, the PSO predicted the adsorptions of Cd(II) well, while the Elovich model could not fit the adsorption well. Moreover, the internal diffusion was not the sole rate-determining factor for the C = 8.6288 in the fitting results of the internal diffusion model. In short, the kinetic studies showed that the PSO could fit the experimental data well, which suggested that the Cd(II) adsorption by MCB was mainly based on chemisorption [42].

The R^2 of the Langmuir model and Freundlich model fitting for the adsorption of Cd(II) were 0.9956 and 0.9910 (Figure S4d). By comparing the R^2 values of the two models, the results showed that the Langmuir model better illustrated the interaction between MCB and Cd(II). The results implied that the absorption of MCB on Cd(II) was unimolecular [42], and the adsorption capacity was 612.43 mg·L^{-1}.

To sum up, the PSO and Langmuir models fit the adsorption process of Cd(II) on MCB better, and adsorption capacity was 612.43 mg·L^{-1}, indicating the Cd(II) adsorption by MCB was mainly based on chemisorption, and was unimolecular.

3.5.2. Adsorption Mechanisms

As shown in Figure S5, pH was also an important index affecting the adsorption effect of Cd(II) by MCB. With the pH of 3 and 5, the adsorption capacity was 62.4 and 85.4 mg·g^{-1}, respectively. When the reaction solution was acidic, the adsorption effect was poor, and the stronger the acid, the worse the adsorption effect. On the contrary, with the pH of 9 and 11, the adsorption capacity was 97.3 and 99.5 mg·g^{-1}, respectively. The form of Cd(II) in different pH reaction solutions may be varied. When pH $\leq$ 6, there were ions in a free state, like Cd^{2+}, and when pH was 6–9, there were ions existed in the form of Cd^{2+} and Cd(OH)$_2$, while when pH $\geq$ 9, there were ions mainly existing in the form of Cd(OH)$_2$. When pH $\leq$ 3, there was electrostatic repulsion between MCB and Cd(II) for the surface of MBC was positively charged when the pH value of the Cd(II) solution was less than 3. Moreover, there was a lot of H$^+$ in the acidic solution, and the H$^+$ would compete with Cd(II) for adsorption sites on the surface of MCB, which reduced the adsorption effect.

The SEM and SEM–EDS imagines of Cd(II)-loaded MCB were shown in Figure S6, and the XRD spectra and FTIR spectra of Cd(II)-loaded MCB were shown in Figure 7. Compared with the MCB (Figures S2a and S6a), some pore structures of Cd(II)-loaded MCB were blocked with small solid agglomerates, which were also attached to the biochar skeleton. It was speculated that these solid agglomerates were the chemical precipitation of Cd(II). The EDS confirmed that the particles on biological carbon were Cd(II)-contained substances. These results indicated that MCB may adsorb Cd(II) via pore filling. As shown in Figure S6c,f, compared with the MCB, the presence of Cd(II) on the MCB surface suggests that the MCB had successfully loaded Cd(II) after absorption. According to the SEM–EDS imagines, Cd(II) was evenly distributed on the surface of Cd(II)-loaded MCB, and the distribution of Cd(II) was related to the distribution of Fe on the surface, indicating that the iron oxide on the surface of MCB may play a part in the adsorption on Cd(II).

As shown in Figure 7c, comparing the biochar before adsorption, the XRD spectra of Cd(II)-loaded MCB showed new peaks, such as CdCO$_3$ (PDF#42–1342), Cd$_3$(PO$_4$)$_2$ (PDF#72–1959), and Cd(OH)$_2$ (PDF#20–0179), determining that Cd(II) may be removed via co–precipitation and ion-exchange. Moreover, iron oxides loaded on MCB also reacted with Cd(II) to form $CdFe_2O_4$ (PDF#22–1063). Figure 7d showed the FTIR spectra of Cd(II)-loaded MCB. The broad peak band from 630 to 750 cm^{-1} was mainly metal oxygen-containing functional groups, such as Fe–O, while after Cd(II) adsorption, these broad peak bands mentioned above decreased, indicating that Cd(II) reacted with the oxygenic groups. After Cd(II) adsorption, the intensity of peaks at 1573, 1398, 1029, 981, and 800–700 cm^{-1} changed slightly, which indicated that C=C, C=O, C–O, and other groups participated in the adsorption reaction, which determining qualitatively that cation–π interaction existed in the adsorption [43].

The XPS spectra were displayed in Figure S7. The C 1s fine spectrum of the sample consisted of three peaks, C–C (284 eV), C–O–C (286 eV), and O–C=O/C=O (289 eV). While after adsorption, the peaks of C 1 shifted, which indicated the structure changed after adsorption, and oxygen-contained functional groups made a contribution to the Cd(II) adsorption. The peak positions for Cd 3d were 406 and 414 eV, whose percentages were 47.6 and 52.4%, respectively. There was no REDOX reaction involved for the valence of Cd(II) did not change. Combined with the XRD analysis mentioned above, Fe(II) and Fe(III) compounds were loaded on the surface and their contents were 73.4 and 26.6%, respectively. As shown in Figure S7b, the peak positions for Fe 2p were 710, 711, 717, 723, and 725 eV. The Fe 2p peaks were changed after absorption, showing that presence of Cd(II) caused the variation in Fe 2p. Besides, compared with Fe 2p before adsorption (Figure S8), the contents of Fe^0 and Fe^{2+} loaded on Cd(II)-loaded MCB decreased, thus, Fe^0 and Fe^{2+} may were oxidized during the absorption. The Fe(III) loaded on MCB interacted with the anions (such as carbonate and phosphate) exited from MCB. The hydrolysis of Fe(III) made the solution acidic, and H$^+$ replaced Mg(II) in MCB, resulting in that a Si–O–Mg–O–Si group broke down into two Si–OH groups, so a large number of Si–OH groups were generated on the surface of MCB [43]. Moreover, the Si–OH group can have the following surface coordination adsorption with Cd(II) [44]. The reaction process was shown in Equation (6) to Equation (8).

$$SiOH + Cd^{2+} \rightarrow SiOCd^+ + H^+ \tag{6}$$

$$2SiOH + Cd^{2+} \rightarrow (SiO)_2Cd + 2H^+ \tag{7}$$

$$SiOH + Cd^{2+} + H_2O \rightarrow SiOCdOH + 2H^+ \tag{8}$$

After adsorption, the characteristic peak of Fe_3O_4 (30.1°) was still observed, while the characteristic peak of Fe^0 disappeared, indicating that Fe loaded on MCB was oxidized [45,46], which was shown as Equation (9) to Equation (12).

$$Fe^0 \rightarrow Fe(OH)_2 \rightarrow Fe_3O_4 \rightarrow (\gamma\text{-}Fe_2O_3) \tag{9}$$

$$2Fe^0 + 4H^+ + O_2 \rightarrow 2Fe^{2+} + 2H_2O \tag{10}$$

$$Fe^{2+} + H_2O \rightarrow Fe^{3+} + 1/2H_2 + OH^- \tag{11}$$

$$Fe^{3+} + 3OH^- \rightarrow Fe(OH)_3 \tag{12}$$

To sum up, the adsorption was dominated via chemisorption with the mechanisms of ion-exchange, electrostatic attraction, pore-filling, co-precipitation, and formation of complexations (Figure 8).

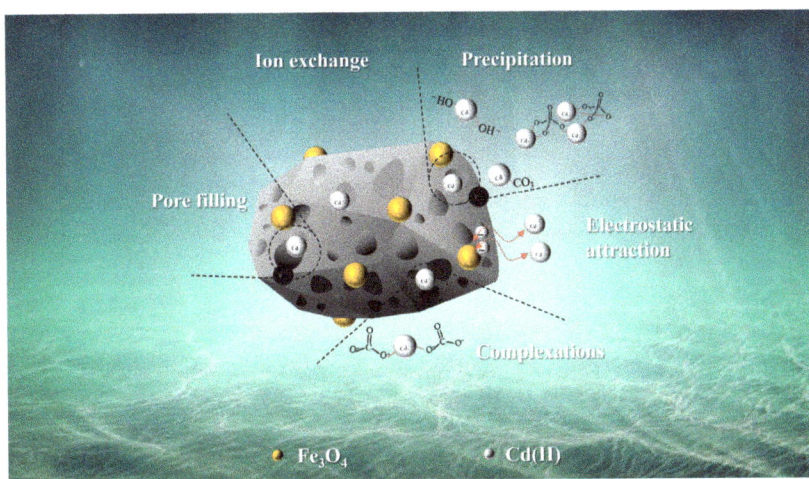

Figure 8. Surface interaction between Cd(II) and MCB functional materials.

4. Conclusions

In this work, MCB was prepared via cow manure, and employed for Cd(II) removal. PT, HT, and iron content are the key factors of MCB preparation, while the optimal conditions were obtained by the RSM method as 160 °C for HT, 600 °C for pyrolysis, and Fe loading with 10 wt%, and experimental parameters were optimized as 1 g·L^{-1} for dosage, 7 for pH, and 100 mg·L^{-1} for initial concentration. The BP–ANN model fit better than the RSM model, with lower error values. The constructed BP–ANN model overcame the shortage of large amounts of data training to achieve high precision, demonstrating better contingency and stability. As for thermodynamics, the PSO and the Langmuir model could better mimic the surface interactions between Cd(II) and MCB. The adsorption was dominated by chemisorption with the mechanisms of ion-exchange, electrostatic attraction, pore-filling, co-precipitation, and formation of complexations. All these results demonstrated the potential of cow manure as feasible biomass feedstock for functional materials preparation and wastewater purification.

Supplementary Materials: The following supporting information can be downloaded at: https://www.mdpi.com/article/10.3390/pr11082295/s1, Text S1: Calculation of adsorption capacity; Table S1: Variables and experimental conditions for RSM; Table S2: BBD experimental design with the actual values of adsorption quantity with different biochar; Table S3: ANOVA for response surface quadratic model; Table S4: BBD experimental design with the actual values of adsorption quantity under different reaction conditions; Table S5: ANOVA for response surface quadratic model; Table S6: Comparison of the statistical parameters of ANN and RSM models for the various responses; Table S7: Fitting parameters of kinetic model and isotherm model of Cd(II) absorption; Figure S1: N$_2$ adsorption–desorption isotherms and pore size distribution curve; Figure S2: SEM (a and b) and SEM–EDS imagines of MCB (c – f); Figure S3: Magnetic hysteresis loop of MCB; Figure S4: Fitting-figures of PFO, PSO (a), internal diffusion model (b), Elovich model (c), and isotherms (d) for the adsorption of Cd(II) on MCB. Experimental conditions: [initial concentration of Cd(II)] = 100 mg·L^{-1}, [dosage] = 1 g·L^{-1}, [time] = 24 h, [temperature] = 25 °C; Figure S5: The effect of pH for adsorption Cd(II) by MCB. Experimental conditions: [initial concentration of Cd(II)] = 100 mg·L^{-1}, [dosage] = 1 g·L^{-1}, [time] = 24 h, [temperature] = 25 °C; Figure S6: SEM–EDS imagines of Cd(II)–loaded MCB; Figure S7: XPS spectrums of Cd(II)–loaded MCB; Figure S8: XPS Fe 2p spectrum of MCB.

Author Contributions: Y.W.: formal analysis, original draft; D.C.: data curation, software, methodology; Y.Z.: data curation, investigation, methodology; H.W.: supervision, writing—review and editing; R.X.: funding acquisition, resources, supervision. All authors have read and agreed to the published version of the manuscript.

Funding: This work was supported by the National Natural Science Foundation of China [22264025, 52100147], Applied Basic Research Foundation of Yunnan Province [202201AS070020, 202201AU070061] and the Science and Technology Research Project of Education Department of Jiangxi Province [DA202102159].

Data Availability Statement: Not applicable.

Conflicts of Interest: The authors declare no conflict of interest.

References

1. Ammar, N.S.; Fathy, N.A.; Ibrahim, H.S.; Mousa, S.M. Micro–mesoporous modified activated carbon from corn husks for removal of hexavalent chromium ions. *Appl. Water Sci.* **2021**, *11*, 154. [CrossRef]
2. Chin, J.F.; Heng, Z.W.; Teoh, H.C.; Chong, W.C.; Pang, Y.L. Recent development of magnetic biochar crosslinked chitosan on heavy metal removal from wastewater–Modification, application and mechanism. *Chemosphere* **2022**, *291*, 133035. [CrossRef]
3. Cuellar, A.D.; Webber, M.E. Cow power: The energy and emissions benefits of converting manure to biogas. *Environ. Res. Lett.* **2008**, *3*, 034002. [CrossRef]
4. Deng, Y.; Zhou, X.; Shen, J.; Xiao, G.; Hong, H.; Lin, H.; Wu, F.; Liao, B.-Q. New methods based on back propagation (BP) and radial basis function (RBF) artificial neural networks (ANNs) for predicting the occurrence of haloketones in tap water. *Sci. Total Environ.* **2021**, *772*, 145334. [CrossRef] [PubMed]
5. Du, Y.; Dai, M.; Cao, J.; Peng, C.; Ali, I.; Naz, I.; Li, J. Efficient removal of acid orange using a porous adsorbent–supported zero–valent iron as a synergistic catalyst in advanced oxidation process. *Chemosphere* **2020**, *244*, 125522. [CrossRef]
6. Erdem, H. The effects of biochars produced in different pyrolsis temperatures from agricultural wastes on cadmium uptake of tobacco plant. *Saudi J. Biol. Sci.* **2021**, *28*, 3965–3971. [CrossRef] [PubMed]
7. Gadekar, M.R.; Ahammed, M.M. Modelling dye removal by adsorption onto water treatment residuals using combined response surface methodology–artificial neural network approach. *J. Environ. Manag.* **2019**, *231*, 241–248. [CrossRef]
8. Guo, X.; Zhu, L.; Xu, X.; Ma, M.; Zou, G.; Wei, D. Competitive or synergistic? Adsorption mechanism of phosphate and oxytetracycline on chestnut shell–derived biochar. *J. Clean. Prod.* **2022**, *370*, 133526. [CrossRef]
9. Hassan, M.A.; Mahmoud, Y.K.; Elnabtiti, A.A.S.; El-Hawy, A.S.; El-Bassiony, M.F.; Abdelrazek, H.M.A. Evaluation of Cadmium or Lead exposure with nannochloropsis oculata mitigation on productive performance, biochemical, and oxidative stress biomarkers in barki rams. *Biol. Trace Elem. Res.* **2023**, *201*, 2341–2354. [CrossRef]
10. He, X.L.; Fan, S.K.; Zhu, J.; Guan, M.Y.; Liu, X.X.; Zhang, Y.S.; Jin, C.W. Iron supply prevents Cd uptake in *Arabidopsis* by inhibiting IRT1 expression and favoring competition between Fe and Cd uptake. *Plant Soil* **2017**, *416*, 453–462. [CrossRef]
11. He, Y.; Lin, H.; Dong, Y.; Li, B.; Wang, L.; Chu, S.; Luo, M.; Liu, J. Zeolite supported Fe/Ni bimetallic nanoparticles for simultaneous removal of nitrate and phosphate: Synergistic effect and mechanism. *Chem. Eng. J.* **2018**, *347*, 669–681. [CrossRef]
12. Hu, Y.B.; Du, T.; Ma, L.; Feng, X.; Xie, Y.; Fan, X.; Fu, M.L.; Yuan, B.; Li, X.Y. Insights into the mechanisms of aqueous Cd(II) reduction and adsorption by nanoscale zerovalent iron under different atmosphere conditions. *J. Hazard. Mater.* **2022**, *440*, 129766. [CrossRef] [PubMed]
13. Huang, Q.; Yu, Y.; Wan, Y.; Wang, Q.; Luo, Z.; Qiao, Y.; Su, D.; Li, H. Effects of continuous fertilization on bioavailability and fractionation of cadmium in soil and its uptake by rice (*Oryza sativa* L.). *J. Environ. Manag.* **2018**, *215*, 13–21. [CrossRef] [PubMed]
14. Karki, R.; Chuencharn, W.; Surendra, K.; Sung, S.; Raskin, L.; Khanal, S.K. Anaerobic co–digestion of various organic wastes: Kinetic modeling and synergistic impact evaluation. *Bioresour. Technol.* **2022**, *343*, 126063. [CrossRef]
15. Khetib, Y.; Sedraoui, K.; Gari, A. Improving thermal conductivity of a ferrofluid–based nanofluid using Fe_3O_4—Challenging of RSM and ANN methodologies. *Chem. Eng. Commun.* **2022**, *209*, 1070–1081. [CrossRef]
16. Klasson, K.T. Biochar characterization and a method for estimating biochar quality from proximate analysis results. *Biomass Bioenergy* **2017**, *96*, 50–58. [CrossRef]
17. Li, X.; Yang, B.; Yang, J.; Fan, Y.; Qian, X.; Li, H. Magnetic properties and its application in the prediction of potentially toxic elements in aquatic products by machine learning. *Sci. Total Environ.* **2021**, *783*, 147083. [CrossRef]
18. Liu, X.; Lee, C.; Kim, J.Y. Comparison of mesophilic and thermophilic anaerobic digestions of thermal hydrolysis pretreated swine manure: Process performance, microbial communities and energy balance. *J. Environ. Sci.* **2023**, *126*, 222–233. [CrossRef]
19. Liu, Y.; Li, Y.; Xia, Y.; Liu, K.; Ren, L.; Ji, Y. The Dysbiosis of Gut Microbiota Caused by Low–Dose Cadmium Aggravate the Injury of Mice Liver through Increasing Intestinal Permeability. *Microorganisms* **2020**, *8*, 211. [CrossRef]
20. Liu, Y.; Wang, L.; Gu, K.; Li, M. Artificial Neural Network (ANN)–Bayesian Probability Framework (BPF) based method of dynamic force reconstruction under multi–source uncertainties. *Knowl.-Based Syst.* **2022**, *237*, 107796. [CrossRef]

21. Mariuzza, D.; Lin, J.C.; Volpe, M.; Fiori, L.; Ceylan, S.; Goldfarb, J.L. Impact of Co–Hydrothermal carbonization of animal and agricultural waste on hydrochars' soil amendment and solid fuel properties. *Biomass Bioenergy* **2022**, *157*, 106329. [CrossRef]
22. Medellín-Castillo, N.A.; Cruz-Briano, S.A.; Leyva-Ramos, R.; Moreno-Piraján, J.C.; Torres-Dosal, A.; Giraldo-Gutiérrez, L.; Labrada-Delgado, G.J.; Pérez, R.O.; Rodriguez-Estupiñan, J.P.; Lopez, S.Y.R.; et al. Use of bone char prepared from an invasive species, pleco fish (*Pterygoplichthys* spp.), to remove fluoride and Cadmium in water. *J. Environ. Manag.* **2020**, *256*, 109956. [CrossRef] [PubMed]
23. Mei, Y.; Xu, J.; Zhang, Y.; Li, B.; Fan, S.; Xu, H. Effect of Fe–N modification on the properties of biochars and their adsorption behavior on tetracycline removal from aqueous solution. *Bioresour. Technol.* **2021**, *325*, 124732. [CrossRef] [PubMed]
24. Muhammad, N.; Ge, L.; Chan, W.P.; Khan, A.; Nafees, M.; Lisak, G. Impacts of pyrolysis temperatures on physicochemical and structural properties of green waste derived biochars for adsorption of potentially toxic elements. *J. Environ. Manag.* **2022**, *317*, 115385. [CrossRef] [PubMed]
25. Obi, C.C.; Nwabanne, J.T.; Igwegbe, C.A.; Ohale, P.E.; Okpala, C.O.R. Multi–characteristic optimization and modeling analysis of electrocoagulation treatment of abattoir wastewater using iron electrode pairs. *J. Water Process Eng.* **2022**, *49*, 103136. [CrossRef]
26. Peng, B.; Liu, Q.; Li, X.; Zhou, Z.; Wu, C.; Zhang, H. Co-pyrolysis of industrial sludge and rice straw: Synergistic effects of biomass on reaction characteristics, biochar properties and heavy metals solidification. *Fuel Process. Technol.* **2022**, *230*, 107211. [CrossRef]
27. Peng, H.; Guo, J.; Qiu, H.; Wang, C.; Zhang, C.; Hao, Z.; Rao, Y.; Gong, Y. Efficient removal of Cr(VI) with biochar and optimized parameters by response surface methodology. *Processes* **2021**, *9*, 889. [CrossRef]
28. Plattes, M.; Bertrand, A.; Schmitt, B.; Sinner, J.; Verstraeten, F.; Welfring, J. Removal of tungsten oxyanions from industrial wastewater by precipitation, coagulation and flocculation processes. *J. Hazard. Mater.* **2007**, *148*, 613–615. [CrossRef]
29. Qiu, B.; Tao, X.; Wang, H.; Li, W.; Ding, X.; Chu, H. Biochar as a low–cost adsorbent for aqueous heavy metal removal: A review. *J. Anal. Appl. Pyrolysis* **2021**, *155*, 105081. [CrossRef]
30. Sun, H.; Zhou, Q.; Zhao, L.; Wu, W. Enhanced simultaneous removal of nitrate and phosphate using novel solid carbon source/zero–valent iron composite. *J. Clean. Prod.* **2021**, *289*, 125757. [CrossRef]
31. Sun, L.; Gong, P.; Sun, Y.; Qin, Q.; Song, K.; Ye, J.; Zhang, H.; Zhou, B.; Xue, Y. Modified chicken manure biochar enhanced the adsorption for Cd^{2+} in aqueous and immobilization of Cd in contaminated agricultural soil. *Sci. Total Environ.* **2022**, *851*, 158252. [CrossRef] [PubMed]
32. Szogi, A.A.; Takata, V.H.; Shumaker, P.D. Chemical extraction of phosphorus from dairy manure and utilization of recovered manure solids. *Agronomy* **2020**, *10*, 1725. [CrossRef]
33. Tan, W.T.; Zhou, H.; Tang, S.F.; Zeng, P.; Gu, J.F.; Liao, B.H. Enhancing Cd(II) adsorption on rice straw biochar by modification of iron and manganese oxides. *Environ. Pollut.* **2022**, *300*, 118899. [CrossRef]
34. Wang, H.; Chen, D.; Wen, Y.; Zhang, Y.; Liu, Y.; Xu, R. Iron–rich red mud and iron oxide–modified biochars: A comparative study on the removal of Cd(II) and influence of natural aging processes. *Chemosphere* **2023**, *330*, 138626. [CrossRef] [PubMed]
35. Wang, H.; Cui, T.; Chen, D.; Luo, Q.; Xu, J.; Sun, R.; Zi, W.; Xu, R.; Liu, Y.; Zhang, Y. Hexavalent chromium elimination from wastewater by integrated micro–electrolysis composites synthesized from red mud and rice straw via a facile one–pot method. *Sci. Rep.* **2022**, *12*, 14242. [CrossRef] [PubMed]
36. Wang, H.; Liu, Y.; Ifthikar, J.; Shi, L.; Khan, A.; Chen, Z.; Chen, Z. Towards a better understanding on mercury adsorption by magnetic bio–adsorbents with gamma–Fe_2O_3 from pinewood sawdust derived hydrochar: Influence of atmosphere in heat treatment. *Bioresour. Technol.* **2018**, *256*, 269–276. [CrossRef]
37. Wang, H.; Wang, S.; Chen, Z.; Zhou, X.; Wang, J.; Chen, Z. Engineered biochar with anisotropic layered double hydroxide nanosheets to simultaneously and efficiently capture Pb^{2+} and CrO_4^{2-} from electroplating wastewater. *Bioresour. Technol.* **2020**, *306*, 123118. [CrossRef]
38. Wang, N.; Sun, Q.; Zhang, T.; Mayoral, A.; Li, L.; Zhou, X.; Xu, J.; Zhang, P.; Yu, J. Impregnating subnanometer metallic nanocatalysts into self-pillared zeolite nanosheets. *J. Am. Chem. Soc.* **2021**, *143*, 6905–6914. [CrossRef]
39. Witek-Krowiak, A.; Chojnacka, K.; Podstawczyk, D.; Dawiec, A.; Pokomeda, K. Application of response surface methodology and artificial neural network methods in modelling and optimization of biosorption process. *Bioresour. Technol.* **2014**, *160*, 150–160. [CrossRef]
40. Xu, L.; Tian, S.; Hu, Y.; Zhao, J.; Ge, J.; Lu, L. Cadmium contributes to heat tolerance of a hyperaccumulator plant species Sedum alfredii. *J. Hazard. Mater.* **2023**, *441*, 129840. [CrossRef]
41. Yan, R.; Liao, J.; Yang, J.; Sun, W.; Nong, M.; Li, F. Multi–hour and multi–site air quality index forecasting in Beijing using CNN, LSTM, CNN–LSTM, and spatiotemporal clustering. *Expert Syst. Appl.* **2021**, *169*, 114513. [CrossRef]
42. Yang, R.; Liang, X.; Strawn, D.G. Variability in Cadmium uptake in common wheat under Cadmium stress: Impact of genetic variation and silicon supplementation. *Agriculture* **2022**, *12*, 848. [CrossRef]
43. Yu, X.; Zhou, H.; Ye, X.; Wang, H. From hazardous agriculture waste to hazardous metal scavenger: Tobacco stalk biochar–mediated sequestration of Cd leads to enhanced tobacco productivity. *J. Hazard. Mater.* **2021**, *413*, 125303. [CrossRef] [PubMed]
44. Zhang, A.; Li, X.; Xing, J.; Xu, G. Adsorption of potentially toxic elements in water by modified biochar: A review. *J. Environ. Chem. Eng.* **2020**, *8*, 104196. [CrossRef]

45. Zhou, X.; Xu, D.; Yang, J.; Yan, Z.; Zhang, Z.; Zhong, B.; Wang, X. Treatment of distiller grain with wet–process phosphoric acid leads to biochar for the sustained release of nutrients and adsorption of Cr(VI). *J. Hazard. Mater.* **2023**, *441*, 129949. [CrossRef] [PubMed]
46. Zhou, X.; Zhu, Y.; Niu, Q.; Zeng, G.; Lai, C.; Liu, S.; Huang, D.; Qin, L.; Liu, X.; Li, B.; et al. New notion of biochar: A review on the mechanism of biochar applications in advannced oxidation processes. *Chem. Eng. J.* **2021**, *416*, 129027. [CrossRef]

Disclaimer/Publisher's Note: The statements, opinions and data contained in all publications are solely those of the individual author(s) and contributor(s) and not of MDPI and/or the editor(s). MDPI and/or the editor(s) disclaim responsibility for any injury to people or property resulting from any ideas, methods, instructions or products referred to in the content.

Article

Alkali Etching Hydrochar-Based Adsorbent Preparation Using Chinese Medicine Industry Waste and Its Application in Efficient Removal of Multiple Pollutants

Xinyan Zhang [1,*], Shanshan Liu [1], Qingyu Qin [2], Guifang Chen [1] and Wenlong Wang [1]

[1] National Engineering Laboratory for Reducing Emissions from Coal Combustion, Engineering Research Center of Environmental Thermal Technology of Ministry of Education, Shandong Key Laboratory of Energy Carbon Reduction and Resource Utilization, School of Energy and Power Engineering, Shandong University, Jinan 250061, China

[2] Laboratory of Biomass and Bioprocessing Engineering, College of Engineering, China Agricultural University, Beijing 100083, China

* Correspondence: sddxzxy2020@sdu.edu.cn

Abstract: The annual discharge (6–7 million tons per year) of Chinese medicine industry waste (CMIW) is large and harmful. CMIW with a high moisture content can be effectively treated by hydrothermal carbonization (HTC) technology. Compared with CMIW, the volume and number of pores of the prepared hydrochar increased significantly after alkali etching (AE), and they had abundant oxygen-containing functional groups. These properties provide physical and chemical adsorption sites, improving the adsorbent activity of the alkaline etching of Chinese medicine industry waste hydrochar (AE-CMIW hydrochar). However, few studies have investigated the adsorption of organic dyes and heavy metals in mixed solutions. This study proposed a method of coupling HTC with AE to treat CMIW and explored the potential of AE-CMIW hydrochar to remove metal ions and organic dyes from mixed solution. We analyzed the removal rates of metal ions and organic dyes by the adsorbents and investigated their differences. The results showed that the lead ion, cadmium ion, and methylene blue could be efficiently removed by AE-CMIW hydrochar in a mixed solution, with removal rates of more than 98%, 20–57%, and 60–80%, respectively. The removal rates were different mainly due to the various electrostatic interactions, physical adsorption, differences in the hydrating ion radius of the metal ions, and functional group interactions between the AE-CMIW hydrochar and the lead ion, cadmium ion, and methylene blue. This study provides a technical method for preparing multi-pollutant adsorbents from CMIW, which enables efficient utilization of organic solid waste and achieves the purpose of treating waste with waste.

Keywords: chinese medicine industry waste; hydrothermal carbonization; alkali etching; multi-pollutant adsorbent; treating waste with waste

1. Introduction

The annual output of Chinese medicine industry waste (CMIW), a type of organic solid waste containing many bacteria, is huge [1]. It can reach six to seven million tons per year [2–4]. The unreasonable disposal of CMIW causes environmental pollution and ecological damage and threatens human health [5]. The Ministry of Ecology and Environment of the People's Republic of China promulgated the Law of the People's Republic of China on the Prevention and Control of Environmental Pollution by Solid Waste, which stated that the prevention and control of environmental pollution by solid waste should adhere to the principles of reduction, recycling and harmless treatment [6]. In the European Union countries, the waste management system presupposes an integrated system of various aspects: social, economic, regulatory, managerial, technical. The new edition of the Environmental Code of the Republic of Kazakhstan on 2 January 2021 indicated that the level of this country's waste management is closer to international standards [7]. The microbes in CMIW can be killed by heat treatment. The CMIW can be converted into carbon material after high-temperature treatment, which is a method for efficient resource utilization of CMIW. The heat treatment of organic solid waste usually includes pyrolysis, baking, and hydrothermal carbonization (HTC). The moisture content of CMIW is high, and pre-drying is required if CMIW is treated by pyrolysis and baking, which can increase the overall operational cost. HTC is a thermochemical process in a wet medium that can effectively treat organic solid waste with high moisture content, such as CMIW [8,9]. Studies have shown that HTC is a promising technique for converting organic solid waste into hydrochar, which can be used as a fuel, adsorbent, catalyst, and others [10–12]. Hydrochar has porous structures and abundant surface functional groups, which also depend on the feed-stock and treatment conditions [13,14]. Studies have shown that hydrochar can be effectively applied to use in wastewater treatment in a wide variety of pollutants including heavy metals and dyes [15–17]. Studies have shown that alkali etching (AE) can increase the porosity of hydrochar [18]. This phenomenon happens because the energized element particles (K^+ and Na^+) of basic activators (e.g., KOH and NaOH) generated by absorbing heat energy can penetrate the carbon structure of the hydrochar and further promote the development of pores [19].

Organic dyes and heavy metals are common toxic and harmful pollutants in the wastewater of the printing, paper-making, leather, textile, and cosmetic industries [20], mainly because many heavy metals (particularly cadmium) are used as mordants in the dyeing process in these industries. Organic dyes have complex and special structural properties and are difficult to degrade using strong oxidants [21]. The accumulation of organic dyes in organisms can easily cause gene mutations and accumulation in the environment, leading to the deterioration of water quality [22]. Heavy metals are highly toxic and carcinogenic [23], can accumulate in living organisms through the food chain, and are difficult to be eliminated from the body through metabolism [24]. These multiple pollutants in wastewater cause great harm to the ecosystem and human health. Therefore, there is an urgent need to develop efficient pollutant removal methods. Removal methods include ultra-filtration, photo-catalytic reduction, reverse osmosis, and adsorption [25–29], among which adsorption is the most convenient, efficient, and low-cost method [30]. Scholars have focused on removing a single pollutant in wastewater using the adsorption method [31–34]. Little research has been conducted on organic dyes and heavy metals simultaneously. Studies have shown that organic dyes or heavy metals can combine with the functional groups of adsorbents through electrostatic, chelation, and π–π interactions [23,35]. Therefore, competitive adsorption relationships exist between organic dyes and heavy metals in the same environment, leading to challenges in wastewater treatment due to the coexistence of heavy metals and organic dyes. Therefore, exploring an environmentally friendly adsorbent that can efficiently adsorb both organic dyes and heavy metals is necessary.

This study proposed an innovative method of coupling HTC with AE to treat CMIW. After this treatment, the CMIW formed porous structures and rich functional groups, making AE-CMIW hydrochar with both physical and chemical adsorption sites. Thus, CMIW might be advantageous as a raw material for the adsorbents of multiple pollutants. This study investigated the potential of the prepared AE-CMIW hydrochar as an adsorbent for multiple pollutants and the differences in the adsorption of multiple pollutants. The surface morphology and functional groups were analyzed using the Brunauer–Emmet–Teller (BET) model, scanning electron microscopy (SEM), and Fourier transform infrared spectroscopy (FTIR). The adsorption properties of the AE-CMIW hydrochar on lead ion (Pb^{2+}), cadmium ion (Cd^{2+}) and methylene blue (MB) were measured. Subsequently, the differences in the adsorption characteristics and kinetics of the AE-CMIW hydrochar on multiple pollutants in the mixed solution were investigated. This study made an organic solid waste such as CMIW achieve efficient resource utilization and treat multiple pollutants in wastewater. Therefore, this research is a resource utilization model for organic solid waste treatment.

2. Materials and Methods

2.1. Experimental Material

CMIW was obtained from pharmaceutical factories in Jinan, Shandong Province, China. They consist of a variety of traditional Chinese medicine residues, which mainly included Radix Glycyrrhizae residues, Panax notoginseng residues, Radix Isatidis residues and residues of Huoxiang Zhengqi Liquid. Their main components were crude fibre, crude fat, crude protein, amino acids, polysaccharide, starch, saponins, inorganic elements, total organic matter and total organic carbon. A multiple-component solution was prepared using methylene blue (MB; $C_{16}H_{18}N_3SCl$), $Pb(NO_3)_2$ and $Cd(NO_3)_2 \cdot 4H_2O$ for MB macromolecules, Pb^{2+} and Cd^{2+}, respectively. All chemicals were of analytical grade, and no further purification was required. Deionized water was used for solution preparation and dilution to the desired concentration of 100 mg/L to mimic pollutant concentrations in industrial effluents.

2.2. Hydrothermal Carbonization and Alkali Etching

HTC of CMIW was performed using an autoclave reaction kettle (Parr 4848; Parr Instrument Co., Moline, IL, USA). Nitrogen was pumped into the reactor five times to drain air into the kettle. The temperature of the reactor was controlled using a proportional integral derivative (PID) controller [36]. The pressure was generated by self-boosting, and it could reach 2–4 Mpa in the kettle. In the HTC experiments, 100 g of CMIW (~70% w.t.) and 300 mL of deionized water were mixed in a reaction kettle and auto-stirred at a rate of 400 rpm [37]. The temperature in the reaction kettle was set to 200, 230 or 260 °C at the rate of 3–5 °C/min and then maintained at a constant temperature for 1, 2, and 4 h, respectively. After HTC, the hydrochar was separated from the mixture via vacuum filtration. The hydrochar was dried at 40 °C for 24 h for further analysis. Subsequently, a concentration gradient series of aqueous KOH solutions was prepared, and AE was conducted at a 1:15 (w/v) hydrochar/KOH solution ratio. The conditions of HTC coupled with AE are denoted as AEHTC A-B-C in the figures and tables, where A is the temperature (°C), B is the duration (h), and C is the alkali concentration (mol/L).

2.3. Analytical Method

Hydrochar surface morphologies were determined with a ZEISS SUPRA 55 field emission scanning electron microscope (SEM, Carl Zeiss Co., Ltd., Oberkochen, Germany) at 30 kV acceleration voltage. Prior to analysis, the samples were mounted on a stub and sputter coated with gold [38]. The specific surface area (S_{BFT}) was calculated using the BET method, whereas the micro-pore and mesopore volumes were calculated using the Barrett–Joyner–Halenda model. The N_2 adsorption–desorption isotherms were plotted with a BK100 specific surface area and aperture synchronous analyzer (Beijing Jingwei Gaobo Science and Technology Co. Ltd., Beijing, China) at 77 K and using N_2 as the analytical gas. Before the measurement, all samples were degassed at 100 °C for 12 h to remove physisorbed water and impurities. The powder sample was uniformly adhered to the conductive tape. The acceleration voltage of the instrument was 0.02–30 kV, and the amplification factor was 20 K [39]. The surface element compositions of the hydrochar samples were analyzed using a matching Energy Dispersive Spectrometer (EDS, Carl Zeiss Co., Ltd., Oberkochen, Germany). The scans with Cu Kα radiation source were in the range of 10–70 (2θ).

The functional groups were analyzed using a Spectrum 400 Fourier infrared spectrometer (Thermo Nicolet Corp., Waltham, MA, USA). The sample and KBr were dried and mixed in a ratio of 1:100. Then, they were ground with a mortar and mixed thoroughly. The mixture was laminated for infrared spectral analysis. The infrared spectrum range was set as 4000–400 cm^{-1}, and the resolution was set as 4 cm^{-1}. The infrared spectral image of the sample was taken by a Merlin small cold field emission scanning electron microscope at 1 kV [37].

The isopotential point (pH$_{IEP}$) of the hydrochar in an aqueous environment was tested using a zeta potential analyzer (Malvern Zetasizer Nano ZS90, Arkansas, USA). The pH$_{IEP}$ of hydrochar samples was determined by measuring the initial and final solution pH values. The initial solution pH was varied in the range of 2–10 in intervals of 2 by using 0.1 M HCl and NaOH solutions. The solution (50 mL) with sample (50 mg) was vibrated for 24 h. The pH$_{IEP}$ was obtained by plotting the of initial solution pH against final solution pH [40].

Hydrochar samples (0.5 g) were put into the polyfluoroethylene tube; then, they were digested by microwave digestion (MDS-6C, SINEO) using HNO$_3$-HCl for 1.5 h. The residual concentrations of the heavy metals were determined using inductively coupled plasma–optical emission spectrometry (ICP-OES, Optima 7000D, Waltham, MA, USA). Inductively coupled plasma is a high-frequency electromagnetic field generated by a high-frequency current through an induction coil, which ionizes the working gas (Ar) to form a high temperature plasma with flame discharge. The default flow rate of Ar is 0.8 L/min, and the partial pressure of Ar is maintained between 0.6 and 0.8 Mpa during operation. The maximum temperature of the plasma is 10,000 K. The sample solution passes through the injection capillary and then enters into the atomizer by a peristaltic pump to form aerosol. The high temperature plasma is introduced through the carrier gas and then evaporates, atomizes, excites, ionizes, and produces radiation. The light source passes through the daylighting tube into the slit, the reflector, the prism, the middle step grating and the collimator, and then forms a two-dimensional spectrum. The spectral lines fall on the CID detector of 540 × 540 pixels in the form of light spots, and each spot covers several pixels. A spectrometer measures elemental concentrations by measuring the number of quantal light falling onto a pixel.

2.4. Adsorption Experiment

A mixed solution of MB, Pb^{2+} and Cd^{2+} was used to simulate the industrial wastewater. The concentrations of the three solutes were all configured at 100 mg/L, and then, the AE-CMIW hydrochar treated under different conditions was placed in 40 mL of the mixed solution and oscillated at 40 rpm/min in the oscillator for 24 h until the adsorption reached equilibrium. After adsorption, the supernatant was filtered and collected. The absorbance of the filtrate was measured using a UV–vis spectrophotometer (Thermo Fisher Scientific, Waltham, MA, USA) at a wavelength of 665 nm. Three parallel tests were performed for each group. The removal rates of the three pollutants were calculated using Equation (1)

$$R = \frac{C_0 - C_e}{C_0} \times 100 \tag{1}$$

where C_0 (mg/L) is the initial concentration, and C_e (mg/L) is the equilibrium of the mixed solution.

2.5. Adsorption Kinetics Experiment

In the adsorption kinetics experiment, 100 mg of the AE-CMIW hydrochar sample was added to separate vials containing 40 mL of a 100 mg/L mixed solution. The vials oscillated at 40 rpm/min for 1, 2, 4, 6, 8, 10, 12, 18, and 24 h. The adsorption capacities for MB and metal ions were measured and calculated. Pseudo-first-order (PFO) and pseudo-second-order (PSO) kinetic models were employed to describe the adsorption behavior of AE-CMIW hydrochar [39]. The PFO and PSO models are expressed by Equations (2) and (3), respectively:

$$qt = qe(1 - \exp(-k1t)) \tag{2}$$

$$qt = \frac{k2qe^2 t}{1 + k2qet} \tag{3}$$

where q_e (mg/g) is the adsorption capacity at equilibrium, and q_t (mg/g) is the adsorption capacity at time t (h). k_1 (h^{-1}) and k_2 (h^{-1}) represent the kinetic adsorption rate constants of the PFO and PSO models, respectively.

3. Results and Discussion

3.1. Pore Structure Analyses

The pore structure analyses of the AE-CMIW hydrochar are presented in Table 1. Studies have shown that the S_{BET}, pore size and pore volume are the indicators of the available absorption inter-phase on the hydrochar, and they have a significant influence on the adsorptive application and solute binding capacity of hydrochar [15,41]. The S_{BET} and total pore volume (V_{total}) of the AE-CMIW hydrochar increased significantly ($p < 0.05$) compared with those of CMIW. The S_{BET} increased by approximately two to three times. The total pore volume increased approximately three to four times. The average pore diameter of hydrochar also increased compared to that of CMIW. The increases in S_{BET} and V_{total} indicated that these hydrochars could be used as adsorbents to provide more adsorption sites. With the increase in HTC temperature, the S_{BET} decreased, and the V_{total} and average pore diameter increased. With the increase in HTC duration, the S_{BET} decreased, and the average pore diameter increased. The static adsorption isotherms and pore size distributions of the AE-CMIW hydrochar are shown in Figure 1. The static adsorption and desorption isotherms were H3-type hysteresis rings, which showed that the hydrochars were typical mesoporous material, and the overall hysteresis ring was relatively narrow, indicating that the pore structure of hydrochar was the accumulation of flake particles, with layered structure characteristics [39]. The pore size distribution diagram showed that the pore size of hydrochar was mostly concentrated in the range of 2~10 nm. With the increase in HTC temperature, the S_{BET} decreased. The S_{BET} of AEHTC 200-2-3 was 53.47 m^2/g, while that of AEHTC 260-2-3 was 42.61 m^2/g. When HTC temperature rose, CMIW had more violent depolymerization reaction. Therefore, a large number of small molecules accumulated and precipitated on the surface of hydrochars, resulting in a part of the pore structure being blocked.

Table 1. Porous structure properties of CMIW and AE-CMIW hydrochar.

Sample	S_{BET} (m^2/g)	V_{total} (cm^3/g)	Average Pore Diameter (nm)
CMIW	19.121 ± 0.05 [f]	0.055 ± 0.01 [c]	11.445 ± 0.11 [bc]
AEHTC 200-2-3	53.466 ± 0.04 [c]	0.213 ± 0.03 [b]	14.729 ± 0.15 [d]
AEHTC 230-1-3	44.633 ± 0.06 [e]	0.225 ± 0.02 [b]	18.437 ± 0.13 [bc]
AEHTC 230-2-3	43.381 ± 0.07 [b]	0.226 ± 0.02 [b]	18.977 ± 0.10 [bc]
AEHTC 230-4-3	42.023 ± 0.05 [a]	0.225 ± 0.04 [a]	19.527 ± 0.09 [a]
AEHTC 260-2-3	42.611 ± 0.06 [d]	0.236 ± 0.05 [a]	20.056 ± 0.12 [bc]

[a–f] Means followed by different superscripts in the same column are significantly different at $p < 0.05$. [a–f] Means followed by same superscripts in the same column are not significantly different at $p > 0.05$.

3.2. Surface Microstructures Analyses

Scanning electron microscopy (SEM) images of the CMIW and AE-CMIW hydrochars are shown in Figure 2. The SEM image of CMIW showed a smooth surface. The AE-CMIW hydrochar had rough and uneven surfaces with a coral-like morphology and abundant pores. When the HTC temperature rose, the pressure in the reactor increased accordingly. Then, the organic matter of CMIW would undergo a series of degradation and depolymerization reactions. Part of the organic matter would be converted into gas or tar and precipitated, part would be dissolved in the liquid phase in the form of small organic molecules, and the rest would reaggregate and precipitate on the surface of solid materials. With the assistance of AE, the hydrochars had loose porous surface structure and rough surface morphology [42,43]. After the reactions of HTC coupled with alkaline etching, AE-CMIW hydrochar formed new morphological feature compared with CMIW. The number of pores on the surface of hydrochar increased, and some spherical particles were formed. These pore structures could provide attachment sites for heavy metals or organic dyes. Therefore, these morphological characteristics are conducive to the adsorption of AE-CMIW hydrochar [42].

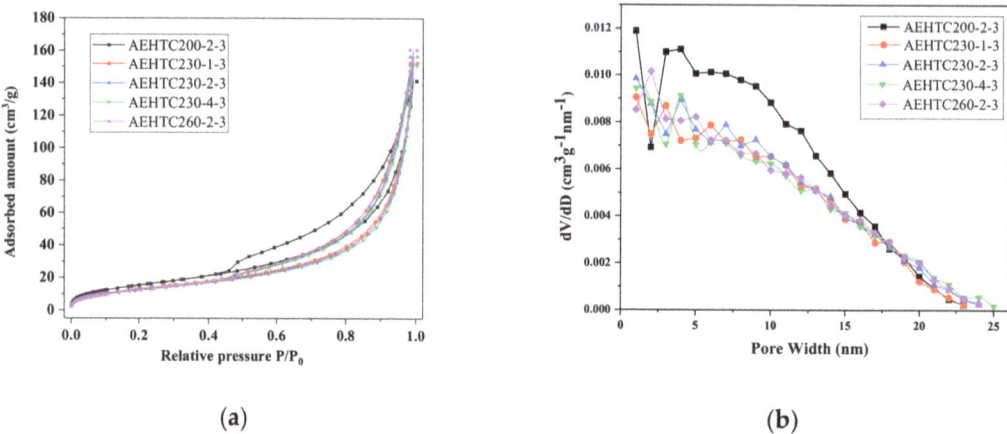

Figure 1. N_2 adsorption–desorption isotherms (a) and pore size distributions (b) of AE-CMIW hydrochar.

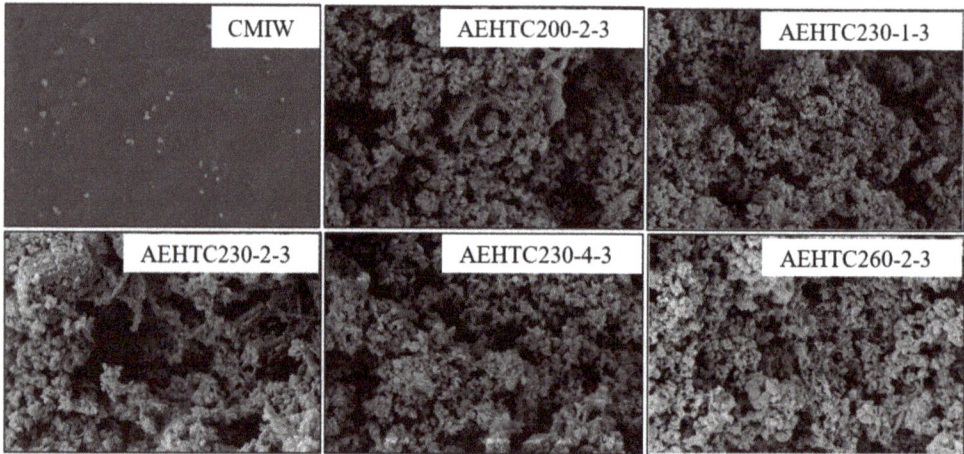

Figure 2. Contrast analysis diagrams of surface microstructures of CMIW and AE-CMIW hydrochar.

3.3. Functional Groups Analyses

The surface functional groups of AE-CMIW hydrochar were determined by the FTIR (Figure 3). During the HTC reaction process, hydronium ions could effectively promote the hydrolysis of macromolecular materials to generate various low polymers, which would further decompose into acids, aldehydes and phenolic substances, forming various functional groups on the surface of hydrochar materials. The peaks at 3430 and 1030 cm^{-1} were attributed to the hydroxyl groups (-OH) in the hydroxyl or carboxyl groups [44,45]. This peak was commonly found in alcohols and acids [46] and could also be characteristic of cellulose [47]. The characteristic peak at 1626 cm^{-1} was associated with C=O stretching vibration in the carboxyl or carbonyl groups [48]. These peaks were caused by HTC and AE treatment. The fluctuation of 1407 cm^{-1} was related to O-CH$_3$ in lignin [49]. These results showed that AE-CMIW hydrochar had many oxygen-containing functional groups. They were conducive to the adsorption property of AE-CMIW hydrochar.

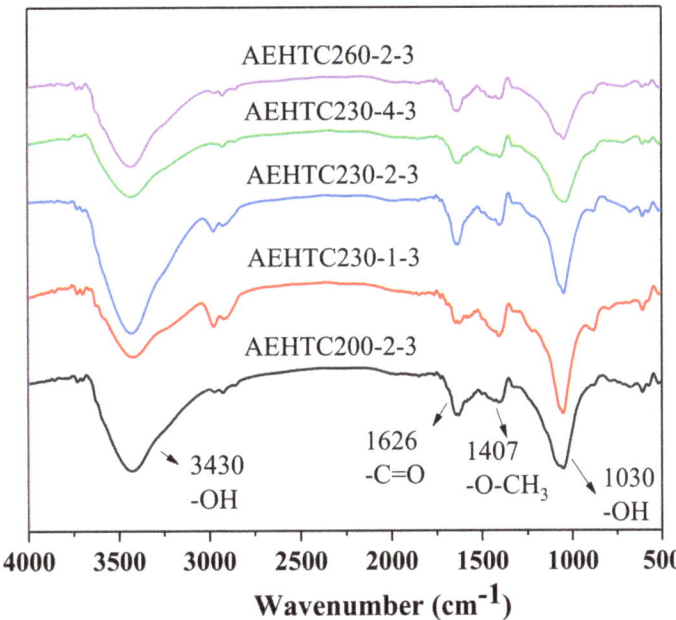

Figure 3. FTIR spectra of AE-CMIW hydrochar.

3.4. Adsorption Properties Analyses of AE-CMIW Hydrochar for Multiple Pollutants

The CMIW was treated by the technique of HTC coupled with AE. The effects of the AE-CMIW hydrochar on the removal rates of MB, Pb^{2+}, and Cd^{2+} in the mixed solution are shown in Figure 4. The AE-CMIW hydrochar exhibited adsorption capacities for MB, Pb^{2+}, and Cd^{2+} in the mixed solution. The surface of the AE-CMIW hydrochar was rough and had abundant pore structures (SEM images), which provided many sites for MB, Pb^{2+}, and Cd^{2+} to attach to the hydrochar. The S_{BET} data of AE-CMIW hydrochar and the removal rate of MB, Pb^{2+} and Cd^{2+} were analyzed and correlated. This has shown that the removal rate of MB was greatly affected by the S_{BET} of AE-CMIW hydrochar, and they showed a relatively positive correlation. However, the removal rates of Pb^{2+} and Cd^{2+} were less affected by the changes of S_{BET}. Other factors may mainly affect the adsorption of Pb^{2+} and Cd^{2+} by AE-CMIW hydrochar, for example, electrostatic attraction and functional group interactions. FTIR analysis revealed that the surface of the AE-CMIW hydrochar contained abundant oxygen-containing functional groups [50]. These functional groups could form complexes with MB and form coordination with heavy metal ions. In addition, there were π–π stacking interactions, electrostatic interactions, ion exchange, and physical functions between hydrochar and MB [51]. In contrast, the possible mechanisms of Pb^{2+} and Cd^{2+} adsorption by hydrochar involved electrostatic attraction, metal–π interactions, ion exchange, precipitation, and complexation [52,53]. In addition, the electrostatic attraction was the main mechanism of Pb^{2+} adsorption by hydrochar. Taken together, these factors lead to adsorption between the AE-CMIW hydrochar and MB, Pb^{2+}, and Cd^{2+}.

In the mixed solution, the removal rates of MB, Pb^{2+}, and Cd^{2+} by the AE-CMIW hydrochar were significantly different ($p < 0.05$). The removal rates of Pb^{2+} by the AE-CMIW hydrochar were the highest, reaching more than 98%, and their removal rates were stable in the different HTC coupled with AE conditions. With the increase in HTC temperature, the removal rates of MB decreased gradually, and the removal rates of Cd^{2+} decreased first and then increased. With the increase in HTC time, the removal rates of MB increased first and then decreased, and the removal rates of Cd^{2+} decreased first and then increased. The removal rates of MB reach 60–80%. In contrast, the removal rates of Cd^{2+} by the AE-CMIW hydrochar were the lowest (20–57%) among the three. In addition, the removal rates of Cd^{2+} by different hydrochars were significantly different ($p < 0.05$). These results indicated varied adsorption strengths of MB, Pb^{2+} and Cd^{2+} by the AE-CMIW hydrochar. That is, the interaction force was different, and hydrochar materials have certain selectivity for the adsorption of multiple pollutants. We explained the reasons for the adsorption differences in detail in Sections 3.5 and 3.6.

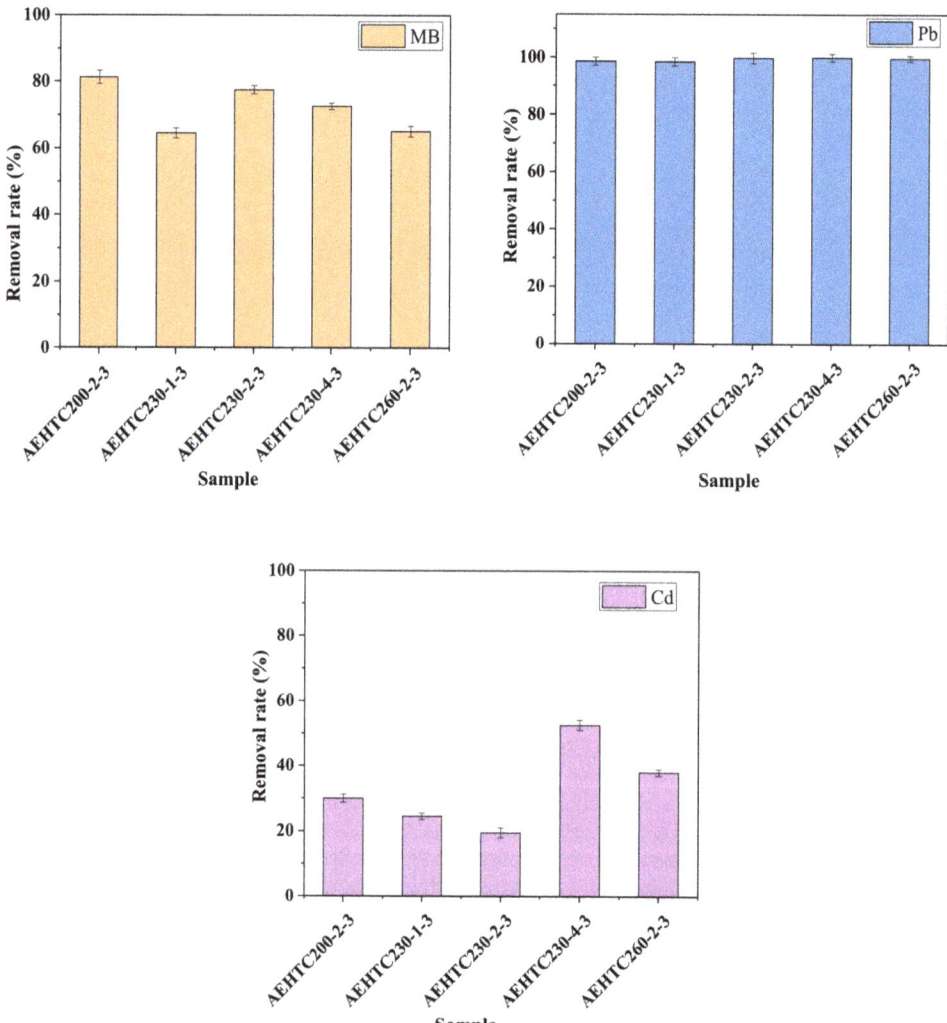

Figure 4. Effects of AE-CMIW hydrochar on the removal rates of MB, Pb^{2+}, Cd^{2+} in mixed solution.

3.5. Adsorption Difference Analyses of AE-CMIW Hydrochar for Multiple Pollutants

The adsorption strength of adsorbents is affected by many factors, among which electrostatic attraction, physical adsorption, and functional group interactions (including ion exchange, hydrogen bonding, and π–π stacking interactions) are the main factors [54–56]. The pH significantly affects the removal efficiency of adsorbents for dyes and metal ions [57,58]. The pH of the isoelectric point (pH_{IEP}) is the pH at which the net surface charge of the solid material is zero [59]. The pH_{IEP} of the AE-CMIW hydrochar is shown in Figure 5. The pH_{IEP} of AE-CMIW hydrochar was approximately pH 3.63 ($p < 0.05$). When pH < pH_{IEP}, the surface of the hydrochar was protonated and positively charged. While pH > pH_{IEP}, the surface of the hydrochar was negatively charged, which was conducive to the adsorption of cations [60]. In other words, the AE-CMIW hydrochar was negatively charged at the experimental pH (~6.5) of the adsorbate solutions. Thus, the AE-CMIW hydrochar could attract MB, Pb^{2+} and Cd^{2+} with a positive charge by electrostatic interactions. For metal ions, the electronegativities of Pb^{2+} and Cd^{2+} were 1.87 and 1.69, respectively; thus, the adsorption of Pb^{2+} by the AE-CMIW hydrochar was stronger than that of Cd^{2+}. In addition, the hydrated ionic radii of the metal ions affected the rate and amount of metal ions entering the pores of the AE-CMIW hydrochar. The results showed that the hydrated ionic radii of Pb^{2+} and Cd^{2+} were 0.401 and 0.426 nm, respectively. Cations with small hydrated ionic radii are more likely to penetrate the pores and channels of the hydrochar. Therefore, the adsorption amount and speed of Pb^{2+} by the AE-CMIW hydrochar were higher than that of Cd^{2+}. Pb^{2+} and Cd^{2+} competed for the adsorption sites on hydrochar, which was mainly determined by the different properties of these metal cations. Metal–π interactions, electrostatic attraction, and precipitation are the mechanisms of Pb^{2+} adsorption by AE-CMIW hydrochar [61–63]. The mechanisms of Cd^{2+} adsorption by AE-CMIW hydrochar involve electrostatic attraction and ion exchange [64,65], indicating that the adsorption between Pb^{2+} and AE-CMIW hydrochar was due to both chemical and physical interactions, whereas it was mainly due to the chemical interaction between Cd^{2+} and AE-CMIW hydrochar. Therefore, the adsorption rate and strength of Pb^{2+} and Cd^{2+} by AE-CMIW hydrochar varied, and the removal rates of Pb^{2+} by the AE-CMIW hydrochar were higher than those of Cd^{2+}.

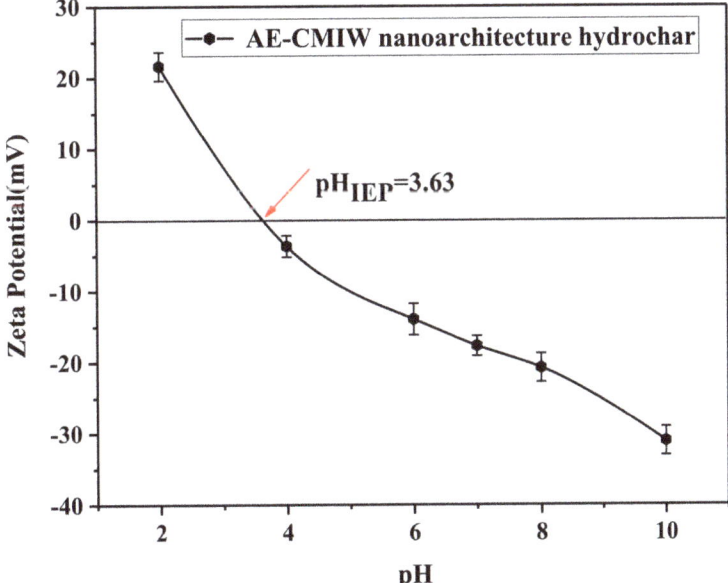

Figure 5. Plot for the pH_{IEP} determination of AE-CMIW hydrochar.

The distribution of main elements (SEM-EDS images) on the surface and near-surface of the AE-CMIW hydrochar after the adsorption reaction is shown in Table 2. The contents of Pb^{2+} on the surface and near-surface of AE-CMIW hydrochar were significantly higher than those of Cd^{2+} ($p < 0.05$) (Table 2). MB was an organic substance, which could interact with the functional groups of AE-CMIW hydrochar in addition to electrostatic attraction, indicating that MB could combine with AE-CMIW hydrochar in various ways to produce adsorption. In this experiment, electrostatic interaction was the main factor for the adsorption of MB, Pb^{2+}, and Cd^{2+} by the AE-CMIW hydrochar, followed by the interaction between the functional groups [66]. Therefore, the adsorption order of the three pollutants by the AE-CMIW hydrochar was Pb^{2+} > MB > Cd^{2+}.

Table 2. Distribution of main elements on the surface and near-surface of AE-CMIW hydrochar after the adsorption reaction.

Element Contents (w.t.%)	C	O	Fe	Cd	Pb
AEHTC 200-2-3	11.87 ± 0.01 [a]	31.31 ± 0.02 [a]	35.92 ± 0.01 [a]	3.20 ± 0.07 [a]	17.69 ± 0.02 [a]
AEHTC 230-1-3	5.03 ± 0.05 [b]	22.69 ± 0.03 [b]	42.38 ± 0.04 [b]	2.54 ± 0.04 [b]	27.37 ± 0.05 [b]
AEHTC 230-2-3	5.95 ± 0.02 [a]	24.56 ± 0.05 [a]	41.44 ± 0.06 [a]	2.17 ± 0.02 [a]	25.89 ± 0.04 [a]
AEHTC 230-4-3	6.44 ± 0.01 [b]	24.73 ± 0.03 [b]	44.78 ± 0.01 [b]	5.21 ± 0.01 [b]	18.84 ± 0.05 [b]
AEHTC 260-2-3	7.90 ± 0.01 [a]	27.82 ± 0.02 [a]	39.13 ± 0.02 [a]	4.59 ± 0.01 [a]	20.55 ± 0.02 [a]

The figures are the mean value ± standard deviation in the table. [a–b] Means followed by different superscripts in the same column are significantly different at $p < 0.05$. [a–b] Means followed by same superscripts in the same column are not significantly different at $p > 0.05$.

3.6. Adsorption Kinetics Analyses

The adsorption of MB, Pb^{2+}, and Cd^{2+} by the AE-CMIW hydrochar in the mixed solution was complex. The MB, Pb^{2+}, and Cd^{2+} kinetics curves are illustrated in Figure 6. The fitting parameters of MB, Pb^{2+}, and Cd^{2+} adsorption kinetics are shown in Table 3. The PFO kinetic model mainly describes the simple physical adsorption or diffusion of boundary-layer particles. The PSO kinetic model mainly describes chemical surface adsorption, which involves electron sharing or transfer between the adsorbent and adsorbate [67].

By comparing the adsorption kinetics parameters of MB, Pb^{2+} and Cd^{2+} on AE-CMIW mixed solution, the following results can be obtained. The adsorption of MB on the AE-CMIW hydrochar was consistent with the PSO kinetic model (Figure 6a,b, Table 3), indicating that the adsorption process was multiple composite adsorptions dominated by chemical adsorption. The adsorption rate of MB by the AE-CMIW hydrochar was high for the first 2.5 h, and then, the adsorption rate decreased. When the adsorption time of MB was 5 h, the adsorption capacity reached 80% of the saturated adsorption capacity, and the highest equilibrium adsorption capacity was 31.1 mg/g (AEHTC 200-2-3 hydrochar). Both the PFO kinetic model and the PSO kinetic model could well fit the adsorption behavior of Pb^{2+} by the AE-CMIW hydrochar (Figure 6c,d, Table 3), which indicated that the adsorption of Pb^{2+} by the AE-CMIW hydrochar was the chemical and physical adsorptions. In the mixed solution, Pb^{2+} was the least affected by the other adsorbates, and its adsorption rates by the AE-CMIW hydrochar were the fastest. Thus, the removal rates of Pb^{2+} by the AE-CMIW hydrochar were the highest. The adsorption of Pb^{2+} reached equilibrium for less than 2.5 h, and the equilibrium adsorption capacity was approximately 43 mg/g. The PSO kinetic model could more accurately describe the adsorption of Cd^{2+} by the AE-CMIW hydrochar (Figure 6e,f, Table 3). The adsorption rates of Cd^{2+} by AE-CMIW hydrochar were significantly lower than those of Pb^{2+}, and their adsorption equilibriums were reached at approximately 20 h. The equilibrium adsorption capacities of Cd^{2+} by different AE-CMIW hydrochars were significantly lower than those of Pb^{2+}. Therefore, the removal rates of Cd^{2+} by AE-CMIW hydrochar were significantly lower than those of Pb^{2+}.

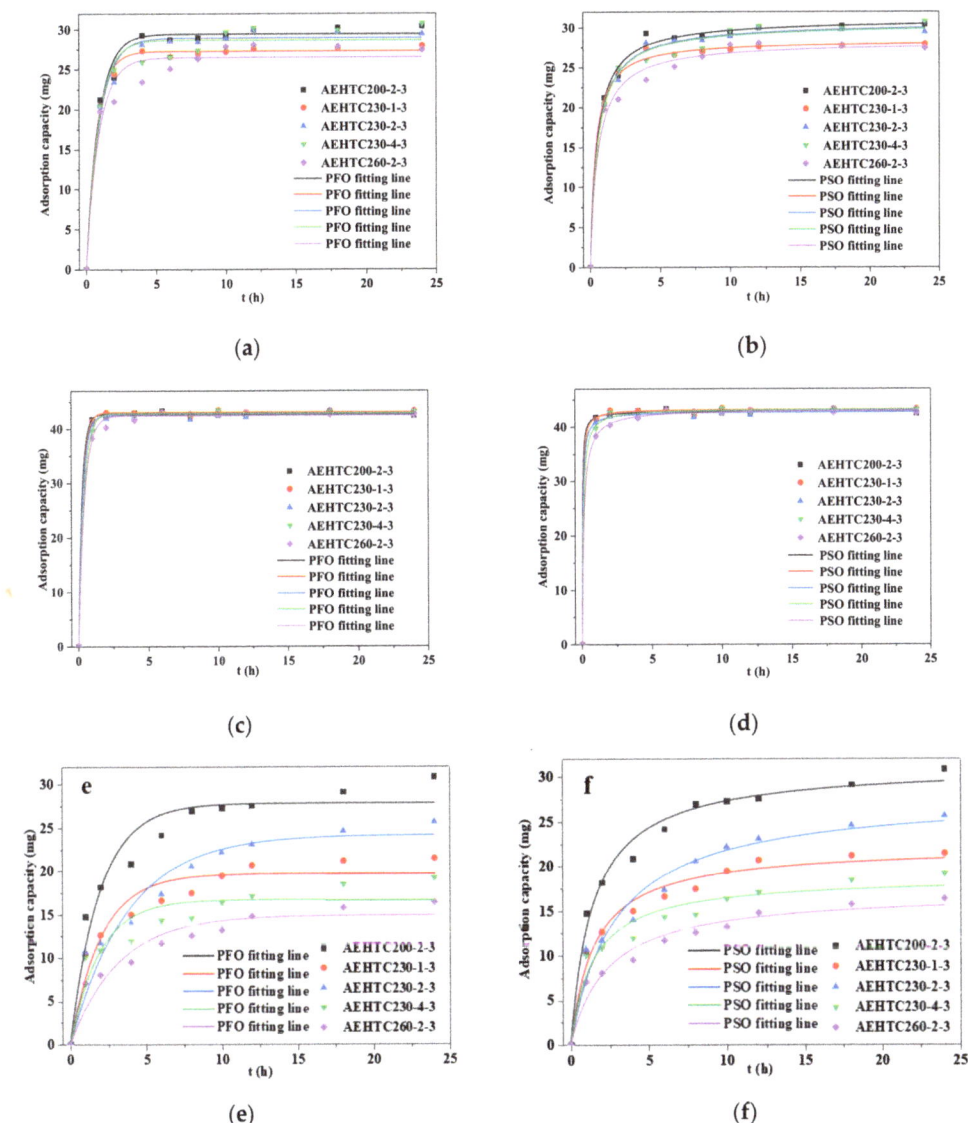

Figure 6. Adsorption experimental (symbols) and fitting (lines) kinetics curves of MB (**a**,**b**), Pb^{2+} (**c**,**d**), Cd^{2+} (**e**,**f**) on AE-CMIW hydrochar in mixed solution.

Table 3. Adsorption kinetics parameters of MB, Pb^{2+}, Cd^{2+} on AE-CMIW hydrochar in mixed solution.

Sample		Pseudo-First-Order			Pseudo-Second-Order		
		q_e (mg/g)	k_1 (h^{-1})	R^2	q_e (mg/g)	k_2 (g/mg·h)	R^2
AEHTC 200-2-3	MB	29.471	1.109	0.986	31.076	0.067	0.995
	Pb^{2+}	42.738	3.772	0.999	42.922	0.858	0.999
	Cd^{2+}	27.823	0.510	0.939	31.207	0.022	0.983

Table 3. Cont.

Sample		Pseudo-First-Order			Pseudo-Second-Order		
		q_e (mg/g)	k_1 (h^{-1})	R^2	q_e (mg/g)	k_2 (g/mg · h)	R^2
AEHTC 230-1-3	MB	27.232	1.328	0.996	28.380	0.100	0.997
	Pb^{2+}	43.028	3.321	0.999	43.255	0.622	0.999
	Cd^{2+}	19.661	0.493	0.924	22.122	0.030	0.975
AEHTC 230-2-3	MB	28.949	1.056	0.987	30.622	0.063	0.996
	Pb^{2+}	42.465	3.238	0.998	42.786	0.490	0.999
	Cd^{2+}	24.169	0.267	0.924	27.949	0.013	0.959
AEHTC 230-4-3	MB	28.638	1.149	0.966	30.470	0.104	0.988
	Pb^{2+}	42.860	2.626	0.999	43.315	0.299	0.999
	Cd^{2+}	16.662	0.524	0.866	18.760	0.038	0.941
AEHTC 260-2-3	MB	26.482	1.097	0.953	28.204	0.066	0.985
	Pb^{2+}	42.421	2.253	0.997	43.223	0.178	0.999
	Cd^{2+}	14.889	0.888	0.907	17.059	0.025	0.959

4. Conclusions

This study proposed a method of coupling HTC with AE for treating CMIW. The adsorption characteristics of metal ions and organic dyes by the AE-CMIW hydrochar in the mixed solution were investigated. The reasons for adsorption differences and the adsorption kinetics were analyzed. The conditions of coupling HTC with AE had significant effects on the removal rates of MB and Cd^{2+} ($p < 0.05$), while the removal rates of Pb^{2+} were stable under different conditions. The removal rates of Pb^{2+}, MB, and Cd^{2+} by AE-CMIW hydrochar were ≥98%, 60–80%, 20–57%, respectively.

According to the pH$_{IEP}$ (pH 3.63), the AE-CMIW hydrochar was negatively charged at experimental pH (~6.5), and then, these hydrochar could attract MB, Pb^{2+} and Cd^{2+} with a positive charge by electrostatic interactions. The electronegativity of Pb^{2+} (1.87) was higher than that of Cd^{2+} (1.69), indicating that the adsorption of Pb^{2+} by the AE-CMIW hydrochar was stronger than that of Cd^{2+}. The hydrated ionic radii of Pb^{2+} (0.401 nm) was smaller than that of Cd^{2+} (0.426 nm), indicating that the adsorption amount and speed of Pb^{2+} were higher than those of Cd^{2+}. The AE-CMIW hydrochar had abundant pore structures and functional groups, providing physical and chemical binding sites for MB, Pb^{2+} and Cd^{2+} on the absorbents to achieve pollutant removal. The adsorption kinetics analyses showed that the adsorptions of Cd^{2+} and MB by AE-CMIW hydrochar were consistent with the PSO kinetic model, indicating that their adsorption processes were multiple composite adsorptions dominated by chemical adsorption. The adsorption of Pb^{2+} conformed to both the PFO and PSO kinetic models, indicating that it was dominated by the chemical and physical adsorptions. Therefore, the removal rate of Pb^{2+} by the AE-CMIW hydrochar was the highest.

Author Contributions: Conceptualization, X.Z. and W.W.; methodology, S.L., Q.Q. and G.C.; software, S.L.; validation, X.Z. and S.L.; formal analysis, X.Z. and W.W.; investigation, X.Z., S.L., Q.Q. and G.C.; resources, W.W.; data curation, X.Z.; writing—original draft preparation, X.Z., S.L. and W.W.; writing—review and editing, X.Z; visualization, S.L.; supervision, X.Z. and W.W.; project administration, X.Z.; funding acquisition, X.Z and W.W. All authors have read and agreed to the published version of the manuscript.

Funding: This work was supported by the National Natural Science Foundation of China (51976110), the Postdoctoral Innovation Project of Shandong Province (202101002), and the Major Science and Technology Innovation Project of Tai'an City (2021ZDZX031).

Institutional Review Board Statement: Not applicable.

Informed Consent Statement: Not applicable.

Data Availability Statement: Not applicable.

Conflicts of Interest: The authors declare no conflict of interest.

References

1. Wang, R.K.; Jin, Q.Z.; Ye, X.M.; Lei, H.Y.; Jia, J.D.; Zhao, Z.H. Effect of process wastewater recycling on the chemical evolution and formation mechanism of hydrochar from herbaceous biomass during hydrothermal carbonization. *J. Clean. Prod.* **2020**, *277*, 123281. [CrossRef]
2. Zhu, W.Q.; Wang, L.J.; Liang, p.; He, Z.L.; Yan, G.H.; Liang, P.; He, Z.L.; Yan, G.H. Analysis of the status quo and future prospects of Chinese medicinal resources sustainable development. *World Chin. Med.* **2018**, *13*, 1752–1755.
3. Chen, Z.Y.; Chen, H.S.; Wu, X.Y.; Zhang, J.H.; Evrendilek, D.E.; Liu, J.Y. Temperatureand heating rate-dependent pyrolysis mechanisms and emissions of Chinese medicine residues and numerical reconstruction and optimization of their nonlinear dynamics. *Renew. Energ.* **2021**, *164*, 1408–1423. [CrossRef]
4. Shen, Q.B.; Wang, Z.Y.; Yu, Q.; Cheng, Y.; Liu, Z.D.; Zhang, T.P. Removal of tetracycline from an aqueous solution using manganese dioxide modified biochar derived from Chinese herbal medicine residues. *Environ. Res.* **2020**, *183*, 109195. [CrossRef] [PubMed]
5. Li, K.; Zhang, D.; Niu, X.; Guo, H.; Yu, Y.; Tang, Z.; Lin, Z.; Fu, M. Insights into CO_2 adsorption on KOH-activated biochars derived from the mixed sewage sludge and pine sawdust. *Sci. Total Environ.* **2022**, *826*, 154133. [CrossRef]
6. Ministry of Ecology and Environment of the People's Republic of China. The Law of the People's Republic of China on the Prevention and Control of Environmental Pollution by Solid Waste. 2020. Available online: https://www.mee.gov.cn/ywgz/fgbz/fl/202004/t20200430_777580.shtml (accessed on 30 April 2020).
7. Zhamiyeva, R.; Sultanbekova, G.; Balgimbekova, G.; Mussin, K.; Abzalbekova, M.; Kozhanov, M. Problems of the effectiveness of the implementation of international agreements in the field of waste management: The study of the experience of Kazakhstan in the context of the applicability of European legal practices. *Int. Environ. Agreem.* **2022**, *22*, 177–199. [CrossRef]
8. Liu, L.M.; Zhai, Y.B.; Liu, X.M.; Liu, X.p.; Wang, Z.X.; Zhu, Y.P.; Wang, Z.X.; Zhu, Y.; Xu, M. Features and mechanisms of sewage sludge hydrothermal carbonization aqueous phase after ferrous/persulfate-based process: The selective effect of oxidation and coagulation. *J. Clean. Prod.* **2022**, *366*, 132831. [CrossRef]
9. Zhai, Y.; Peng, C.; Xu, B.; Wang, T.; Li, C.; Zeng, G.; Zhu, Y. Hydrothermal carbonisation of sewage sludge for char production with different waste biomass: Effects of reaction temperature and energy recycling. *Energy* **2017**, *127*, 167–174. [CrossRef]
10. Ponnusamy, V.K.; Nagappan, S.; Bhosale, R.R.; Lay, C.H.; Duc Nguyen, D.; Pugazhendhi, A.; Chang, S.W.; Kumar, G. Review on sustainable production of biochar through hydrothermal liquefaction: Physico-chemical properties and applications. *Bioresour. Technol.* **2020**, *310*, 123414. [CrossRef]
11. Zhang, X.Y.; Qin, Q.Y.; Liu, X.; Wang, W.L. Effects of stepwise microwave synergistic process water recirculation during hydrothermal carbonization on properties of wheat straw. *Front. Energy Res.* **2022**, *10*, 846752. [CrossRef]
12. Zhang, X.Y.; Qin, Q.Y.; Sun, X.; Wang, W.L. Hydrothermal Treatment: An Efficient Food Waste Disposal Technology. *Front. Nutr.* **2022**, *9*, 986705. [CrossRef] [PubMed]
13. Panequea, M.; De la Rosaa, J.M.; Kernb, J.; Rezac, M.T.; Knicker, H. Hydrothermal carbonization and pyrolysis of sewage sludges: What happen to carbon and nitrogen? *J. Anal. Appl. Pyrolysis* **2017**, *128*, 314–323. [CrossRef]
14. Peng, C.; Zhai, Y.B.; Zhu, Y.; Xu, B.B.; Wang, T.F.; Li, C.T.; Zeng, G.M. Production of char from sewage sludge employing hydrothermal carbonization: Char properties, combustion behavior and thermal characteristics. *Fuel* **2016**, *176*, 110–118. [CrossRef]
15. Ighalo, J.O.; Rangabhashiyam, S.; Dulta, K.; Umeh, C.T.; Iwuozor, K.O.; Aniagor, C.O.; Eshiemogie, S.O.; Iwuchukwu, F.U.; Igwegbe, C.A. Recent advances in hydrochar application for the adsorptive removal of wastewater pollutants. *Chem. Eng. Res. Des.* **2022**, *184*, 419–456. [CrossRef]
16. Ding, Z.; Zhang, L.; Mo, H.; Chen, Y.; Hu, X. Microwaveassisted catalytic hydrothermal carbonization of Laminaria japonica for hydrochars catalyzed and activated by potassium compounds. *Bioresour. Technol.* **2021**, *341*, 125835. [CrossRef] [PubMed]
17. Khoshbouy, R.; Takahashi, F.; Yoshikawa, K. Preparation of high surface area sludge-based activated hydrochar via hydrothermal carbonization and application in the removal of basic dye. *Environ. Res.* **2019**, *175*, 457–467. [CrossRef]
18. Liu, Z.Y.; Wang, Z.H.; Chen, H.G.; Cai, T.; Liu, Z.D. Hydrochar and pyrochar for sorption of pollutants in wastewater and exhaust gas: A critical review. *Environ. Pollut.* **2021**, *268*, 115910. [CrossRef]
19. Ge, S.; Foong, S.Y.; Ma, N.L.; Liew, R.K.; Mahari, W.A.W.; Xia, C.; Yek, p.N.Y.; Peng, W.; Nam, W.L.; Lim, X.Y.; et al. Vacuum pyrolysis incorporating microwave heating and base mixture modification: An integrated approach to transform biowaste into eco-friendly bioenergy products. *Renew. Sustain. Energy Rev.* **2020**, *127*, 109871. [CrossRef]
20. Verma, M.; Lee, I.Y.; Hong, Y.M.; Kumar, V.; Kim, H. Multifunctional β-Cyclodextrin-EDTA-Chitosan polymer adsorbent synthesis for simultaneous removal of heavy metals and organic dyes from wastewater. *Environ. Pollut.* **2022**, *292*, 118447. [CrossRef]
21. Hu, L.; Yang, Z.; Cui, L.; Li, Y.; Ngo, H.H.; Wang, Y.; Wei, Q.; Ma, H.; Yan, L.; Du, B. Fabrication of hyperbranched polyamine functionalized graphene for high-efficiency removal of Pb(II) and methylene blue. *Chem. Eng. J.* **2016**, *287*, 545–556. [CrossRef]
22. Deng, J.H.; Zhang, G.M.; Gong, J.L.; Niu, Q.Y.; Liang, J. Simultaneous removal of Cd (II) and ionic dyes from aqueous solution using magnetic graphene oxide nanocomposite as an adsorbent. *Chem. Eng. J.* **2013**, *226*, 189–200. [CrossRef]
23. Qin, X.M.; Bai, L.; Tan, Y.Z.; Li, L.; Song, F.; Wang, Y.Z. β-Cyclodextrin-crosslinked polymeric adsorbent for simultaneous removal and stepwise recovery of organic dyes and heavy metal ions: Fabrication, performance and mechanisms. *Chem. Eng. J.* **2019**, *372*, 1007–1018. [CrossRef]
24. Liu, M.; Wang, Z.; Zong, S.; Chen, H.; Zhu, D.; Wu, L.; Hu, G.; Cui, Y. SERS detection and removal of mercury(II)/silver(I) using oligonucleotide-Functionalized core/shell magnetic silica sphere@Au nanoparticles. *ACS Appl. Mater. Interfaces* **2014**, *6*, 7371–7379. [CrossRef] [PubMed]

25. Kumar, K.; Kumar, V.; Vlaskin, M.S. Environmental Technology & Innovation Hydropyrolysis of freshwater macroalgal bloom for bio-oil and biochar production: Kinetics and isotherm for removal of multiple heavy metals. *Environ. Technol. Innov.* **2021**, *22*, 101440.
26. Lakard, S.; Magnenet, C.; Mokhter, M.A.; Euvrard, M.; Buron, C.C.; Lakard, B. Retention of Cu(II) and Ni(II) ions by filtration through polymer-modified membranes. *Separ. Purif. Technol.* **2015**, *149*, 1–8. [CrossRef]
27. Luo, S.; Qin, F.; Zhao, H.; Liu, Y.; Chen, R. Fabrication uniform hollow Bi_2S_3 nanospheres via Kirkendall effect for photocatalytic reduction of Cr (VI) in electroplating industry wastewater. *J. Hazard. Mater.* **2017**, *340*, 253–262. [CrossRef]
28. Verma, M.; Mitan, M.; Kim, H.; Vaya, D. Efficient photocatalytic degradation of Malachite green dye using facilely synthesized cobalt oxide nanomaterials using citric acid and oleic acid. *J. Phys. Chem. Solid.* **2021**, *155*, 110125. [CrossRef]
29. Zhang, X.; Qian, J.; Pan, B. Fabrication of novel magnetic nanoparticles of multifunctionality for water decontamination. *Environ. Sci. Technol.* **2016**, *50*, 881–889. [CrossRef]
30. Guo, D.M.; An, Q.D.; Xiao, Z.Y.; Zhai, S.R.; Yang, D.J. Efficient removal of Pb(II), Cr(VI) and organic dyes by polydopamine modified chitosan aerogels. *Carbohydr. Polym.* **2018**, *202*, 306–314. [CrossRef]
31. Hadi, p.; Guo, J.; Barford, J.; Mckay, G. Multilayer dye adsorption in activated carbons-facile approach to exploit vacant sites and interlayer charge interaction. *Environ. Sci. Technol.* **2016**, *50*, 5041–5049. [CrossRef]
32. Li, C.; Yan, Y.; Zhang, Q.; Zhang, Z.; Huang, L.; Zhang, J.; Xiong, Y.; Tan, S. Adsorption of Cd^{2+} and Ni^{2+} from aqueous single-metal solutions on graphene oxide-chitosan-poly (vinyl alcohol) hydrogels. *Langmuir* **2019**, *35*, 4481–4490. [CrossRef] [PubMed]
33. Pap, S.; Kirk, C.; Bremner, B.; Turk Sekulic, M.; Shearer, L.; Gibb, S.W.; Taggart, M.A. Low-cost chitosan-calcite adsorbent development for potential phosphate removal and recovery from wastewater effluent. *Water Res.* **2020**, *173*, 115573. [CrossRef]
34. Verma, M.; Singh, K.P.; Kumar, A. Reactive magnetron sputtering based synthesis of WO_3 nanoparticles and their use for the photocatalytic degradation of dyes. *Solid State Sci.* **2020**, *99*, 105847. [CrossRef]
35. Hou, Z.; Wen, Z.; Wang, D.; Wang, J.; François-Xavier, C.p.; Wintgens, T.P.; Wintgens, T. Bipolar jet electrospinning bi-functional nanofibrous membrane for simultaneous and sequential filtration of Cd^{2+} and BPA from water: Competition and synergistic effect. *Chem. Eng. J.* **2018**, *332*, 118–130. [CrossRef]
36. Zhang, X.Y.; Gao, B.; Zhao, S.N.; Wu, p.F.; Han, L.J.; Liu, X.; Wu, P.F.; Han, L.J.; Liu, X. Optimization of a "coal-like" pelletization technique based on the sustainable biomass fuel of hydrothermal carbonization of wheat straw. *J. Clean. Prod.* **2020**, *242*, 118426. [CrossRef]
37. Zhang, X.Y.; Li, Y.F.; Wang, M.Y.; Han, L.J.; Liu, X. Effects of hydrothermal carbonization conditions on the combustion and kinetics of wheat straw hydrochar pellets and efficiency improvement analyses. *Energ. Fuel.* **2020**, *34*, 587–598. [CrossRef]
38. Zhang, X.Y.; Peng, W.Q.; Han, L.J.; Xiao, W.H.; Liu, X. Effects of different pretreatments on compression molding of wheat straw and mechanism analysis. *Bioresour. Technol.* **2018**, *251*, 210–217. [CrossRef]
39. Zhang, X.Y.; Liu, S.S.; Wang, M.M.; Ma, X.L.; Sun, X.; Liu, X.; Wang, L.S.; Wang, W.L. Hydrochar magnetic adsorbent derived from Chinese medicine industry waste via one-step hydrothermal route: Mechanism analyses of magnetism and adsorption. *Fuel* **2022**, *326*, 125110. [CrossRef]
40. Li, Y.F.; Zhang, X.Y.; Zhang, p.Z.; Liu, X.; Han, L.J.; Zhang, P.Z.; Liu, X.; Han, L.J. Facile fabrication of magnetic bio-derived chars by co-mixing with Fe_3O_4 nanoparticles for effective Pb^{2+} adsorption: Properties and mechanism. *J. Clean. Prod.* **2020**, *262*, 121350. [CrossRef]
41. Ighalo, J.O.; Kurniawan, S.B.; Iwuozor, K.O.; Aniagor, C.O.; Ajala, O.J.; Oba, S.N.; Iwuchukwu, F.U.; Ahmadi, S.; Igwegbe, C.A. A review of treatment Technologies for the Mitigation of the toxic environmental effects of acid mine drainage (AMD). *Process Saf. Environ. Prot.* **2022**, *157*, 37–58. [CrossRef]
42. Mahmoud, M.E.; Mohamed, A.K.; Salam, M.A. Self-decoration of N-doped graphene oxide 3-D hydrogel onto magnetic shrimp shell biochar for enhanced removal of hexavalent chromium. *J. Hazard. Mater.* **2021**, *408*, 124951. [CrossRef] [PubMed]
43. Zhu, X.D.; Liu, Y.C.; Qian, F.; Zhou, C.; Zhang, S.C.; Chen, J.M. Preparation of magnetic porous carbon from waste hydrochar by simultaneous activation and magnetization for tetracycline removal. *Bioresour. Technol.* **2014**, *154*, 209–214. [CrossRef] [PubMed]
44. Mihajlović, M.; Petrović, J.; Maletić, S.; Isakovski, M.K.; Stojanović, M.; Lopičić, Z.; Trifunović, S. Hydrothermal carbonization of Miscanthus × giganteus: Structural and fuel properties of hydrochars and organic profile with the ecotox-icological assessment of the liquid phase. *Energy Convers. Manag.* **2018**, *159*, 254–263. [CrossRef]
45. Zheng, C.; Ma, X.; Yao, Z.; Chen, X. The properties and combustion behaviors of hydrochars derived from co-hydrothermal carbonization of sewage sludge and food waste. *Bioresour. Technol.* **2019**, *285*, 121347. [CrossRef]
46. Wang, Z.; Zhai, Y.; Wang, T.; Peng, C.; Li, S.; Wang, B.; Liu, X.; Li, C. Effect of temperature on the sulfur fate during hydrothermal carbonization of sewage sludge. *Environ. Pollut.* **2020**, *260*, 114067. [CrossRef]
47. Liu, X.; Zhai, Y.; Li, S.; Wang, B.; Wang, T.; Liu, Y.; Qiu, Z.; Li, C. Hydrothermal carbonization of sewage sludge: Effect of feed-water pH on hydrochar's physicochemical properties, organic component and thermal behavior. *J. Hazard. Mater.* **2020**, *388*, 122084. [CrossRef]
48. Wu, J.; Yang, J.; Huang, G.; Xu, C.; Lin, B. Hydrothermal carbonization synthesis of cassava slag biochar with excellent adsorption performance for Rhodamine B. *J. Clean. Prod.* **2020**, *251*, 119717. [CrossRef]
49. Tu, R.; Jiang, E.C.; Yan, S.; Xu, X.W.; Rao, S. The pelletization and combustion properties of terrified Camellia shell via dry and hydrothermal torrefaction: A comparative evaluation. *Bioresour. Technol.* **2018**, *264*, 78–89. [CrossRef]

50. Zhu, G.; Yang, L.; Gao, Y.; Xu, J.; Chen, H.; Zhu, Y.; Zhu, C. Characterization and pelletization of cotton stalk hydrochar from HTC and combustion kinetics of hydrochar pellets by TGA. *Fuel* **2019**, *244*, 479–491. [CrossRef]
51. Fan, S.; Tang, J.; Wang, Y.; Li, H.; Zhang, H.; Tang, J.; Wang, Z.; Li, X. Biochar prepared from co-pyrolysis of municipal sewage sludge and tea waste for the adsorption of methylene blue from aqueous solutions: Kinetics, isotherm, thermodynamic and mechanism. *J. Mol. Liq.* **2016**, *220*, 432–441. [CrossRef]
52. Deng, R.; Huang, D.; Wan, J.; Xue, W.; Wen, X.; Liu, X.; Chen, S.; Lei, L.; Zhang, Q. Recent advances of biochar materials for typical potentially toxic elements management in aquatic environments: A review. *J. Clean. Prod.* **2020**, *255*, 119523. [CrossRef]
53. Peng, H.; Gao, p.; Chu, G.; Pan, B.; Peng, J.; Xing, B.; Gao, P.; Chu, G.; Pan, B.; Peng, J.; et al. Enhanced adsorption of Cu(II) and Cd(II) by phosphoric acid-modified biochars. *Environ. Pollut.* **2017**, *229*, 846–853. [CrossRef] [PubMed]
54. Karami, K.; Beram, S.M.; Bayat, p.; Siadatnasab, F.; Ramezanpour, A.; Bayat, P.; Siadatnasab, F.; Ramezanpour, A. A novel nanohybrid based on metal–organic framework MIL101−Cr/PANI/Ag for the adsorption of cationic methylene blue dye from aqueous solution. *J. Mol. Struct.* **2022**, *1247*, 131352. [CrossRef]
55. Siruru, H.; Syafii, W.; Wistara, I.N.J.; Pari, G.; Budiman, I. Properties of sago waste charcoal using hydrothermal and pyrolysis carbonization. *Biomass. Convers. Bior.* **2020**, *12*, 5543–5554. [CrossRef]
56. Zhang, W.; Yang, J.M.; Yang, R.N.; Yang, B.C.; Quan, S.; Jiang, X. Effect of free carboxylic acid groups in UiO-66 analogues on the adsorption of dyes from water: Plausible mechanisms for adsorption and gate-opening behavior. *J. Mol. Liq.* **2019**, *283*, 160–166. [CrossRef]
57. Ramalingam, B.; Parandhaman, T.; Choudhary, p.; Das, S.K.; Choudhary, P.; Das, S.K. Biomaterial functionalized graphene-magnetite nanocomposite: A novel approach for simultaneous removal of anionic dyes and heavy-metal ions. *ACS Sustain. Chem. Eng.* **2018**, *6*, 6328–6341. [CrossRef]
58. Usman, M.; Ahmed, A.; Yu, B.; Wang, S.; Shen, Y.; Cong, H. Simultaneous adsorption of heavy metals and organic dyes by β-Cyclodextrin-Chitosan based cross-linked adsorbent. *Carbohydr. Polym.* **2021**, *255*, 117486. [CrossRef]
59. Sumalinog, D.A.G.; Capareda, S.C.; de Luna, M.D.G. Evaluation of the effectiveness and mechanisms of acetaminophen and methylene blue dye adsorption on activated biochar derived from municipal solid wastes. *J. Environ. Manag.* **2018**, *210*, 255–262. [CrossRef]
60. Essandoh, M.; Kunwar, B.; Pittman, C.U.; Mohan, D.; Mlsna, T. Sorptive removal of salicylic acid and ibuprofen from aqueous solutions using pine wood fast pyrolysis biochar. *Chem. Eng. J.* **2015**, *265*, 219–227. [CrossRef]
61. Li, B.; Guo, J.Z.; Liu, J.L.; Fang, L.; Lv, J.Q.; Lv, K. Removal of aqueous-phase lead ions by dithiocarbamate-modified hydrochar. *Sci. Total Environ.* **2020**, *714*, 136897. [CrossRef]
62. Xia, Y.; Liu, H.; Guo, Y.; Liu, Z.; Jiao, W. Immobilization of heavy metals in contaminated soils by modified hydrochar: Efficiency, risk assessment and potential mechanisms. *Sci. Total Environ.* **2019**, *685*, 1201–1208. [CrossRef] [PubMed]
63. Zhou, N.; Chen, H.; Feng, Q.; Yao, D.; Chen, H.; Wang, H.; Zhou, Z.; Li, H.; Tian, Y.; Lu, X. Effect of phosphoric acid on the surface properties and Pb(II) adsorption mechanisms of hydrochars prepared from fresh banana peels. *J. Clean. Prod.* **2017**, *165*, 221–230. [CrossRef]
64. Elaigwu, S.E.; Rocher, V.; Kyriakou, G.; Greenway, G.M. Removal of Pb^{2+} and Cd^{2+} from aqueous solution using chars from pyrolysis and microwave-assisted hydrothermal carbonization of prosopis africana shell. *J. Ind. Eng. Chem.* **2014**, *20*, 3467–3473. [CrossRef]
65. Sun, K.; Tang, J.; Gong, Y.; Zhang, H. Characterization of potassium hydroxide (KOH) modified hydrochars from different feedstocks for enhanced removal of heavy metals from water. *Environ. Sci. Pollut. Res.* **2015**, *22*, 16640–16651. [CrossRef]
66. Manjunath, S.V.; Baghel, R.S.; Kumar, M. Antagonistic and synergistic analysis of antibiotic adsorption on Prosopis juliflora activated carbon in multicomponent systems. *Chem. Eng. J.* **2020**, *381*, 122713. [CrossRef]
67. Suteu, D.; Zaharia, C.; Badeanu, M. Kinetic modeling of dye sorption from aqueous solutions onto apple seed powder. *Cell. Chem. Technol.* **2016**, *50*, 1085–1091.

Disclaimer/Publisher's Note: The statements, opinions and data contained in all publications are solely those of the individual author(s) and contributor(s) and not of MDPI and/or the editor(s). MDPI and/or the editor(s) disclaim responsibility for any injury to people or property resulting from any ideas, methods, instructions or products referred to in the content.

Article

Adsorption Characteristics and Mechanism of Methylene Blue in Water by NaOH-Modified Areca Residue Biochar

Yixin Lu [1,2], Yujie Liu [2], Chunlin Li [3], Haolin Liu [1], Huan Liu [1], Yi Tang [1], Chenghan Tang [2], Aojie Wang [2] and Chun Wang [4,*]

[1] School of Materials and Environmental Engineering, Chengdu Technological University, Chengdu 611730, China
[2] Faculty of Geosciences and Environmental Engineering, Southwest Jiaotong University, Chengdu 611756, China
[3] Sichuan Development Environmental Science and Technology Research Institute Co., Ltd., Chengdu 610094, China
[4] Sichuan Academy of Eco-Environmental Sciences, Chengdu 610046, China
* Correspondence: wangc2007@163.com; Tel.:+86-188-4831-0198

Citation: Lu, Y.; Liu, Y.; Li, C.; Liu, H.; Liu, H.; Tang, Y.; Tang, C.; Wang, A.; Wang, C. Adsorption Characteristics and Mechanism of Methylene Blue in Water by NaOH-Modified Areca Residue Biochar. *Processes* 2022, 10, 2729. https://doi.org/10.3390/pr10122729

Academic Editors: Shicheng Zhang, Gang Luo, Abdul-Sattar Nizami, Andrzej Bialowiec and Yan Shi

Received: 25 November 2022
Accepted: 13 December 2022
Published: 17 December 2022

Publisher's Note: MDPI stays neutral with regard to jurisdictional claims in published maps and institutional affiliations.

Copyright: © 2022 by the authors. Licensee MDPI, Basel, Switzerland. This article is an open access article distributed under the terms and conditions of the Creative Commons Attribution (CC BY) license (https://creativecommons.org/licenses/by/4.0/).

Abstract: To solve the water pollution problem caused by methylene blue (MB), areca residue biochar (ARB) was prepared by pyrolysis at 600 °C, and modified areca residue biochar (M-ARB) was obtained by modifying ARB with 1.5 mol/L NaOH, and they were utilized to adsorb and eliminate MB from water. The structural characteristics of ARB and M-ARB were examined, and the main influencing factors and adsorption mechanism of MB adsorption process were investigated. The outcomes demonstrated an increase in M-ARB's specific surface area and total pore volume of 66.67% and 79.61%, respectively, compared with ARB, and the pore structure was more abundant, and the content of oxygen element was also significantly increased. When the reaction temperature was 25 °C, starting pH of the mixture was 10, the initial MB concentration was 50 mg/L, the ARB and M-ARB dosages were 0.07 g/L and 0.04 g/L, respectively, the adsorption equilibrium was achieved at about 210 min, and the elimination rate for MB exceeded 94%. The adsorption behaviors of ARB and M-ARB on MB were more in line with the Langmuir isotherm model ($R^2 > 0.95$) and the quasi-secondary kinetic model ($R^2 > 0.97$), which was characterized by single-molecule layer chemisorption. The highest amount of MB that may theoretically be absorbed by M-ARB in water ranging from 136.81 to 152.72 mg/g was 74.99–76.59% higher than that of ARB. The adsorption process was a spontaneous heat absorption reaction driven by entropy increase, and the adsorption mechanism mainly involved electrostatic gravitational force, pore filling, hydrogen bonding, and π–π bonding, which was a complex process containing multiple mechanisms of action. NaOH modification can make the ARB have more perfect surface properties and more functional group structures that can participate in the adsorption reaction, which can be used as an advantageous adsorption material for MB removal in water.

Keywords: areca residue; methylene blue; biochar; modification; adsorption

1. Introduction

Areca as a leisure food in the market that the majority of consumers love, and its market demand is increasing day-by-day. Along with the economic development and the growth of areca demand, there is also the growth of the waste areca residue produced by people after consuming areca fruits [1]. Areca residue is a solid waste contains a large amount of coarse fiber components, which lacks specialized treatment and disposal methods and is usually discarded directly in the natural environment [2]. The random disposal of areca residue not only affects environmental hygiene, but also causes different degrees of damage to the water, soil, and its surrounding environment.

A type of cationic organic dyes, called methylene blue (MB), are employed not only in textile, paper, and leather industries, but also has a large market in rubber and pharmaceutical production [3,4]. MB, as one of the most prevalent contaminants in dyeing water, is frequently detected in the natural environment [5]. The untreated MB wastewater discharged directly into the water will cause adverse effects on the aquatic ecological environment and may also directly or indirectly pose a threat to human health [6]. Therefore, the search for effective removal methods of MB in water has recently emerged as a hub for water treatment at home and abroad in recent years.

To lessen the negative impacts of MB on the aquatic environment, the effectiveness of adsorption [7], photocatalysis [8], electrocatalysis [9], nanofilms [10], and ultrasonic irradiation [11] for the treatment of MB in water has been reported extensively in the literature previously. Biochar, a new class of adsorbent emerged from adsorption method, is an environmental functional material with rich carbon content, well-developed pore structure, and high stability generated by high-temperature pyrolysis of biomass raw materials under anaerobic or oxygen-limited conditions, which shows a very broad application prospect in the field of MB treatment because of its cheap cost and numerous feedstock sources [12,13]. Common waste biomass feedstocks include plant waste [14–16], animal manure waste [17], and residual sludge [18]. Among them, the conversion of plant residues into carbon adsorbents is considered to be an economic, efficient, and environmentally friendly way to remove pollutants from wastewater. Ezz et al. [19] used rice straw digestate to prepare biochar, with a total cost of 0.3022 US $/kg and a maximum adsorption capacity of 18.90 mg/g for MB, which was a feasible economic adsorbent. Chen et al. [20] prepared biochar with rape straw residue, which can achieve the adsorption capacity of 148.94 mg/g of MB in water. Sawalha et al. [21] used eight kinds of plant biomass residues (coffee granules, almond shells, sunflower shells, pistachio shells, peanut shells, jujube seeds, jute stalks, and grape vine stalks) available locally in Palestine to prepare biochar for adsorption of MB in water and found that jute stalks provide the highest removal efficiency. The performance of biochar prepared from different types of residues varies widely. To improve the performance of biochar, a series of physical, chemical, or biological methods are usually used to modify it to have greater adsorption potential [22–26]. Among the many modification methods, alkali modification can effectively reduce the ash content of biochar, change the elemental composition, functional group distribution, specific surface area, and pore structure of biochar surface [27]. Thus, it becomes a promising modification pathway. Areca residue is a waste biomass that contains a lot of organic material and can be processed to yield biochar. However, the conventional treatment of areca residue is still mainly to treat it as solid waste, and there are few reports on the exploration of making it into biochar or modified biochar. What is more, the research on the application of modified areca residue biochar in the removal of MB from wastewater are even less, and its adsorption feasibility and mechanisms are still unclear.

Therefore, in this study, waste areca residue was selected for the preparation of biochar, and NaOH was used as a modifier to modify them and investigate their feasibility for application in the removal of MB from water. The primary goals of this research were (1) to look into how NaOH modification affects the structural features of areca residue biochar; (2) to investigate the factors influencing the adsorption effect of areca residue biochar on MB before and after NaOH modification; and (3) to resolve the kinetics, isotherms, thermodynamic characteristics, and adsorption mechanism of the adsorption process, as well as to address the issue of waste areca residue pollution, while also reusing it to treat MB in water, and realize the reuse of waste biomass.

2. Materials and Methods

2.1. Reagents and Raw Materials

The areca residue was gathered in the streets of Zhuzhou City, Hunan Province, China, from an areca residue collection column, a portable device specifically designed to collect areca residue discarded by the public. After removing the impurities from the waste areca residue and drying it, the residue was crushed by a crusher. After being crushed, the powdered areca residue was sieved using a 60 mesh sieve to remove any particles larger than 0.25 mm before being used as a raw material for the subsequent preparation of biochar.

The main chemical reagents used in the experiments included analytically pure MB ($C_{16}H_{18}ClN_3S$), hydrochloric acid (HCl), caustic soda (NaOH), and ethanol (C_2H_5OH), which was purchased from Chengdu Kelong Chemical Reagent Factory, and $C_{16}H_{18}ClN_3S$ was used to prepare the MB solution. The water used in the experiment for solution preparation, sample analysis, and vessel rinsing referred to deionized water, unless otherwise specified.

2.2. Preparation and Modification of Biochar

After drying, the powdered areca residue was packed in a quartz boat, put into the chamber of a tube furnace, warmed to 600 °C with a 20 °C/min temperature gradient, and then set 150 min at a steady temperature. The quartz boat was chilled to room temperature after the pyrolysis. Then, the item was split into two sections. The pH of the supernatant was stabilized by washing one portion of the pyrolysis product with deionized water, dried, and sieved through a 100-mesh sieve to obtain a sieve out product with particle size less than 0.15 mm, which was areca residue biochar labeled as ARB and stored for backup. Another part of the pyrolysis product was modified with NaOH solution at 1.5 mol/L. The following was the course of treatment: 50 mL NaOH solution was added to each 1 g of pyrolysis product, and the product was shaken at a frequency of 150 rpm one day at 25 °C. The residue was sorted out and repeatedly rinsed with deionized water to stabilize the supernatant's pH, dried and sieved through 100-mesh sieve to obtain a sieve product with particles size of no more than than 0.15 mm, which was the modified areca residue biochar labeled as M-ARB and kept as a reserve.

2.3. Characterization of Biochar

The structures of ARB and M-ARB were characterized using BET, SEM-EDS, and FTIR techniques to analyze their basic physicochemical properties. Among them, the total pore volume, average pore size, and specific surface area of ARB and M-ARB were analyzed by NOVA 4000e-type specific surface area analyzer (Quantachrome, Beijing, China). The surface morphology was examined by Gemini 300 field emission scanning electron microscope (ZEISS, Oberkochen, Germany), and the EDS (line scan) was analyzed by Xplore energy spectrometer (OXFORD, Oxfordshire, UK). Surface functional groups were analyzed by a Nicolet 670 FTIR spectrometer (Thermo Fisher, Waltham, MA, USA).

2.4. Sequential Batch Adsorption Experiments

In a batch of 250 mL conical flasks, 100 mL of MB solution and an appropriate amount of biochar were added to each batch, and the sequential batch tests were run at various initial pH of the solutions (2–11), biochar dosing amounts (0.02–0.11 g), contact times (30–400 min), and initial MB concentrations (40–150 mg/L), in order to study the effects of ARB and M-ARB on the adsorption of MB. The main factors influencing the adsorption of ARB and M-ARB were investigated. By adding 1 mol/L of HCl and NaOH, the solution's original pH was changed. To investigate the effect of biochar dosing, the pH, reaction time, and starting MB concentration were set to 10, 400 min, and 50 mg/L, respectively. To study the influence of reaction time, the initial pH, ARB dosing, M-ARB dosing, and initial MB concentration were set to 10, 0.07 g, 0.04 g, and 50 mg/L, respectively. To exam the impact of the starting MB concentration, the pH, ARB dosage, M-ARB dosage, and contact time were set to 10, 0.07 g, 0.04 g, and 400 min, respectively. Experiments were conducted in a THZ-82 air-bath thermostat shaker (Changzhou Yineng Experimental Instrument Factory, Changzhou, China) with the oscillation frequency set to 150 rpm and the temperature set to 25 °C.

After the shaking, the mixture was put into a 2–16P centrifuge (SIGMA, Osterode, Germany) and spun for five minutes at 4000 rpm. After centrifugation, the supernatant was loaded into a cuvette, and the remaining MB concentration was tested by a 722S visible spectrophotometer (Shanghai Jingke, Shanghai, China) at 665 nm. Equations (1) and (2) were used to calculate the adsorption amount (q_t, mg/g) and removal rate (w, %) of MB, respectively.

$$w = \frac{\rho_0 - \rho_t}{\rho_0} \times 100\% \tag{1}$$

$$q_t = \frac{(\rho_0 - \rho_t)V}{m} \tag{2}$$

where ρ_0 refers to the MB starting mass concentration in solution, mg/L; ρ_t refers to the MB mass concentration at moment t, mg/L; V refers to the water sample volume, mL; m refers to the dosage of ARB and M-ARB, g.

During the analysis of regeneration performance, ethanol was used as the desorption agent. ARB and M-ARB saturated with MB were added to ethanol, respectively, and the constant temperature oscillation was conducted for 90 min at 25 °C and 150 r/min. After the oscillation, the filter residue was washed with deionized water three times. After drying the filter residue, the removal rates of MB by ARB and M-ARB after desorption were tested. The same batch of ARB and M-ARB was repeatedly operated five times to investigate the changes in the adsorption properties of ARB and M-ARB.

2.5. Adsorption Kinetics

In order to better understand the kinetics of adsorption behavior of MB on ARB and M-ARB, the experimental data of the adsorption amount at different adsorption times were selected and fitted with the quasi-primary kinetic model, quasi-secondary kinetic model, Elovich model, and Weber–Morris model, as shown in Equations (3), (4), (5), and (6), respectively.

i. Quasi-primary kinetic model

$$\ln(q_e - q_t) = \ln q_e - k_1 t \tag{3}$$

ii. Quasi-secondary dynamical model

$$\frac{t}{q_t} = \frac{t}{q_e} + \frac{1}{k_2 q_e^2} \tag{4}$$

iii. Elovich model

$$q_t = k_3 \ln t + a \tag{5}$$

iv. Weber–Morris model

$$q_t = k_{4i} t^{0.5} + b_i \tag{6}$$

where the MB adsorption at equilibrium and at time t are represented by q_e and q_t, respectively, mg/g; t is the contact time, min. k_1 is the quasi-primary kinetic model's rate constant, min^{-1}; k_2 is the quasi-secondary kinetic model's rate constant, g/(mg × min); k_3 is the Elovich model's rate constant, mg/(g × min); k_{4i} is the Weber–Morris model rate constant, mg/(g × min$^{0.5}$); ⓘ is the i-th stage of adsorption; a and b_i are the empirical constants of the Elovich and Weber–Morris models, respectively.

2.6. Adsorption Isotherm

To investigate the relationship between MB concentration and MB adsorption at adsorption equilibrium, the experimental data of MB adsorption at reaction temperatures of 25–45 °C and initial MB concentrations of 40–150 mg/L were selected and fitted with the Langmuir model, Freundlich model, and Temkin model, given by Equations (7), (8), and (9), respectively.

i. Langmuir model.

$$q_e = \frac{q_m K_L \rho_e}{1 + K_L \rho_e} \quad (7)$$

ii. Freundlich model.

$$q_e = K_F \rho_e^{1/n} \quad (8)$$

iii. Temkin model.

$$q_e = K_T \ln \rho_e + c \quad (9)$$

where q_e and q_m are the equilibrium adsorption and theoretical maximum adsorption of MB, respectively, mg/g; ρ_e refers the MB mass concentration at adsorption equilibrium, mg/L; K_L is the Langmuir constant, L/mg; k_F and n are the correlation constants of the Freundlich model; K_T and c are the correlation constants of the Temkin model.

2.7. Adsorption Thermodynamics

A thermodynamic model, as shown in Equations (10) and (11), was used to make the experimental results more suitable for the adsorption of MB in water by ARB and M-ARB at different reaction temperatures to analyze the thermodynamic characteristics of the adsorption process.

$$\Delta G^\theta = -RT \ln \frac{q_e}{\rho_e} \quad (10)$$

$$\ln \frac{q_e}{\rho_e} = -\frac{\Delta H^\theta}{RT} + \frac{\Delta S^\theta}{R} \quad (11)$$

where ΔG^θ refers the Gibbs free energy change, kJ/mol; R refers the ideal gas constant, 8.314 J/(mol × K); T refers the thermodynamic temperature, K; q_e refers the equilibrium adsorption amount of MB, mg/g; ρ_e refers the mass concentration of MB at adsorption equilibrium, mg/L; ΔH_θ refers the enthalpy change, kJ/mol; ΔS^θ is the entropy change, kJ/(mol × K).

3. Results and Discussion

3.1. Structural Characterization of ARB

Table 1 reflects the results regarding the BET analysis of ARB and M-ARB. It demonstrates that, as compared to ARB, M-ARB's specific surface area and total pore volume increased, but the average pore size decreased, which meant the pore structure of areca residue biochar could be altered by NaOH modification. The specific surface area of M-ARB reached 105 m^2/g, which increased by 66.67%, compared with ARB. The total pore volume of M-ARB increased by 79.61%, compared with ARB and reached 0.282 cm^3/g, while the average pore size decreased by 38.92%, compared with ARB. Mu et al. [28] prepared tea residue biochar (TRB) by NaOH-assisted pyrolysis, and compared to unmodified TRB, its specific surface area and total pore volume lessened by 59.64% and 24.77%, while the average pore size increased by 85.35%, contrary to the change pattern obtained in this study. This may be because the specific surface area and total pore volume of TRB decreased because of the inhibition of biomass pyrolysis by the addition of NaOH during the pyrolysis process in this study. In contrast, in this study, NaOH was used to modify the products after pyrolysis, and during the modification process, NaOH was in full contact with the areca residue biochar and chemically reacted with the impurities produced by the thermal decomposition of the biochar, which led to the solubilization of the pores, causing more pore structures and bringing about the pores' enlargement. The investigation by Choudhary et al. [29] confirmed a related occurrence using NaOH impregnation to treat *Opuntia ficus-indica* biochar. The increase in the specific surface area and total pore volume of

M-ARB can give more adsorption places for the attachment of more places for contaminants to attach themselves to adsorb, which makes the adsorption potential of M-ARB better than that of ARB. Zhou et al. [30] prepared nitrogen-doped biochar nanomaterials for the adsorption of cationic dyes from seaweed at 600 °C pyrolysis temperature and found that a greater specific surface area and appropriate pore structure can minimize the resistance to the diffusion process of pollutants to biochar and provide more active adsorption sites, which was beneficial for improving the adsorption capacity of organic dyes.

Table 1. Results of BET analysis of ARB and M-ARB.

Biochar	Specific Surface Area (m^2/g)	Total Hole Volume (cm^3/g)	Average Pore Size (nm)
ARB	63	0.157	5.593
M-ARB	105	0.282	3.416

The SEM analysis results of ARB and M-ARB can be seen in Figure 1. As can be observed, ARB's surface was comparatively smooth, and the pore structure was not immediately apparent. Compared with ARB, the porosity of M-ARB was significantly higher, the diameter of pores was reduced, the pores were more densely arranged, and the surface was more concave and uneven, showing a honeycomb shape. The analysis suggested that, during the pyrolysis process, water, organic matter, and other volatile substances inside the areca residue were released, which prompted the formation of pore structure. However, some substances were not volatile, and their blockage inside the biochar or attachment to the biochar surface resulted in the insufficient opening of pores [31]. The etching effect of NaOH cleared the impurities attached to the biochar surface and blocked inside the biochar, which led to the formation of more pore structures. The etching pore-making effect of NaOH led to the appearance of more microporous structures in the biochar, which can provide more subsequent pollutant adsorption. The pore-forming effect of NaOH made biochar appear to have more microporous structures, which can provide more adsorption sites and larger adsorption space for subsequent pollutant adsorption [32].

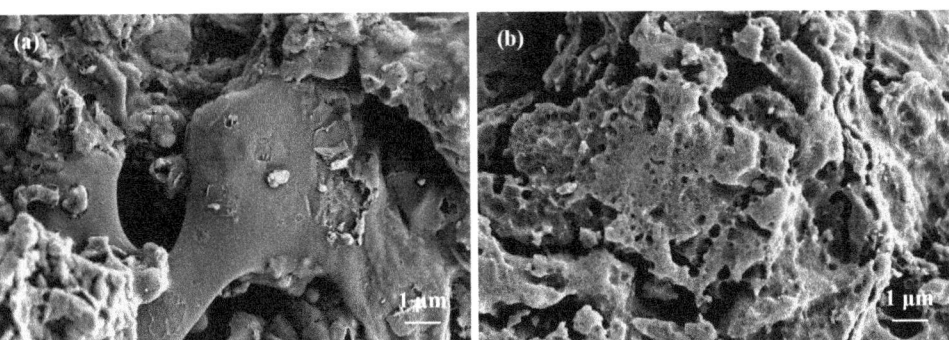

Figure 1. Scanning electron microscope of ARB (a) and M-ARB (b).

To understand the elemental composition of ARB and M-ARB, they were analyzed by EDS, and the results can be seen in Figure 2. Figure 2 shows that the areca residue biochar prepared by pyrolysis was very rich in elemental species. The highest atomic mass fraction (At) of C element was 88.59% in ARB and 79.32% in M-ARB, indicating that both had a more stable aromatic carbon structure. However, the elemental C content in the modified ARB was reduced, compared with the unmodified ARB, which may be due to two reasons. One is that the NaOH modification makes some of the C elements react with it and be carried out with the dissolution of other substances, resulting in the reduction of C content [33]; the other is that the NaOH modification makes the content of other elements increase, which indirectly results in the relative reduction of C elements in M-ARB [34]. Compared with ARB, the atomic mass fraction of O element in M-ARB showed the most significant increase by 4.89 percentage points, and the content of elements N, Na, Mg, Al, S, Cl, and Ca also increased. NaOH modification exposed the oxygen-containing functional groups masked by areca residue biochar, and more types of elements formed chemical bonds with or attached to the biochar, which was more conducive to the adsorption of pollutants. In a study by Cai et al. [35] on the modification of Oiltea camellia shells biochar with NaOH, it was also found that the alkali modification improved the functional group structure on biochar surface, added the number of oxygen-containing functional groups, and consequently significantly improved its ability to absorb and remove contaminants.

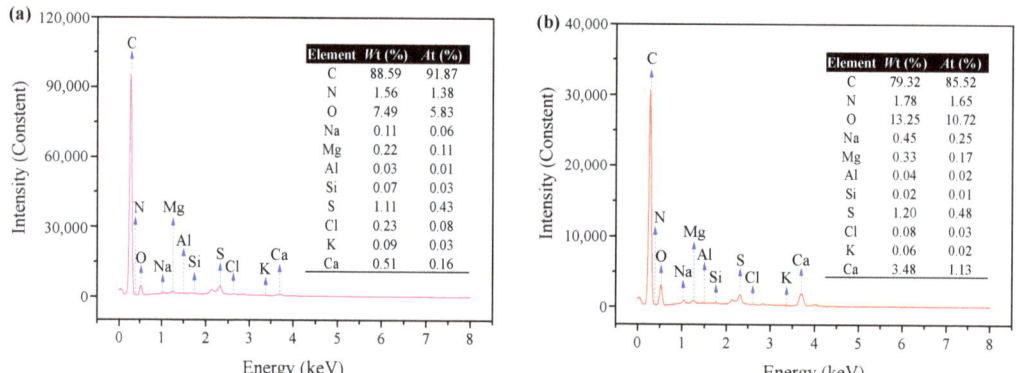

Figure 2. EDS analysis results of ARB (a) and M-ARB (b).

3.2. Analysis of Adsorption Influencing Factors

3.2.1. Effect of Initial pH of the Solution

The effect of the initial pH upon the effect of ARB and M-ARB on the adsorption of MB can be seen in Figure 3. The adsorption and removal rate of MB by ARB and M-ARB increased rapidly in the interval of pH range from 2 to 4. In the interval of the pH range from 4 to 8, the increasing trend of the MB removal effect became slower. In the interval of pH range from 8 to 10, the increasing rate of MB removal became fast again, but showed a decreasing trend after the pH exceeded 10. When the solution's original pH was 10, the adsorption value of MB by ARB and M-ARB were 86.23 mg/g and 120.05 mg/g, respectively, reaching the maximum adsorption value at this pH condition, while the adsorption of M-ARB was significantly higher than that of ARB, and its corresponding MB removal rate was increased by 27.06%, compared with that of ARB. This showed that the alkaline water environment was more favorable for the adsorption of MB by areca residue biochar, and the optimal solution initial pH was 10.

It was analyzed that, under acidic conditions, the higher concentration of hydrogen ions in water not only competed with MB cations for adsorption, but also caused electrostatic repulsion between the surface of biochar and MB cations, due to protonation; thus, the more acidic the conditions, the less effective the MB removal [36]. As the starting pH added, this competitive adsorption and electrostatic repulsion gradually decreased; thus, the adsorption effect of ARB and M-ARB on MB gradually picked up with the increase of pH [37]. However, after pH exceeded 10, the presence of too many anions in water increased the attraction for MB, resulting in MB that is not easily diffused from the liquid phase to biochar surface; thus, the removal of MB began to deteriorate. In examining the pH impact from 2.5 to 11 on the removal of MB by rice husk, cow dung, and sludge biochar, Ahmad et al. [38] found that the adsorption behavior depended mainly upon the electrostatic gravitational force between biochar and MB. MB existed in cationic form in aqueous solution, and higher pH biochar particles accumulated more negative charges on the surface, thus promoting the adsorption of MB through electrostatic gravitational force, while lower pH slowed down the efficiency of adsorption occurrence.

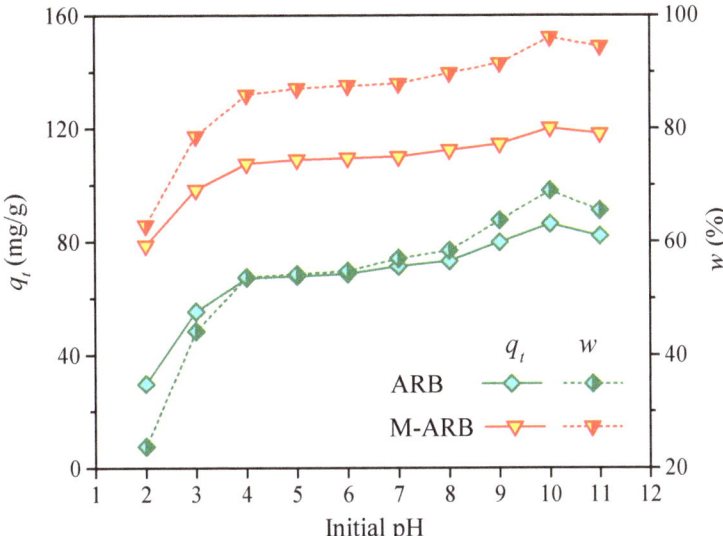

Figure 3. Effect of initial pH on MB adsorption performance.

3.2.2. Effect of Biochar Dosage

The amount of biochar added is the most direct factor for determining the adsorption effect, which determines the number of adsorption sites available for pollutant removal and the size of the adsorption capacity. Generally, the higher the amount of biochar added, the better the adsorption effect, but in practice, all factors should be integrated to select an optimal amount to achieve the best adsorption efficiency, while avoiding the waste of resources [39].

The impacts of diverse ARB and M-ARB dosing amounts on the removal efficiency and adsorption of MB in water were investigated, and the outcomes can be seen in Figure 4. As depicted in Figure 4, the adsorption of MB through ARB and M-ARB showed a rapid decreasing trend when the dosing amount raised to 0.6 g/L from 0.2 g/L, and this trend gradually slowed down after the dosing amount exceeded 0.6 g/L. The removal rates of MB by ARB increased rapidly when the dosage of ARB raised from 0.2 g/L to 0.7 g/L and then slowed down after the dosage of ARB exceeded 0.7 g/L. The removal efficiency of MB by M-ARB increased rapidly when the dosage of M-ARB raised from 0.2 g/L to 0.4 g/L and then slowed down after the M-ARB dosage exceeded 0.4 g/L. It was analyzed that, when the MB starting concentration was certain, with the increase of biochar dosing, the adsorption sites on biochar increased, the total adsorption capacity raised, and the MB clearance rate was higher because more MB might be adsorbed [40]. However, the utilization rate of biochar per unit mass decreased; thus, the MB adsorption capacity decreased [41]. Considering both adsorbent utilization and adsorbate removal, 0.07 g/L was selected as the ideal amount of ARB to take, and the adsorption amount and clearance rate of MB at this dosage were 68.37 mg/g and 95.72%, respectively. In contrast, the optimum dosage of M-ARB was only 0.04 g/L. The adsorption amount and removal rate of MB reached 120.08 mg/g and 96.06%, respectively, at this dosage. It can be seen that, compared with ARB, M-ARB can obtain better MB treatment effect with less dosage.

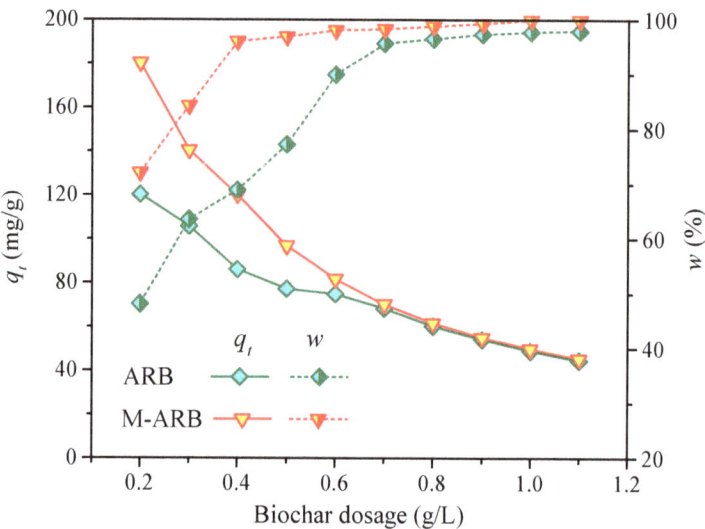

Figure 4. Effect of biochar dosage on MB adsorption performance.

3.2.3. Effect of Contact Time

Adsorption is a dynamic equilibrium process, and when pollutants are adsorbed, they need a certain amount of time to reach adsorption equilibrium; insufficient contact time may lead to incomplete adsorption, and too long contact time may lead to desorption and also waste time [42]. For this reason, this study investigated the adsorption and removal of MB from water by ARB and M-ARB at different contact times, and the results can be seen in Figure 5.

It is evident that, as the reaction time increased, both the adsorption amount and removal efficiency of MB exhibited an upward trend, and this trend, in turn, showed a pattern of fast, then slow, and finally, stable [43]. The rate of adsorption and elimination during the 30–120 min reaction time of MB by these two types of biochar showed a rapid increasing trend and belonged to the rapid adsorption stage (Stage I). At a contact time of 120–210 min, the rising trend became slower and belonged to the slow adsorption stage (Stage II). At a contact time of 210–400 min, the rising trend was very slow, and the overall trend remained the same, which belonged to the equilibrium adsorption stage (Stage III). Stage I was the initial step of the reaction, and the MB concentration in this stage was high. A lot of free adsorption places can be seen on the ARB and M-ARB's surface, and the difference of MB concentration between solid and liquid phases was large, so the driving force for MB to approach ARB and M-ARB was large, and MB can be rapidly adsorbed by ARB and M-ARB in a short period of time. After entering Stage II, the concentration of MB decreased little-by-little, and the available adsorption points on the ARB and M-ARB gradually reduced, too. The concentration difference of MB between the solid and liquid interface decreased, and the driving force for adsorption decreased, so the rate of MB adsorption by ARB and M-ARB became slower. After entering Stage III, the MB concentration decreased to a very low level, and the adsorption sites on the ARB were gradually occupied until saturation was reached and the MB removal effect was maximized. Therefore, the adsorption could reach equilibrium when the contact time was about 210 min. At this time, the adsorption amount and removal rate of MB by ARB were 67.21 mg/g and 94.10%, and those of MB by M-ARB were 118.28 mg/g and 94.62%, respectively. Overall, the NaOH modification could make the areca residue biochar possess a better MB removal effect in a shorter time.

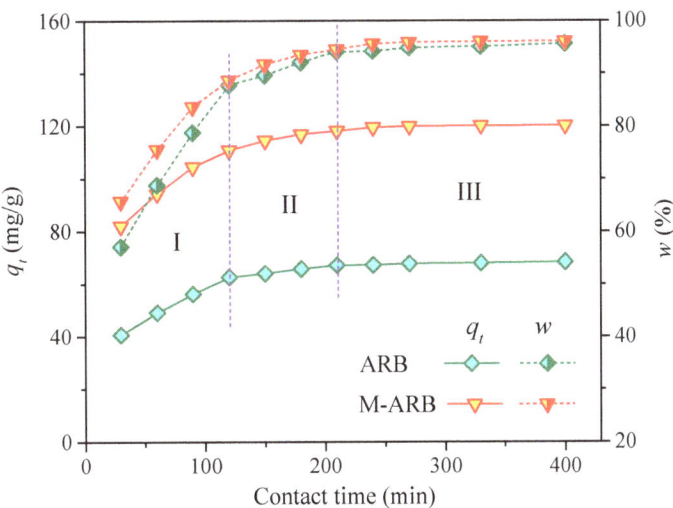

Figure 5. Effect of contact time on MB adsorption performance.

3.2.4. Effect of MB Initial Concentration

The starting concentration of adsorbent is also a vital element in the adsorption behavior. If the initial concentration of adsorbent is too low, the biochar cannot be fully utilized. Too high an initial concentration of adsorbent tends to exceed the adsorption capacity range of biochar [44]. The results of the investigation into the impact of various starting MB concentrations on the adsorption effect of ARB and M-ARB are displayed in Figure 6.

It can be seen that, at the starting MB concentration of 40 mg/L, the adsorption amounts of ARB and M-ARB on MB were 55.53 mg/g and 97.68 mg/g, respectively, and MB was removed at a rate of above 97%. At a 150 mg/L starting MB concentration, the adsorption amounts of ARB and M-ARB to MB increased to 77.33 mg/g and 135.33 mg/g, respectively, but the removal rate of MB decreased to less than 40%. When the initial concentration of MB raised from 40 mg/L to 150 mg/L, the elimination rate of MB by ARB and M-ARB kept decreasing. The adsorption amount increased rapidly at the initial MB concentration of 40–60 mg/L, and then the increase gradually decreased. The analysis suggested that ARB and M-ARB could provide sufficient adsorption sites when the initial MB concentration was low, and thus, efficient MB removal could be achieved. However, due to the low initial concentration of MB, there was still a large amount of unused residual space on the ARB and M-ARB, leading to a low adsorption efficiency of MB per unit mass of ARB and M-ARB. At a fixed amount of ARB and M-ARB dosing, as the initial MB concentration increased, the MB concentration difference between the solid–liquid phases increased, the adsorption driving force was enhanced, and the ARB and M-ARB was more fully utilized, thus the adsorption capacity per unit mass of ARB and M-ARB increased [45]. However, with the initial MB concentration rising even higher, the sites available for adsorption on the ARB and M-ARB were insufficient or had reached saturation, resulting in a decrease of the MB removal rate and more MB remaining in the water that cannot be effectively adsorbed. Thus, it is evident that there was a high removal rate, but the adsorption capacity was low when the initial MB concentration was too low, and the adsorption capacity was high, but the removal rate was low when the initial MB concentration was too high. Therefore, the optimum initial concentration of MB that can be treated with the given amount of biochar was 50 mg/L. Under this condition, the adsorption amounts of ARB and M-ARB for MB were 68.37 mg/g and 120.23 mg/g, respectively, and the elimination rates of MB were 95.72% and 96.18%, respectively.

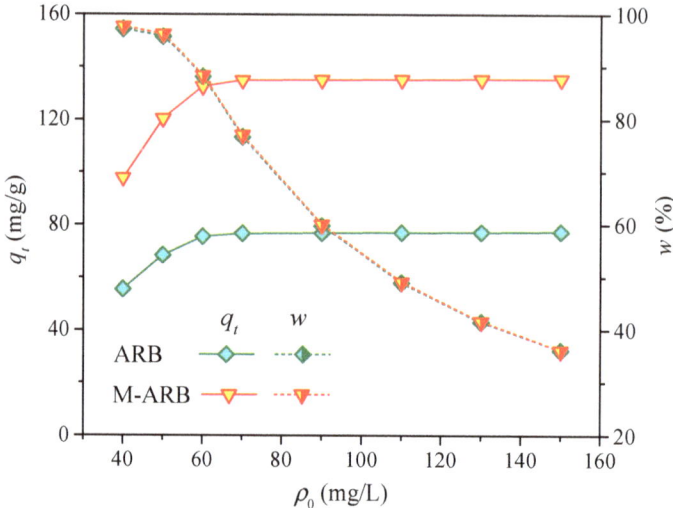

Figure 6. Effect of initial MB concentration on its adsorption performance.

3.3. Adsorption Kinetics

The results of the kinetic fitting for the MB adsorption process on ARB and M-ARB are shown in Figure 7, and the fitted factors can be seen in Table 2. The correlation coefficient R^2 obtained after fitting the adsorption data can be seen using the quasi-secondary kinetic model and exceeded 0.97, which was higher than the quasi-primary kinetic model and the Elovich model. This showed that the adsorption behavior of MB by ARB and M-ARB owned greater conformity to the quasi-secondary kinetic model, reflecting that chemisorption mainly occurred in the removal of MB by these two areca residue biochar [46]. According to the fitted parameters of the quasi-secondary kinetic model, the equilibrium adsorption capacity of M-ARB for MB was 127.23 mg/g and the rate constant k_2 was 0.0005 g/(mg × min), both of which were higher than that of ARB, indicating that the modified areca residue biochar had better adsorption performance for MB.

Table 2. Adsorption kinetics fitting parameters of MB onto ARB and M-ARB.

Model	Parameter	ARB	M-ARB
Quasi-primary kinetic	q_e (mg/g)	66.90	117.22
	k_1 (min^{-1})	0.0249	0.0329
	R^2	0.9220	0.8497
Quasi-secondary kinetic	q_e (mg/g)	74.33	127.23
	k_2 [g/(mg × min)]	0.0004	0.0005
	R^2	0.9761	0.9793
Elovich	a	35.17	74.49
	k_3 [mg/(g × min)]	6.6396	9.1936
	R^2	0.8653	0.8761

To examine the rate-controlling steps of MB adsorption in water by ARB and M-ARB, the segments of the adsorption experiment data were fitted with the Weber–Morris model, and the results are given in Figure 8, along with the fitted parameters for each phase in Table 3. Figure 8 and Table 3 show that the Weber–Morris model may be used to fit the adsorption data into three stages. The correlation coefficient R^2 of the fit for stage I (30–120 min) was the highest, reaching over 0.99, which was higher than that of stage II (120–210 min) and stage III (210–400 min). It is evident that the early stage of adsorption was more consistent with the characteristics of the Weber–Morris model. In terms of the magnitude of the rate constant k_4 values of the Weber–Morris model, stage I had the highest k_4 value, indicating that adsorption can occur rapidly at this stage. The gradually decreasing k_4 values for stages II and III indicated that the adsorption rate gradually became slower, and MB started to travel through the internal pores of biochar and searched for adsorption sites, but the adsorption gradually moved toward saturation as the external driving force decreased and the internal diffusion resistance increased [47]. The results of this analysis were consistent with the previous inferences made when it came to the impact of reaction time upon the adsorption effect of MB. The rate constant k_4 for MB adsorption of M-ARB were 1.33 and 1.16 times higher than those of ARB in stages I and II, respectively, indicating that the MB adsorption process on M-ARB was more likely to reach equilibrium than on ARB. It was also because of this that the remaining MB content in the aqueous solution available for M-ARB adsorption was very small after entering stage III; thus, the rate constant k_4 values for this stage decreased dramatically. A similar three-stage pattern of adsorption and decreasing rate constants for adsorption was also found in the research by Qing et al. upon the adsorption of phosphate in water by sodium alginate/zirconium hydrogels [48]. Moreover, none of the three fitted lines crossed the origin of the coordinate axis, indicating that the ARB and M-ARB adsorption MB process's rate-controlling steps involved other mechanisms besides intraparticle diffusion, which was also controlled by a combination of other factors, such as liquid film diffusion and adsorption reaction [48,49].

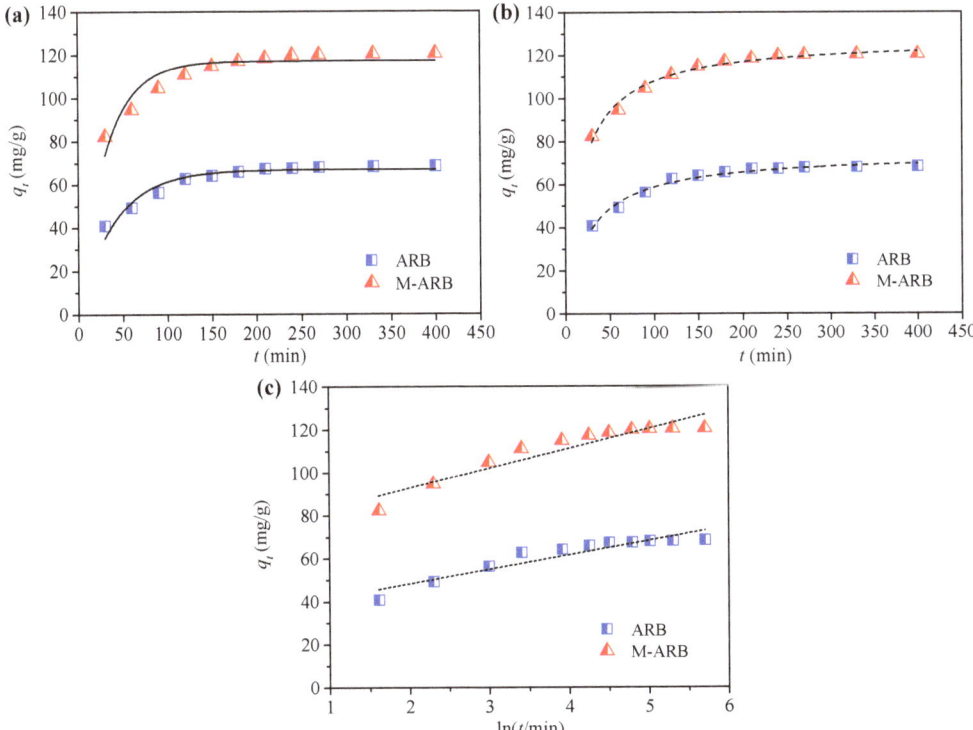

Figure 7. Adsorption kinetics fitting of MB onto ARB and M-ARB: (**a**) quasi-primary kinetic model, (**b**) quasi-secondary kinetic model, and (**c**) Elovich model.

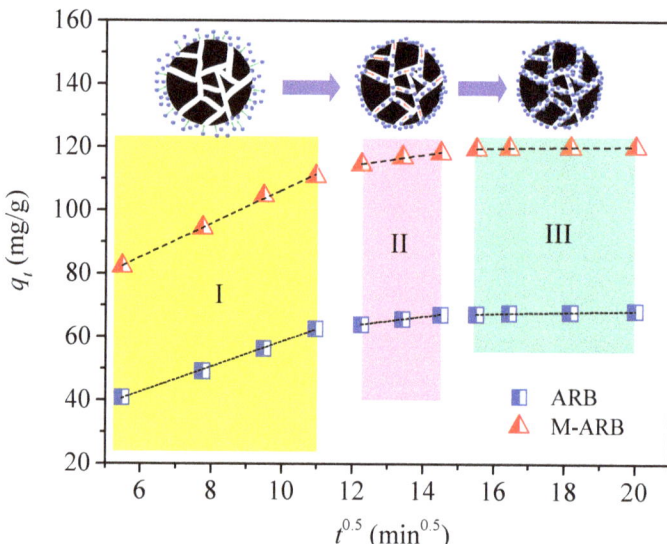

Figure 8. Segmented fit of the Weber–Morris model.

Table 3. Segmented fit parameters of the Weber–Morris model.

Stage	Parameter	ARB	M-ARB
I	b_1	18.65	53.33
	k_{41} (mg/(g × min$^{0.5}$))	3.9917	5.3098
	R^2	0.9979	0.9961
II	b_2	46.84	94.77
	k_{42} (mg/(g × min$^{0.5}$))	1.4079	1.6319
	R^2	0.9981	0.9649
III	b_3	64.19	117.77
	k_{43} (mg/(g × min$^{0.5}$))	0.2093	0.1261
	R^2	0.8924	0.8002

3.4. Adsorption Isotherm

The isothermal adsorption fitted consequences of MB by ARB and M-ARB are shown in Figure 9, and their fitted parameters can be seen in Table 4. From the fitted parameters of the three isothermal adsorption models, the adsorption procedure can be found to be more compatible with the Langmuir model, and its fitted correlation coefficient R^2 was 0.9574–0.9895, which was greater than that of the Freundlich and Temkin models by a large margin. Therefore, the adsorption of ARB and M-ARB on MB was dominated by monolayer molecular adsorption, and there was microporous filling during the adsorption process, and the limit adsorption amount was the filling amount of microporous [50]. According to the Langmuir model fitting parameters, the theoretical maximum adsorption of M-ARB on MB was 136.81–152.72 mg/g, which was 74.99–76.59% higher than that of ARB. The values of the Langmuir constant K_L for M-ARB ranged from 2.9839 to 6.7535, which was 20.06% to 48.81% higher than that of ARB. As seen, the biochar made from modified areca residue displayed increased adsorption ability and can be employed as a more effective adsorbent to remove of MB in water. Additionally showing that heat might encourage the occurrence of adsorption, the values of q_m and K_L rose with the rise in reaction temperature, which led to a more complete removal of MB from aqueous solution [51].

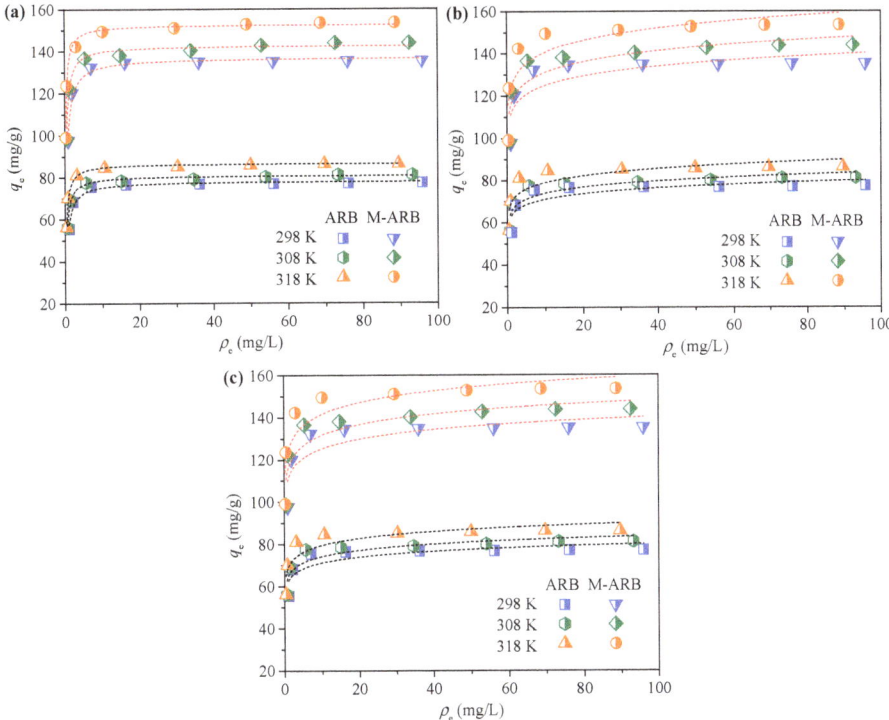

Figure 9. Isothermal adsorption fitting of Langmuir model (**a**), Freundlich model (**b**), and Temkin model (**c**).

Table 4. Isothermal adsorption fitting parameters of MB onto ARB and M-ARB.

Biochar	T/K	Langmuir Model			Freundlich Model			Temkin Model		
		q_m	K_L	R^2	$1/n$	K_F	R^2	c	K_T	R^2
ARB	298	78.18	2.4854	0.9574	0.0522	63.0093	0.6350	62.32	3.8667	0.6636
	308	80.87	3.6369	0.9881	0.0549	65.1281	0.7405	64.63	4.1659	0.7695
	318	86.56	4.5384	0.9895	0.0578	69.1859	0.7192	68.79	4.6546	0.7525
M-ARB	298	136.81	2.9839	0.9716	0.0503	111.4137	0.6477	110.38	6.5314	0.6769
	308	142.81	4.6229	0.9719	0.0539	115.8897	0.7609	115.29	7.1771	0.7871
	318	152.72	6.7535	0.9604	0.0566	123.5513	0.7604	123.30	7.9705	0.7904

In addition, the magnitude of the separation factor R_L can be calculated from $(1 + K_L C_0)^{-1}$ to determine whether the adsorption is beneficial. When the value of R_L is 0, the adsorption process can be considered irreversible. When the R_L value is 1, it can be judged that the adsorption is linear. When the R_L value is 0–1, the adsorption can be judged as favorable. As the R_L value exceeds 1, the adsorption can be judged to be unfavorable [52]. In this study, the R_L values of ARB and M-ARB were 0.0015–0.01 and 0.001–0.0083, respectively, demonstrating that the MB adsorption process by both ARB and M-ARB was favorable, and the appropriate increase of the initial MB concentration was more favorable for promoting the adsorption.

3.5. Adsorption Thermodynamics

The thermodynamic fitting results of the adsorption process of ARB and M-ARB on MB can be seen in Figure 10, and the fitted parameters depicted in Table 5. The ΔG^θ

values of ARB and M-ARB at different temperatures were negative and decreased with raising reaction temperature, demonstrating that the adsorption of these two-areca residue biochar on MB was spontaneous, and as the reaction temperature rose, this spontaneity grew [53]. The absolute value of ΔG^θ for M-ARB was greater than that for ARB at the same reaction temperature. It is, thus, clear that the spontaneous adsorption of MB through the NaOH-modified areca residue biochar was enhanced, and the modification could promote the spontaneous proceeding of the adsorption reaction, thus improving its adsorption performance on MB. Based on the fact that the value of ΔH^θ was greater than 0, it is known that the adsorption process of both ARB and M-ARB on MB was a heat absorption reaction [54]. The adsorption of MB by M-ARB required more heat for the reaction, so raising the reaction temperature can boost its ability to progress in a good way. The main types of adsorption reactions can be determined based on the magnitude of ΔH^θ values. When the value of ΔH^θ is less than 20 kJ/mol, the main form of adsorption could be considered physical adsorption. When the value of ΔH^θ is 20 to 80 kJ/mol, the main type of adsorption can be considered chemisorption [55]. In this study, the ΔH^θ values of ARB and M-ARB exceeded 20 kJ/mol, which indicated that their main reaction type for MB removal from water was chemisorption. The ΔH^θ value of M-ARB was 1.44 times higher than that of ARB, which indicated that chemisorption was more dominant in the removal of MB by M-ARB, which confirmed the results obtained from the previous adsorption kinetic analysis.

Table 5. Adsorption thermodynamic parameters of MB onto ARB and M-ARB.

T (K)	ΔG^θ (kJ/mol)		ΔH^θ (kJ/mol)		ΔS^θ [kJ/(mol × K)]		R^2	
	ARB	M-ARB	ARB	M-ARB	ARB	M-ARB	ARB	M-ARB
298	−8.5827	−10.2628						
308	−9.9041	−12.2349	38.9489	56.2585	0.1592	0.2229	0.9809	0.9937
318	−11.7668	−14.6803						

Additionally, the positive value of ΔS^θ indicated that the adsorption process of MB by both ARB and M-ARB was driven by entropy. The greater instability between the solid–liquid phases during the reaction improved the randomness of the reaction between MB molecules and active sites at the solid–liquid interface, and the greater disorder between the solid–liquid phases after the modification treatment was more favorable for the adsorption process [56].

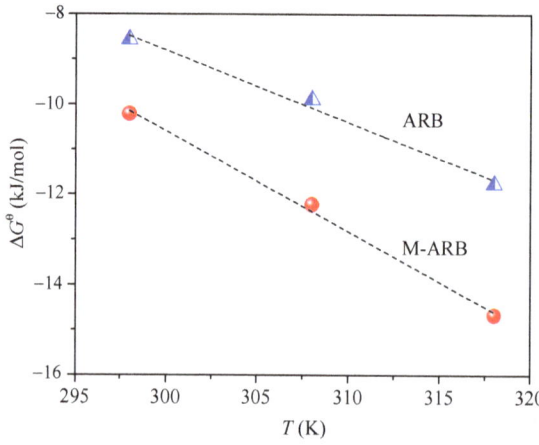

Figure 10. Adsorption thermodynamic fitting of MB onto ARB and M-ARB.

3.6. BET Analysis before and after Adsorption

Table 6 compares the changes of BET of ARB and M-ARB before and after adsorption of MB. As can be seen from Table 6, the specific surface area and total pore volume of ARB decreased by 68.25% and 71.34%, respectively, compared with those before the adsorption of MB. The specific surface area and total pore volume of M-ARB decreased 70.48% and 81.21%, respectively, compared with those before the adsorption of MB. While their average pore diameters increased 2.33 and 5.95 times, respectively, compared with those before the adsorption of MB. It demonstrated that the specific surface area and total pore volume of both areca residue biochar decreased significantly after the adsorption of MB, while the average pore size raised. It might because the decrease in the specific surface area and total pore volume because of the filling of pores on the biochar with MB after the adsorption reaction and the decrease in the number of mesopores and micropores due to the filling, thus increasing the average pore size. It was concluded that pore filling during adsorption was an important mechanism of action for the adsorption of MB by ARB and M-ARB. The role of pore filling in the adsorption of pollutants by biochar has been widely reported [57–59]. Moreover, from the comparison of the data, it can be found that the decrease in the specific surface area and total pore volume and the increase in average pore size of M-ARB were more obvious than those of ARB, showed that the pore filling was more intense during adsorption of MB by M-ARB, and more specific surface area and pores were utilized; thus, a higher amount of pollutants can be adsorbed, which became an important reason why the adsorption effect of M-ARB on MB was better than that of ARB. This is one of the important reasons why M-ARB had a better adsorption effect on MB than ARB.

Table 6. Comparison of BET analysis results before and after adsorption of MB.

Stage	Specific Surface Area (m^2/g)		Total Hole Volume (cm^3/g)		Average Pore Size (nm)	
	ARB	M-ARB	ARB	M-ARB	ARB	M-ARB
Before MB adsorption	63	105	0.157	0.282	5.593	3.416
After MB adsorption	20	31	0.045	0.053	18.611	20.325

3.7. FTIR Comparison before and after Adsorption

ARB and M-ARB surface functional group alterations prior to and during MB adsorption are depicted in Figure 11. It is evident that the surface of the areca residue biochar prepared by pyrolysis at 600 °C contained abundant functional group types. Among them, the ARB had stretching vibration peaks of -OH at 3431 cm^{-1}, C-H at 2926 cm^{-1}, and 2858 cm^{-1}, C=C and C=O at 1633 cm^{-1}, C-C at 1400 cm^{-1}, C-O at 1223 cm^{-1}, and the bending vibration peaks of Si-O-Si at 1105 cm^{-1} and 1034 cm^{-1} [60,61]. M-ARB had the stretching vibration peaks of -OH at 3427 cm^{-1}, C-H at 2918 cm^{-1}, and 2860 cm^{-1}, C=C and C=O at 1595 cm^{-1}, C-C at 1400 cm^{-1}, C-O at 1271 cm^{-1}, and the bending vibration peaks of Si-O-Si at 1099 cm^{-1} and 1026 cm^{-1} and C-H at 796 cm^{-1} [62]. Except for the absorption peak position of C-C, which was the same, after NaOH treatment, the locations of the remaining functional groups' absorption peaks were altered, with the most obvious shift of the absorption peaks at C=O and C=C, and a new C-H bending vibration peak was also generated. This showed that the NaOH modification could change the surface functional group structure of biochar and bring more functional group types to biochar. It was analyzed that NaOH modification could open the pores on biochar that were originally blocked, and at the same time, dissolve some substances on the surface of biochar that masked the functional groups, thus allowing for better exposure of the functional groups, which could provide favorable conditions for the better adsorption of pollutants by areca residue biochar [63].

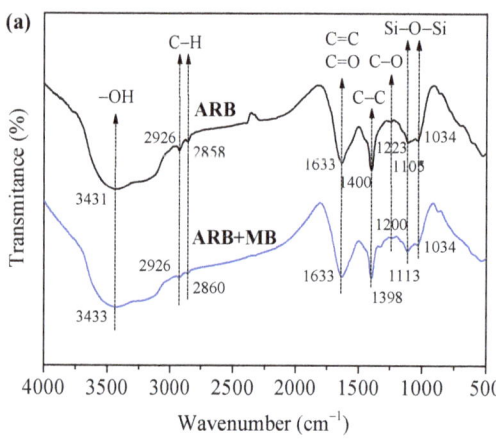
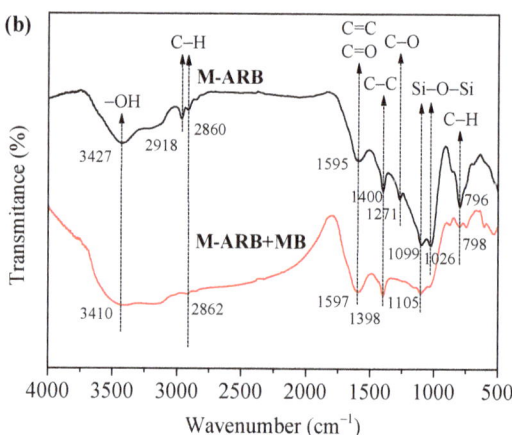

Figure 11. FTIR spectra of ARB (a) and M-ARB (b) before and after MB adsorption.

The locations of the absorbance peak of several functional groups of ARB and M-ARB were changed to varying degrees after MB had been absorbed. Among them, the -OH absorption peaks of ARB and M-ARB were shifted to 3433 cm^{-1} and 3410 cm^{-1}, respectively. This has to do with the creation of hydrogen bonds between the biochar's -OH and the heterocycles on MB that contain nitrogen; thus, hydrogen bonding had a significant impact on the adsorption [64,65]. The C-H stretching vibration peak of ARB at 2858 cm^{-1} had been changed to 2860 cm^{-1}, while the peak at 2926 cm^{-1} had not. The C-H stretching vibration peak of M-ARB at 2860 cm^{-1} was displaced to 2862 cm^{-1}, while the peak at 2918 cm^{-1} vanished. The C=O and C=C absorption peaks of ARB remained in their original locations, whereas M-ARB's C=O and C=C absorption peaks changed to 1597 cm^{-1}, implying the presence of π–π bonding between MB and NaOH-modified areca residue biochar [66]. However, this effect was weaker in the unmodified areca residue biochar and was not sufficient to change its absorption peak position. The C-C absorption peaks of both ARB and M-ARB were shifted to 1398 cm^{-1}, indicating the involvement of C-C functional groups in the adsorption reaction. The C-O absorption peak of ARB moved to 1200 cm^{-1}, while that of M-ARB disappeared. The Si-O-Si bending vibration peak of ARB at 1105 cm^{-1} moved to 1113 cm^{-1}, while the peak at 1034 cm^{-1} was not offset. The Si-O-Si bending vibration peak of M-ARB at 1099 cm^{-1} shifted to 1105 cm^{-1}, while the peak at 1026 cm^{-1} disappeared. In addition, following the adsorption of MB, the newly formed C-H bending vibration peak of M-ARB changed to 798 cm^{-1}, demonstrating that C-H was also engaged in the adsorption reaction. This showed that the NaOH-modified areca residue biochar not only had a richer surface functional group structure, but also the type of functional groups involved in the adsorption reaction and the degree of participation had been enhanced, which enabled it to show better MB removal during adsorption [67].

The adsorption mechanism of ARB and M-ARB for MB in water is depicted in Figure 12.

As can be seen from Figure 12, the adsorption mechanism of ARB on MB mainly involved electrostatic gravitational force, pore filling, and hydrogen bonding, while the adsorption mechanism of M-ARB on MB also involved π–π bonding and the combined effect of other types of functional groups. The adsorption of MB by ARB and M-ARB was a complex process with multiple mechanisms. Wang et al. [68] prepared HNO$_3$-activated and tannic acid-enhanced reed biochar as an adsorbent for MB and showed that the adsorption process was closely related to electrostatic interactions, hydrogen bonding, ion exchange, and n–π/π–π interactions. Zhang et al. [69] prepared biochar from hickory chip and obtained modified biochar for MB adsorption using ball milling and H$_2$O$_2$ treatment. The mechanism involved in this adsorption process includes π–π and van der

Waals attraction mechanism, electrostatic interaction, and ion exchange. Wang et al. [70] used K_2CO_3 modified waste bamboo biochar to remove MB from water, and it was found that electrostatic attraction, hydrogen bonding, and π–π interactions all contributed to MB adsorption. All these studies generally confirmed that the adsorption of MB by biochar was not a single adsorption mechanism, similar to the conclusion obtained in this study. The NaOH modification gave the areca residue biochar a better and more complex structure, thus giving it a greater adsorption potential and creating a valuable and optional pathway for the removal of MB.

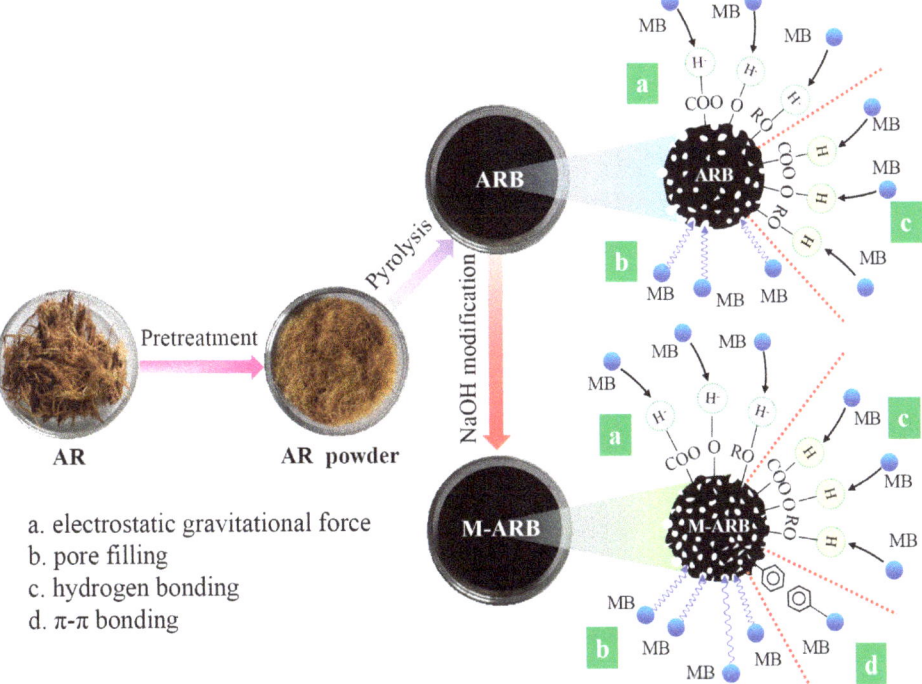

a. electrostatic gravitational force
b. pore filling
c. hydrogen bonding
d. π-π bonding

Figure 12. Adsorption mechanism of ARB and M-ARB for MB.

3.8. Analysis of Regeneration Performance

Figure 13 shows the change of MB removal rate of ARB and M-ARB after five times of regeneration. It can be seen that, with the increase of adsorption-desorption times, the removal rates of MB by ARB and M-ARB showed a downward trend, and the downward trend of ARB was more obvious. After recycling twice, the removal rate of MB by M-ARB was still above 90%. After five times of recycling, the removal rate of MB by M-ARB decreased significantly, but still reached 85.86%, showing good reusability. The removal rate of MB by ARB had dropped to 69.21% by the fifth time of recycling, 16.65 percentage points lower than that of MB by M-ARB. Accordingly, the regeneration performance of areca residue biochar without NaOH modification was relatively poor. NaOH modification enabled areca residue biochar to have a more perfect structure and better physical and chemical properties, so it can still have a higher pollutant removal efficiency after multiple recycling and show better economic and environmental benefits in practical application.

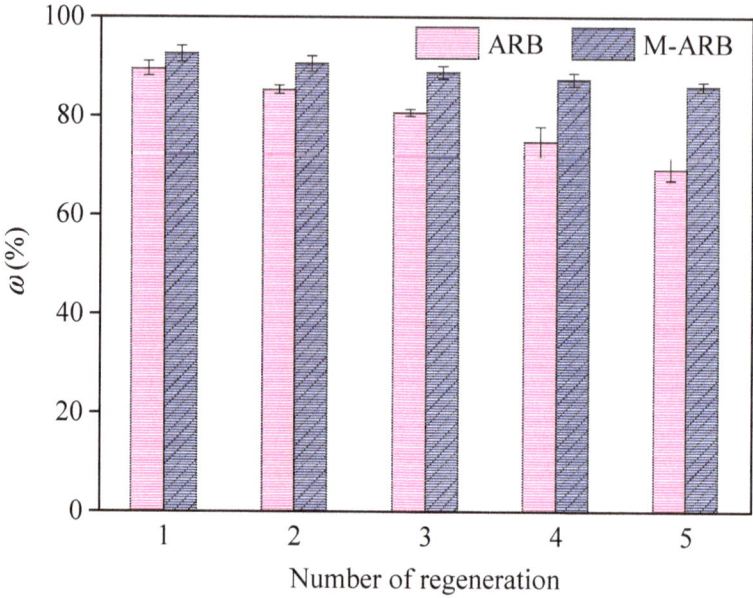

Figure 13. Regeneration performance of ARB and M-ARB for MB adsorption.

3.9. Comparison of Adsorption Capacity

Table 7 lists the adsorption capacity of several biochars for MB in water. It can be seen that, due to the different raw materials and modification methods of biochars, their adsorption capacity for MB also varies greatly. In contrast, under relatively simple preparation and modification conditions, M-ARB in this study, reached a higher MB adsorption capacity in a relatively short contact time, which has great potential to become a high-quality MB adsorbent. The preparation of NaOH modified biochar from areca residue for MB treatment showed good applicability in wastewater treatment and solid waste reuse and helped to promote sustainable ecological development.

Table 7. Comparison of adsorption capacity of different biochar materials for MB.

Raw Materials	Modifying Agent	Pyrolysis Temperature (°C)	Pyrolysis Time (min)	Contact Time (min)	Adsorption Capacity (mg/g)	References
Tea residue	NaOH	700	240	150	105.44	[28]
Reed	HNO$_3$	500	120	720	37.18	[68]
Tea residue	KOH+FeCl$_3$	700	120	360	394.30	[71]
Lychee seed	KOH	700	—	300	124.53	[72]
Lasia rhizome	None	300	180	—	9.58	[73]
Lasia rhizome	HNO$_3$	300	180	—	81.35	[73]
Fallen leaf	None	500	120	1440	78.60	[74]
Cotton residue	None	550	15	720	9.56	[75]
Cotton residue	NaOH	550	15	720	23.82	[75]
Sewage sludge	None	600	60	—	51.10	[76]
ARB	None	600	150	400	86.56	This study
M-ARB	NaOH	600	150	400	152.72	This study

4. Conclusions

This research examined the viability, influencing factors, and mechanisms of adsorption of MB in water via ARB and M-ARB. NaOH modification not only dramatically raised the specific surface area and total pore volume of ARB, but also increased its oxygen content, and enabled it to possess a richer type of surface functional groups and exhibit greater adsorption potential. The theoretical maximum adsorption of MB in water by M-ARB ranged from 136.81 mg/g to 152.72 mg/g, which was increased by 74.99–76.59%, compared with that before modification. The mechanism involved in the adsorption process of M-ARB for MB was more abundant, including the electrostatic gravitational force, pore filling, hydrogen bonding, and π–π bonding, which was a complex process. The conversion of waste areca residue into biochar that can effectively adsorb MB in water provided a preferred pathway for the removal of MB and also provided a new strategy to effectively solve the environmental pollution problems caused by areca residue.

Author Contributions: Conceptualization, Y.L. (Yixin Lu) and C.W.; data curation, Y.L. (Yujie Liu); formal analysis, C.L.; funding acquisition, Y.L. (Yixin Lu) and C.W.; methodology, Y.T.; supervision, C.W.; visualization, A.W.; writing—original draft, H.L. (Haolin Liu) and H.L. (Huan Liu); writing—review and editing, C.T. All authors have read and agreed to the published version of the manuscript.

Funding: This research was funded by the Natural Science Foundation of Sichuan Province (2022NS-FSC0393), Sichuan Science and Technology Program (2022YFG0307), National College Students' Innovation Training Program (202211116025), University's Scientific Research Project of CDTU (2022ZR001), Laboratory Open Fund Project of CDTU (2022CHZH04), and Transverse Project ((2022)82), ((2022)130).

Institutional Review Board Statement: Not applicable.

Informed Consent Statement: Not applicable.

Data Availability Statement: Data are contained within the article.

Conflicts of Interest: The authors declare no conflict of interest.

References

1. Huang, H.; Wang, T.; Han, S.; Bai, Y.; Li, X.Q. Occurrence of areca alkaloids in wastewater of major Chinese cities. *Sci. Total Environ.* **2021**, *783*, 146961. [CrossRef] [PubMed]
2. Subramani, B.; Sheeka, S.S.; Manu, B.; Babunarayan, K.S. Evaluation of pyrolyzed areca husk as a potential adsorbent for the removal of Fe^{2+} ions from aqueous solutions. *J. Environ. Manag.* **2019**, *246*, 345–354. [CrossRef] [PubMed]
3. Rahali, A.; Riazi, A.; Moussaoui, B.; Boucherdoud, A.; Bekta, N. Decolourisation of methylene blue and congo red dye solutions by adsorption using chitosan. *Desalin. Water Treat.* **2020**, *198*, 422–433.
4. Siciliano, A.; Curcio, G.M.; Limonti, C.; Masi, S.; Greco, M. Methylene blue adsorption on thermo plasma expanded graphite in a multilayer column system. *J. Environ. Manag.* **2021**, *296*, 113365. [CrossRef] [PubMed]
5. Ghodbane, I.Z.; Saida, L.; Rim, K.; Rochdi, K. Development of new modified electrode based on beta-cyclodextrin for methylene blue detection. *Sens. Rev.* **2020**, *40*, 477–483. [CrossRef]
6. Punnakkal, V.S.; Jos, B.; Anila, E.I. Polypyrrole-silver nanocomposite for enhanced photocatalytic degradation of methylene blue under sunlight irradiation. *Mater. Lett.* **2021**, *298*, 130014. [CrossRef]
7. Giraldo, S.; Robles, I.; Godínez, L.; Acelas, N.; Flórez, E. Experimental and theoretical insights on methylene blue removal from wastewater using an adsorbent obtained from the residues of the orange industry. *Molecules* **2021**, *26*, 4555. [CrossRef]
8. Perez-Gonzalez, M.; Tomas, S.A. Surface chemistry of TiO_2-ZnO thin films doped with Ag. Its role on the photocatalytic degradation of methylene blue. *Catal. Today* **2021**, *360*, 129–137. [CrossRef]
9. Nwanebu, E.O.; Liu, X.; Pajootan, E.; Yargeau, V.; Omanovic, S. Electrochemical degradation of methylene blue using a Ni-Co-Oxide anode. *Catalysts* **2021**, *11*, 793. [CrossRef]
10. Balushi, K.S.; Devi, G.; Hudaifi, A.; Garibi, A.S.R. Development of chitosan-TiO_2 thin film and its application for methylene blue dye degradation. *Int. J. Environ. Anal. Chem.* **2021**, *2*, 1–14. [CrossRef]
11. Eren, Z.; O'Shea, K. Hydroxyl radical generation and partitioning in degradation of methylene blue and DEET by dual-frequency ultrasonic irradiation. *J. Envior. Eng.* **2019**, *145*, 04019070. [CrossRef]
12. Pan, J.; Zhou, L.; Chen, H.; Liu, X.; Hong, C.; Chen, D.; Pan, B. Mechanistically understanding adsorption of methyl orange, indigo carmine, and methylene blue onto ionic/nonionic polystyrene adsorbents. *J. Hazard. Mater.* **2021**, *418*, 126300. [CrossRef]
13. Mittal, J.; Arora, C.; Mittal, A. Application of biochar for the removal of methylene blue from aquatic environments. In *Biomass-Derived Materials for Environmental Applications*; Elsevier: Amsterdam, The Netherlands, 2022; pp. 29–76.

14. Viegas, C.; Nobre, C.; Correia, R.; Gouveia, L.; Goncalves, M. Optimization of biochar production by co-torrefaction of microalgae and lignocellulosic biomass using response surface methodology. *Energies* **2021**, *14*, 7330. [CrossRef]
15. Prabakaran, E.; Pillay, K.; Brink, H. Hydrothermal synthesis of magnetic-biochar nanocomposite derived from avocado peel and its performance as an adsorbent for the removal of methylene blue from wastewater. *Mater. Today Sustain.* **2022**, *18*, 100123. [CrossRef]
16. Prajapati, A.K.; Mondal, M.K. Green synthesis of Fe_3O_4-onion peel biochar nanocomposites for adsorption of Cr(VI), methylene blue and congo red dye from aqueous solutions. *J. Mol. Liq.* **2022**, *349*, 118161. [CrossRef]
17. Zhu, Y.; Yi, B.; Hu, H.; Zong, Z.; Chen, M.; Yuan, Q. The relationship of structure and organic matter adsorption characteristics by magnetic cattle manure biochar prepared at different pyrolysis temperatures. *J. Environ. Chem. Eng.* **2020**, *8*, 104112. [CrossRef]
18. Yin, Q.; Nie, Y.; Han, Y.; Wang, R.; Zhao, Z. Properties and the application of sludge-based biochar in the removal of phosphate and methylene blue from water: Effects of acid treating. *Langmuir* **2022**, *38*, 1833–1844. [CrossRef]
19. Ezz, H.; Ibrahim, M.G.; Fujii, M.; Nasr, M. Enhanced removal of methylene blue dye by sustainable biochar derived from rice straw digestate. *Key Eng. Mater.* **2022**, *932*, 119–129. [CrossRef]
20. Chen, J.; Tang, C.; Li, X.; Sun, J.; Liu, Y.; Huang, W.; Wang, A.; Lu, Y. Preparation and modification of rape straw biochar and its adsorption characteristics for methylene blue in water. *Water* **2022**, *14*, 3761. [CrossRef]
21. Sawalha, H.; Bader, A.; Sarsour, J.; Al-Jabari, M.; Rene, E.R. Removal of dye (methylene blue) from wastewater using bio-char derived from agricultural residues in palestine: Performance and isotherm analysis. *Processes* **2022**, *10*, 2039. [CrossRef]
22. Xie, Y.; Wang, L.; Li, H.; Westholm, L.J.; Carvalho, L.; Thorin, E.; Yu, Z.; Yu, X.; Skreiberg, O. A critical review on production, modification and utilization of biochar. *J. Anal. Appl. Pyrolysis* **2022**, *161*, 105405. [CrossRef]
23. Zhang, Q.; Wang, J.; Lyu, H.; Zhao, Q.; Lisi, J.; Li, L. Ball-milled biochar for galaxolide removal: Sorption performance and governing mechanisms. *J. Clin. Microbiol.* **2019**, *659*, 1537–1545. [CrossRef] [PubMed]
24. Kwak, J.H.; Islam, M.S.; Wang, S.; Messele, S.A.; Naeth, M.A.; El-Din, M.G.; Chang, S.X. Biochar properties and lead(II) adsorption capacity depend on feedstock type, pyrolysis temperature, and steam activation. *Chemosphere* **2019**, *231*, 393–404. [CrossRef] [PubMed]
25. Miao, Q.; Li, G. Potassium phosphate/magnesium oxide modified biochars: Interfacial chemical behaviours and Pb binding performance. *J. Clin. Microbiol.* **2021**, *759*, 143452. [CrossRef]
26. Zhang, S.; Wang, J. Removal of chlortetracycline from water by immobilized Bacillus subtilis on honeysuckle residue derived biochar. *Water Air Soil Pollut.* **2021**, *232*, 236. [CrossRef]
27. Hussin, M.H.; Chuein, A.L.H.; Idris, N.N.; Hamidon, T.S.; Azani, N.F.S.M.; Abdullah, N.S.; Sharifuddin, S.S.; Ying, A.S. Kinetics and equilibrium studies of methylene blue dye adsorption on oil palm frond adsorbent. *Desalin. Water Treat.* **2021**, *216*, 358–371. [CrossRef]
28. Mu, Y.; Ma, H. NaOH-modified mesoporous biochar derived from tea residue for methylene blue and Orange II removal. *Chem. Eng. Res. Des.* **2021**, *167*, 129–140. [CrossRef]
29. Choudhary, M.; Kumar, R.; Neogi, S. Activated biochar derived from *Opuntia ficus-indica* for the efficient adsorption of malachite green dye, Cu^{+2} and Ni^{+2} from water. *J. Hazard. Mater.* **2020**, *392*, 122441. [CrossRef]
30. Zhou, Y.; Li, Z.; Ji, L.; Wang, Z.; Cai, L.; Guo, J.; Song, W.; Wang, Y.; Piotrowski, A.M. Facile preparation of alveolate biochar derived from seaweed biomass with potential removal performance for cationic dye. *J. Mol. Liq.* **2022**, *353*, 118623. [CrossRef]
31. Vu, M.T.; Le, T.T.; Chao, H.P.; Trinh, T.V.; Lin, C.C.; Tran, H.N. Removal of ammonium from groundwater using NaOH-treated activated carbon derived from corncob wastes: Batch and column experiments. *J. Clean. Prod.* **2018**, *180*, 560–570. [CrossRef]
32. Boguta, P.; Sokolowska, Z.; Skic, K.; Tomczyk, A. Chemically engineered biochar-Effect of concentration and type of modifier on sorption and structural properties of biochar from wood waste. *Fuel* **2019**, *256*, 115893. [CrossRef]
33. Lu, F.; Lu, X.; Li, S.; Zhang, H.; Shao, L.; He, P. Dozens-fold improvement of biochar redox properties by KOH activation. *Chem. Eng. J.* **2022**, *429*, 132203. [CrossRef]
34. Wu, J.; Wang, L.; Ma, H.; Zhou, J. Investigation of element migration characteristics and product properties during biomass pyrolysis: A case study of pine cones rich in nitrogen. *RSC Adv.* **2021**, *11*, 34795–34805. [CrossRef]
35. Cai, T.; Du, H.; Liu, X.; Tie, B.; Zeng, Z. Insights into the removal of Cd and Pb from aqueous solutions by NaOH-EtOH-modified biochar. *Environ. Technol. Innov.* **2021**, *24*, 102031. [CrossRef]
36. Huang, M.; Liao, Z.; Li, Z.; Wen, J.; Zhao, L.; Jin, C.; Tian, D.; Shen, F. Effects of pyrolysis temperature on proton and cadmium binding properties onto biochar-derived dissolved organic matter: Roles of fluorophore and chromophore. *Chemosphere* **2022**, *299*, 134313. [CrossRef]
37. Li, X.; Jiang, X.; Song, Y.; Chang, S. Coexistence of polyethylene microplastics and biochar increases ammonium sorption in an aqueous solution. *J. Hazard. Mater.* **2020**, *405*, 124260. [CrossRef]
38. Ahmad, A.; Khan, N.; Giri, B.S.; Chaturvedi, P.; Chowdhary, P. Removal of methylene blue dye using rice husk, cow dung and sludge biochar: Characterization, application, and kinetic studies. *Bioresour. Technol.* **2020**, *306*, 123202. [CrossRef]
39. Feng, Q.; Chen, M.; Wu, P.; Zhang, X.; Wang, S.; Yu, Z.; Wang, B. Simultaneous reclaiming phosphate and ammonium from aqueous solutions by calcium alginate-biochar composite: Sorption performance and governing mechanisms. *Chem. Eng. J.* **2022**, *429*, 132166. [CrossRef]
40. Wang, B.; Lian, G.; Lee, X.; Gao, B.; Li, L.; Liu, T.; Zhang, X.; Zheng, Y. Phosphogypsum as a novel modifier for distillers grains biochar removal of phosphate from water. *Chemosphere* **2020**, *38*, 124684. [CrossRef] [PubMed]

41. Zhou, H.; Ye, M.; Zhao, Y.; Baig, S.A.; Huang, N.; Ma, M. Sodium citrate and biochar synergistic improvement of nanoscale zero-valent iron composite for the removal of chromium (VI) in aqueous solutions. *J. Environ. Sci.* **2022**, *115*, 227–239. [CrossRef] [PubMed]
42. Akech, S.R.O.; Harrison, O.; Saha, A. Removal of a potentially hazardous chemical, tetrakis (hydroxymethyl) phosphonium chloride from water using biochar as a medium of adsorption. *Environ. Technol. Innov.* **2018**, *12*, 196–210. [CrossRef]
43. Ramola, S.; Belwal, T.; Li, C.; Liu, Y.; Wang, Y.; Yang, S.; Zhou, C. Preparation and application of novel rice husk biochar-calcite composites for phosphate removal from aqueous medium. *J. Clean. Prod.* **2021**, *299*, 126802. [CrossRef]
44. Olu-Owolabi, B.I.; Diagboya, P.N.; Mtunzi, F.M.; During, R.A. Utilizing eco-friendly kaolinite-biochar composite adsorbent for removal of ivermectin in aqueous media. *J. Environ. Manag.* **2021**, *279*, 111619. [CrossRef]
45. Saikia, R.; Goswami, R.; Bordoloi, N.; Senapati, K.K.; Pant, K.K.; Kumar, M.; Kataki, R. Removal of arsenic and fluoride from aqueous solution by biomass based activated biochar: Optimization through response surface methodology. *J. Environ. Chem. Eng.* **2017**, *5*, 5528–5539. [CrossRef]
46. An, Q.; Li, Z.; Zhou, Y.; Meng, F.; Zhao, B.; Miao, Y.; Deng, S. Ammonium removal from groundwater using peanut shell based modified biochar: Mechanism analysis and column experiments. *J. Water Process Eng.* **2021**, *43*, 102219. [CrossRef]
47. Kaith, B.S.; Shanker, U.; Gupta, B. Synergic effect of Guggul gum based hydrogel nanocomposite: An approach towards adsorption-photocatalysis of Magenta-O. *Int. J. Biol. Macromol.* **2020**, *161*, 457–469. [CrossRef]
48. Qing, Z.; Wang, L.; Liu, X.; Song, Z.; Qian, F.; Song, Y. Simply synthesized sodium alginate/zirconium hydrogel as adsorbent for phosphate adsorption from aqueous solution: Performance and mechanisms. *Chemosphere* **2022**, *291*, 133103. [CrossRef]
49. Qiao, H.; Qiao, Y.; Luo, X.; Zhao, B.; Cai, Q. Qualitative and quantitative adsorption mechanisms of zinc ions from aqueous solutions onto dead carp derived biochar. *RSC Adv.* **2021**, *11*, 38273–38282. [CrossRef]
50. Yang, Z.; Hu, W.; Yao, B.; Shen, L.; Jiang, F.; Zhou, Y.; Nunez-Delgado, A. A novel manganese-rich pokeweed biochar for highly efficient adsorption of heavy metals from wastewater: Performance, mechanisms, and potential risk analysis. *Processes* **2021**, *9*, 1209. [CrossRef]
51. Li, S.B.; Dong, L.J.; Wei, Z.F.; Sheng, G.D.; Du, K.; Hu, B.W. Adsorption and mechanistic study of the invasive plant-derived biochar functionalized with CaAl-LDH for Eu(III) in water. *J. Environ. Sci.* **2020**, *96*, 127–137. [CrossRef]
52. Shen, Y.W.; Jiao, S.Y.; Ma, Z.; Gao, W.S.; Chen, J.Q. Humic acid-modified bentonite composite material enhances urea-nitrogen use efficiency. *Chemosphere* **2020**, *255*, 126976. [CrossRef] [PubMed]
53. Liu, Z.F.; Fang, X.; Chen, L.Y.; Tang, B.; Song, F.M.; Li, W.B. Effect of acid-base modified biochar on chlortetracycline adsorption by purple soil. *Sustainability* **2022**, *14*, 5892. [CrossRef]
54. Kaya-Zkiper, K.; Uzun, A.; Soyer-Uzun, S. A novel alkali activated magnesium silicate as an effective and mechanically strong adsorbent for methylene blue removal. *J. Hazard. Mater.* **2022**, *424*, 127256. [CrossRef] [PubMed]
55. Lu, Y.X.; Chen, J.; Zhao, L.; Zhou, Z.; Qiu, C.; Li, Q.L. Adsorption of Rhodamine B from aqueous solution by goat manure biochar: Kinetics, isotherms, and thermodynamic studies. *Pol. J. Environ. Stud.* **2020**, *2*, 2721–2730. [CrossRef] [PubMed]
56. Yan, Y.H.; Chu, Y.T.; Khan, M.A.; Xia, M.Z.; Shi, M.X.; Zhu, S.D.; Lei, W.; Wang, F.Y. Facile immobilization of ethylenediamine tetramethylene-phosphonic acid into UiO-66 for toxic divalent heavy metal ions removal: An experimental and theoretical exploration. *Sci. Total Environ.* **2022**, *806*, 150652. [CrossRef]
57. Cheng, D.L.; Ngo, H.H.; Guo, W.S.; Chang, S.W.; Nguyen, D.D.; Li, J.X.; Ly, Q.V.; Nguyen, T.A.H.; Tran, V.S. Applying a new pomelo peel derived biochar in microbial fell cell for enhancing sulfonamide antibiotics removal in swine wastewater. *Bioresour. Technol.* **2020**, *318*, 123886. [CrossRef]
58. Gao, T.; Shi, W.S.; Zhao, M.X.; Huang, Z.X.; Liu, X.L.; Ruan, W.Q. Preparation of spiramycin fermentation residue derived biochar for effective adsorption of spiramycin from wastewater. *Chemosphere* **2022**, *296*, 133902. [CrossRef]
59. Li, Y.X.; Shang, H.R.; Cao, Y.N.; Yang, C.H.; Feng, Y.J.; Yu, Y.L. High performance removal of sulfamethoxazole using large specific area of biochar derived from corncob xylose residue. *Biochar* **2022**, *1*, 151–161. [CrossRef]
60. Yang, X.D.; Wang, L.L.; Tong, J.; Shao, X.Q.; Chen, R.; Yang, Q.; Li, F.F.; Xue, B.; Li, G.D.; Han, Y.; et al. Synthesis of hickory biochar via one-step acidic ball milling: Characteristics and titan yellow adsorption. *J. Clean. Prod.* **2022**, *338*, 130575. [CrossRef]
61. Wang, T.T.; Zhang, D.; Fang, K.K.; Zhu, W.; Peng, Q.; Xie, Z.G. Enhanced nitrate removal by physical activation and Mg/Al layered double hydroxide modified biochar derived from wood waste: Adsorption characteristics and mechanisms. *J. Environ. Chem. Eng.* **2021**, *9*, 105184. [CrossRef]
62. Zhang, Y.; Tang, J.Y.; Zhang, W.J.; Ai, J.; Liu, Y.Y.; Wang, Q.D.; Wang, D.S. Preparation of ultrahigh-surface-area sludge biopolymers-based carbon using alkali treatment for organic matters recovery coupled to catalytic pyrolysis. *J. Environ. Sci.* **2021**, *106*, 83–96. [CrossRef]
63. Chen, W.; Gong, M.; Li, K.X.; Xia, M.W.; Chen, Z.Q.; Xiao, H.Y.; Fang, Y.; Chen, Y.Q.; Yang, H.P.; Chen, H.P. Insight into KOH activation mechanism during biomass pyrolysis: Chemical reactions between O-containing groups and KOH. *Appl. Energy* **2020**, *278*, 115730. [CrossRef]
64. Hu, M.F.; Liu, L.; Hou, N.; Li, X.S.; Zeng, D.Q.; Tan, H.H. Insight into the adsorption mechanisms of ionizable imidazolinone herbicides in sediments: Kinetics, adsorption model, and influencing factors. *Chemosphere* **2021**, *274*, 129655. [CrossRef]
65. Binh, Q.A.; Kajitvichyanukul, P. Adsorption mechanism of dichlorvos onto coconut fibre biochar: The significant dependence of H-bonding and the pore-filling mechanism. *Water Sci. Technol.* **2019**, *79*, 866–876. [CrossRef]

66. Deng, H.; Zhang, J.Y.; Huang, R.; Wang, W.; Meng, M.W.; Hu, L.N.; Gan, W.X. Adsorption of malachite green and Pb^{2+} by $KMnO_4$-modified biochar: Insights and mechanisms. *Sustainability* **2022**, *14*, 2040. [CrossRef]
67. Zheng, Y.W.; Wang, J.D.; Li, D.H.; Liu, C.; Lu, Y.; Lin, X.; Zheng, Z.F. Insight into the $KOH/KMnO_4$ activation mechanism of oxygen-enriched hierarchical porous biochar derived from biomass waste by in-situ pyrolysis for methylene blue enhanced adsorption. *J. Anal. Appl. Pyrolysis* **2021**, *158*, 105269. [CrossRef]
68. Wang, Y.; Zhang, Y.; Li, S.Y.; Zhong, W.H.; Wei, W. Enhanced methylene blue adsorption onto activated reed-derived biochar by tannic acid. *J. Mol. Liq.* **2018**, *268*, 658–666. [CrossRef]
69. Zhang, Y.; Zheng, Y.L.; Yang, Y.C.; Huang, J.S.; Zimmerman, A.R.; Chen, H.; Hu, X.; Gao, B. Mechanisms and adsorption capacities of hydrogen peroxide modified ball milled biochar for the removal of methylene blue from aqueous solutions. *Bioresour. Technol.* **2021**, *337*, 125432. [CrossRef]
70. Wang, Y.H.; Srinivasakannan, C.; Wang, H.H.; Xue, G.; Wang, L.; Wang, X.; Duan, X. Preparation of novel biochar containing graphene from waste bamboo with high methylene blue adsorption capacity. *Diam. Relat. Mater.* **2022**, *125*, 109034. [CrossRef]
71. Mu, Y.K.; Du, H.X.; He, W.Y.; Ma, H.Z. Functionalized mesoporous magnetic biochar for methylene blue removal: Performance assessment and mechanism exploration. *Diam. Relat. Mater.* **2022**, *121*, 108795. [CrossRef]
72. Sahu, S.; Pahi, S.; Tripathy, S.; Singh, S.K.; Behera, A.; Sahu, U.K.; Patal, R.K. Adsorption of methylene blue on chemically modified lychee seed biochar: Dynamic, equilibrium, and thermodynamic study. *J. Mol. Liq.* **2020**, *315*, 113743. [CrossRef]
73. Fernando, J.C.; Peiris, C.; Navarathna, C.M.; Gunatilake, S.R.; Welikala, U.; Wanasinghe, S.T.; Madduri, S.B.; Jayasinghe, S.; Mlsna, T.E.; Hassan, E.; et al. Nitric acid surface pre-modification of novel Lasia spinosa biochar for enhanced methylene blue remediation. *Groundw. Sustain. Dev.* **2021**, *14*, 100603. [CrossRef]
74. Ji, B.; Wang, J.L.; Song, H.J. Removal of methylene blue from aqueous solutions using biochar derived from a fallen leaf by slow pyrolysis: Behavior and mechanism. *J. Environ. Chem. Eng.* **2019**, *7*, 103036. [CrossRef]
75. Primaz, C.T.; Ribes-Greus, A.; Jacques, R.A. Valorization of cotton residues for production of bio-oil and engineered biochar. *Energy* **2021**, *235*, 121363. [CrossRef]
76. Ribeiro, M.R.; Guimarães, Y.D.; Silva, I.F.; Almeida, C.A.; Silva, M.S.V.; Nascimento, M.A.; da Silva, U.P.; Varejao, E.V.; Renato, N.D.; Teixeira, A.P.D.; et al. Synthesis of value-added materials from the sewage sludge of cosmetics industry effluent treatment plant. *J. Environ. Chem. Eng.* **2021**, *9*, 105367. [CrossRef]

Article

Simulation and Experimental Validation on the Effect of Twin-Screw Pulping Technology upon Straw Pulping Performance Based on Tavares Mathematical Model

Huiting Cheng, Yuanjuan Gong, Nan Zhao, Luji Zhang, Dongqing Lv and Dezhi Ren *

College of Engineering, Shenyang Agricultural University, Shenyang 110866, China
* Correspondence: rdz@syau.edu.cn; Tel.: +86-138-8921-5226

Abstract: Rice straw is waste material from agriculture as a renewable biomass resource, but the black liquor produced by straw pulping causes serious pollution problems. The twin-screw pulping machine was designed by Solidworks software and the straw breakage model was created by the Discrete Element Method (DEM). The model of straw particles breakage process in the Twin-screw pulping machine was built by the Tavares model. The simulation results showed that the highest number of broken straw particles was achieved when the twin-screw spiral casing combination was negative-positive-negative-positive and the tooth groove angle arrangement of the negative spiral casing was 45°−30°−15°. The multi-factor simulation showed that the order of influence of each factor on the pulp yield was screw speed > straw moisture content > tooth groove angle. The Box-Behnken experiment showed that when screw speed was 550 r/min, tooth groove angle was 30°, straw moisture content was 65% and pulping yield achieved up to 92.5%. Twin-screw pulping performance verification experiments were conducted, and the results from the experimental measurements and simulation data from the model showed good agreement.

Keywords: twin-screw pulping; semi-dry pulping; straw breakage; Discrete Element Method; Tavares model

1. Introduction

China's rice planted area and straw yield are large [1], and the utilization rate of straw fertilization, feederization and fuelization are relatively high, while the utilization rate of raw materialization is only 1.0% [2]. Traditional pulping methods have led to a wood shortage and environmental pollution, and straw as a wasted crop can become an alternative method of pulping [3]. Straw pulping has an important role in the Chinese pulping industry, but the complex structure of straw leads to black liquor produced in the pulping process and increases pollution emissions [4]. Therefore, the simulation and experimental study of pulping performance is important to improve the utilization rate of straw raw materialization and promote clean straw pulping.

The Discrete Element Method (DEM) can simulate the motion of individual particles in granular media, and the Tavares model is a practical model to describe the fragmentation of individual particles, which is applied in the fields of particle fragmentation and material wear [5]. The particles under loading assume the following states: unbroken, worn, broken, and separated. DEM solves the particle motion problem based on Newton's equations of motion and uses the contact law to analyze the contact forces between the particles. Tavares proposes a practical model to describe the breakage of a single particle that can be used to describe the breakage of polyhedral particles when exposed to different intensity stresses [6]. The Tavares model was based on the research of Professor L.M. Tavares at the University of Utah and was developed through further research by a research group at the Federal University of Rio de Janeiro [5]. The Tavares model determines whether a particle would crush by calculating the energy displaced in the contact of two elements, and could simulate the breakage of a particle under insufficient stress, explaining the probability of particle breakage through the upper truncated log-normal distribution. The Tavares model extends the breakage distribution function based on simple crushing, enabling breakage prediction to reach a more realistic level, contributing to the description of particle degradation during treatment and particle size reduction under different conditions. It can predict fracture and broken particle percentages for repeated impacts, based on a weakened model of continuum damage mechanics. Considering the special case of materials exhibiting a non-normalized fracture response, it describes the size distribution of particle breakage generations in terms of an incomplete beta function, which predicts the size distribution of fragments produced by stress events by describing experiential expressions for stress energy and fragmentation strength [7]. Currently, the Tavares fracture model is embedded in commercial DEM software (RockyDEM and Altair EDEM) and employed to simulate many complex industrial processes and integrated into micro-scale population balance models describing mill types.

Further, Barrios [8] simulated the fragmentation of the ironstone particle layer shock and compression loading by a DEM-coupled particle replacement model, which was able to describe the force and deformation profile generated by the individual particle loading, deriving a function for calculating the particle size based on the median ratio fracture energy. Carvalho [9] proposed a mathematical model for material grinding, describing the phenomena of fragmentation, wear, and repeated shock caused by the grinding medium, and accurately calculated the wear rate of both materials during the grinding process. Oliveira and Rodriguez [10] simulated the process of material crushing in a vertical stirred mill with a mechanical wear model and predicted the mill power by the Discrete Element Method with an error of less than 3.5% between simulation and experiment, proving that DEM can be used as a tool to predict and analyze the crushing behavior of materials. Tavares and Rodriguez [11] verified the effects of particle fracture energy, particle size, and material properties on the variation in the impact surface by comparing model predictions with single-particle fragmentation, showing that the model could accurately predict the distribution of fragments. Zeng and Mao [12] proposed a numerical method combining the Discrete Element Method (DEM) and the Particle Replacement Model (PRM) to compare the particle crushing process of the simulated vertical mill with the actual grinding process, and concluded that the simulation results were in agreement with the experimental results. Xu [13] used the coupling method of DEM to analyze the contact action between seeds and soil from a microscopic perspective and analyzed the influence of seed displacement during soil cover and compaction processing to verify the feasibility and applicability of the coupling method.

However, the main problems in the straw pulping industry are that it is hard to degrade lignin, chemical pulping is environmentally unfriendly, twin-screw pulping consumes the least energy among mechanical pulping methods, and biological pulping technology is not yet well established [14]. Therefore, the combination of biological and mechanical methods to degrade lignin for pulping is an efficient and environmentally friendly method [15]. At present, the research on twin-screw pulping machines mainly focuses on the optimization of twin-screw structure parameters and process conditions, while there is little research on the simulation of the straw breakage process. Xiaoyang [16] studied the influences of screw speed, inverted thread lead number and groove width of negative spiral casing on specific energy and pulping quality, but failed to consider the influence of tooth groove angle of negative spiral casing on pulping performance. Pradhan [17] conducted material crushing experiments to show that the geometry of the screw element affects the material particle crushing process, modifying the geometry of the screw element to change the size of the largest particle. Shirazian [18] constructed a two-dimensional population balance model (PBM) to find out the twin-screw pelletizing mechanism and carried out experiments on a 12 mm twin-screw with microcrystalline cellulose as the model to predict the GSD values at different spatial locations. Bumm [19] researched the damage to glass fiber of different viscosity and screw speeds, and established a composite modular dynamics model to describe glass fiber fragmentation and compared the simulation results with experimental data to show a consistent relationship. Fleur Rol [20] used a twin-screw extruder to produce CNFs and compared it to an ultrafine grinder (UFG) and found that cellulose pulping with four passes of TSE reduced energy consumption by 4–5 times compared to UFG on the same mass index. Dyna Thenga [21] compared the energy consumption, extrudate and pulping properties of two pretreatment technologies, twin-screw extrusion and steam digestion. It found that twin-screw pulping demanded lower water volume, lower temperature, and lower energy consumption. Thu Ho [22] studied the effect of the fibrillation process on fiber properties in a twin-screw machine, where cellulose fibers were found to exhibit not only fibrillation but also some degree of degradation after several successive passes through the TSE. The method of twin-screw pulping has the advantage of low cost and high fiber content compared to other pulping mechanical methods. Ke pa [23] conducted twin-screw extrusion (TSE), high-energy ball milling (HEBM) and high-pressure homogenization (HPH) of alkali-treated material to convert into cellulose microfiber (MFC) and nanofiber (CNF), and fiber analysis using confocal laser scanning microscopy resulted in the lowest TSE energy consumption. Eduardo Espinosa [24] compared the energy consumption of Twins Screw Extruder (TSE), High Pressure Homogenizer (HPH) and Ultrafine Grinder (UFG) to describe the morphology, crystallinity, thermal stability, chemical structure and mechanical properties of the obtained lignocellulose. Twin-screw pulping was found to consume five times less energy than HPH and UFG. Fangmin Liang [25] used a low-cost Twins Screw Extruder (TSE) for bamboo pulping instead of the high-cost common extrusion Model Screw Device (MSD). TSE extruded material had shorter fibers, higher fines content, lower kink and curl indices compared to MSD, and the absorbance of TSE extruded material (4.50 g/g) was 3 times higher than that of MSD extruded material.

Material characteristics and structural parameters of the twin-screw pulping machine have an important influence on the straw pulping performance. In this work, the Discrete Element Model of straw crushing is created for the pulping process of rice straw in a twin-screw pulping machine based on the Tavares mathematical model, and the number of straw particles broken was analyzed. Ternary quadratic orthogonal center-of-rotation combination tests were conducted with rice straw pretreated using white rot fungi liquid as the experimental object, the straw pulping yield as evaluation indexes, and tooth groove angle, screw speed and straw moisture content as influencing factors, to improve the twin-screw pulp yield and the quality of the pulp without producing black liquor, to ensure a clean and efficient straw process.

2. Materials and Methods

The designed geometric model of the twin-screw pulping machine was imported into Rocky, the particle model of rice straw was set, the Tavares model was selected as the breakage model, and the sum of all forces and moments acting on the particles was calculated. The straw particle was moved to the next position until the end of the simulation, the position, velocity, and time step of the particle were calculated by the Rocky programs, and the simulation results were derived and verified. Based on the results of the simulation analysis, the twin-screw pulping machine was continually optimized until an optimal design was obtained. Multi-factor test with tooth groove angle, screw speed and straw moisture content as influencing factors, to explore the effect of each parameter on the pulping performance of the twin-screw machine, as shown in Figure 1.

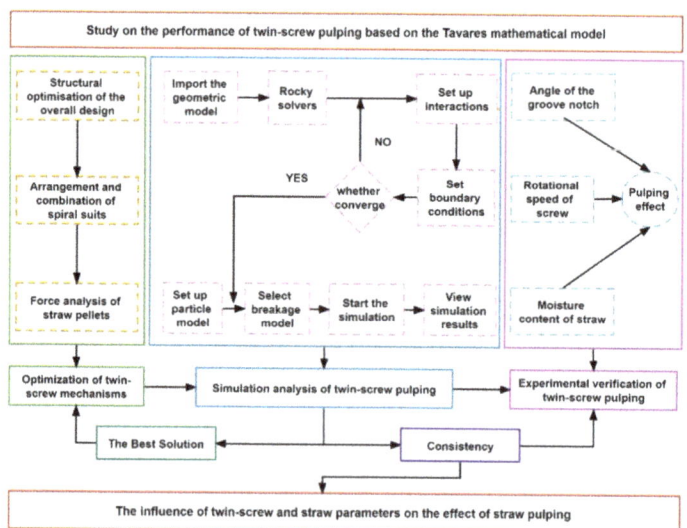

Figure 1. Process of twin-screw pulping simulation and experimental study.

2.1. Experimental Setup

2.1.1. Establishing a Mathematical Model of Twin-Screw Pulping Machine

The 3D model software named Solidworks was used to design and optimize the twin-screw pulping machine, which included screw discharge device, twin-screw device, screw feed device, rigid bearing chock, decelerator, coupling, drive motor, and body frame, as shown in Figure 2.

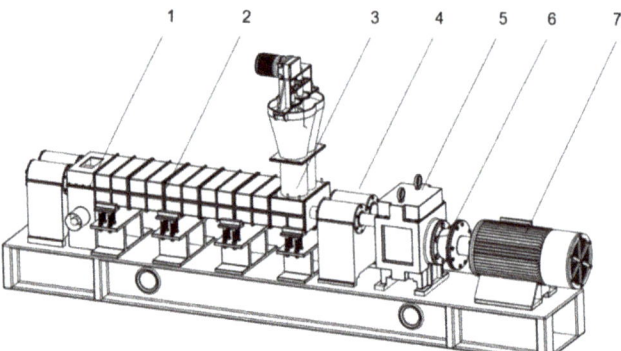

Figure 2. Structure of the twin-screw pulping machine. 1. screw discharge device; 2. twin-screw device; 3. screw feed device; 4. rigid bearing chock; 5. decelerator; 6. coupling; 7. drive motor.

The key component of pulping machinery, the twin-screw device, comprised of the positive spiral casing, negative spiral casing, conveying spiral casing and rotating shaft, as shown in Figure 3.

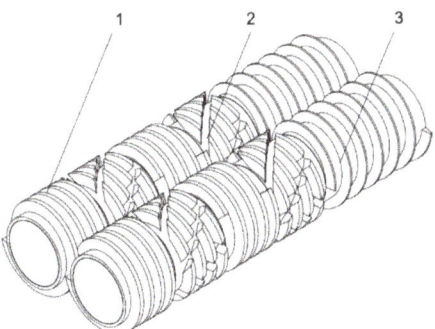

Figure 3. Twin-screw device. 1. positive spiral casing; 2. negative spiral casing; 3. conveying spiral casing.

The force on the straw through the positive spiral casing and the negative spiral casing formed a balanced force system to achieve the purpose of extrusion and shear [26]. The negative spiral casing was designed with the multi-headed reversal spiral groove, and the straw generated shear force and squeezing pressure when passing through the groove of the tooth. The equation of force on the tooth groove is:

$$\begin{cases} F_n = F_n \times \cos \alpha_n \times \sin \gamma \\ F_{r1} = F_n \times \sin_n \\ F_{a1} = F_n \times \cos \alpha_n \times \cos \gamma \end{cases} \quad (1)$$

2.1.2. Twin-Screw Structure Arrangement

In order to optimize the pulping performance of the twin-screw pulping device, the positive spiral casing and negative spiral casing were arranged in different combinations, which made the shearing rate, cumulative strain and specific mechanical energy of the straw material achieved the best value [27], as shown in Figure 4.

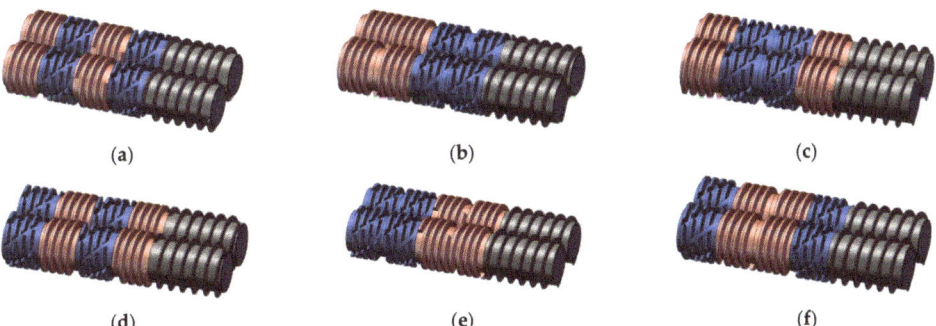

Figure 4. Combination of twin-screw mechanism spiral casing. (**a**) Positive-Negative-Positive-Negative (PNPN); (**b**) Positive-Positive-Negative-Negative (PPNN); (**c**) Positive-Negative-Negative-Positive (PNNP); (**d**) Negative-Positive-Negative-Positive (NPNP); (**e**) Negative-Negative-Positive-Positive (NNPP); (**f**) Negative-Positive-Positive-Negative (NPPN).

The specific mechanical energy (*SEM*) is calculated as follows:

$$SEM\left[\frac{KWh}{t}\right] = \frac{N * C * P_{max}}{N_{max} * C_{max} * Q} \quad (2)$$

where N (r/min) and N_{max} (r/min) are the rotational speed of the screw and the maximum speed of the screw; P_{max} (7 kW) is the maximum power of the motor; C (N·m) and C_{max} (130 N·m) are torque and maximum torque; Q (t/h) is total flow.

2.1.3. Force Analysis of Straw Particles

1. Force analysis of twin-screw on straw particles;

Due to the complex force interaction of straw particles in the extrusion-shear section of the negative spiral casing [28], assuming a single screw as the object of study and straw particles as a whole, when the straw particles moved in the twin-screw pulping device, the force on each micro-element could be decomposed into 8 partial forces, as shown in Figure 5.

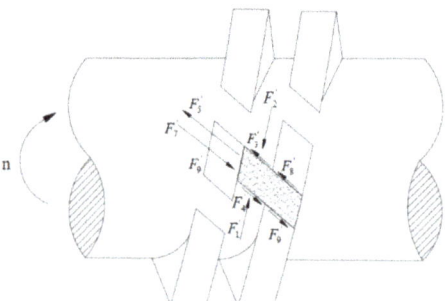

Figure 5. Force Analysis on the straw particles in the negative spiral casing. Where F'_1, F'_3, F'_4, F'_5 are friction forces; F'_2, F'_7 are extrusion forces; F'_8, F'_9 are shear forces.

A is the friction force of the screw surface on the straw particles, which is the driving force of the movement of the straw material flow in the screw teeth grooves, and is calculated as follows:

$$F'_1 = f_b P S_b dL_b \tag{3}$$

where f_b is friction coefficient between screw surface and straw particles; P (Pa) is the pressure of straw particles on screw shaft; dL_b (mm) is the micro increments of the screw surface along the helix direction; S_b (mm^2) is the contact area between straw particles and screw.

2. Force analysis between straw particles;

The action model between the straw particles is the BPM bonding model, where the contact part between the particles of rice straw produces parallel bonding, which is equivalent to setting a set of springs on the circular section of the straw particles [29]. The force analysis between the particles is shown in Figure 6.

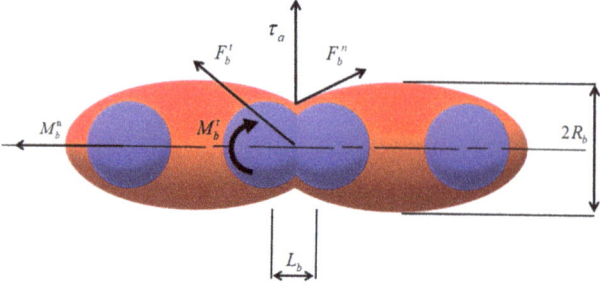

Figure 6. Force Analysis on the straw particles.

The torque of the rice straw particles is determined by Newton's second law, and the contact force between the particles due to relative motion is calculated by the force-displacement law [30]. The force equation of the straw particles is:

$$\begin{aligned} \delta F_b^n &= -v_n S_n A \delta t \\ \delta F_b^t &= -v_t S_t A \delta t \\ \delta M_b^n &= -\omega_n S_t J \delta t \\ \delta M_b^t &= -\omega_n S_n \tfrac{I}{2} \delta t \end{aligned} \quad (4)$$

where δt (s) is time step; v_n (m/s) and v_t (m/s) are normal and tangential velocities of particles; ω_n (r/min) and ω_t (r/min) are normal and tangential angular velocities of particles; J (kg·m^2) is moment of inertia; A (mm^2) is contact area; S_n (N/m) and S_t (N/m) are normal stiffness and tangential stiffness per unit area.

2.2. Simulation Test
2.2.1. Simulation Model

In the Tavares breakage model, the fragmentation probability is based on the upper truncated log-normal distribution [31]. When a particle collision event occurs, the breakage ratio energy decreases owing to the accumulated damage to the particle during loading. When the particle is broken, the geometry of the fragments is generated by the Voronoi fracture equation and distributed according to the Gaudin-Schumann function.

The breakage probability expression is:

$$P_o(e) = \frac{1}{2}\left[1 + erf\left(\frac{\ln e^* - \ln e_{50}}{\sqrt{2\delta^2}}\right)\right] \quad (5)$$

where e^* (J) is relative breakage energy; e_{50} (J) is average breakage energy; δ^2 is variance of the log-normal distribution of the breakage energy.

The relative breakage energy is calculated as follows:

$$e^* = \frac{e_{max} e}{e_{max} - e} \quad (6)$$

The average breakage energy is calculated as follows:

$$e_{50} = e_\infty \left[1 + \left(\frac{d_0}{L}\right)^\phi\right] \quad (7)$$

Rice straw was used as the experimental material, mainly composed of cellulose, hemicellulose and lignin, with a moisture content of 40–70%, a thickness of 5–10 mm and a length of 20–30 mm [32]. The stress-strain relationship of rice straw is following the generalized Hooke's law, and the breakage process is the same as that of the linear elastic material model. Assuming that the rice straw material parameters are linearly elastic, the Hertz-Milling bond contact model is adopted to analyze the rice straw particle fragmentation process, and energy consumption was calculated based on the forces and displacements of particle interactions during bond breakage [33]. The rice straw particle model usually is regarded as an axisymmetric ellipsoid in DEM simulations and is represented by a multi-sphere overlap connection method [34]. Based on the Tavares mathematical model, the rice particle model was simplified to an ellipsoid with a long axis of 52 mm and a short axis of 16 mm by measuring and calculating the dimensions of 100 rice straw particles. Further, regarding the irregular patterns of the straw particles, the straw particles were simulated as irregular polyhedra of 52 mm in length, 16 mm in width and 8 mm in height, as shown in Figure 7.

 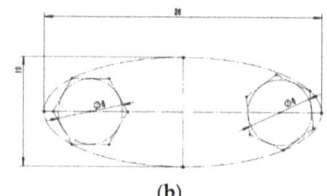 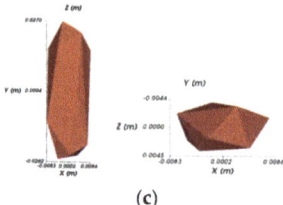

(a) (b) (c)

Figure 7. Modeling of rice straw particles. (**a**) Real straw particles; (**b**) straw particle simplification; (**c**) straw particle model.

Currently, only the JKR model (Johnson-Kendall-Roberts) can describe the surface interaction of elliptical contacts, where the shape of the JKR contact region changes throughout the contact [35]. In the JKR model, the line contact of different objects can be replaced by a cylindrical contact with an equivalent radius of curvature and a cylindrical contact with a bump. The contact between the straw particles can be replaced by line contact between two cylindrical surfaces with an equivalent radius of curvature. The equivalent model is shown in Figure 8. Based on the JKR model, the equivalent work was calculated according to the adhesion force between the straw particles, and the calculated value of the equivalent work was less than the ideal value, while the rough straw particle appearance caused the actual contact surface to be larger than the ideal contact surface [36].

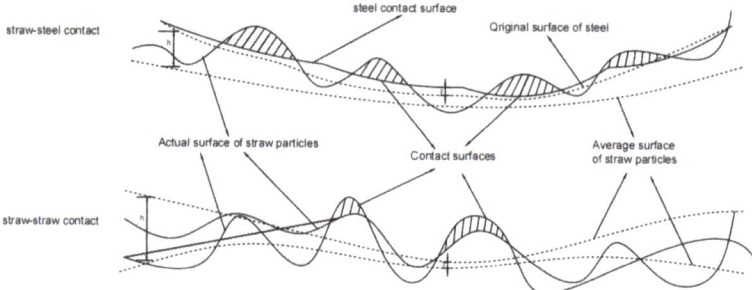

Figure 8. Surface contact equivalent model.

2.2.2. Parameter Setting

The parameters such as density, Poisson's ratio and shear elastic modulus of the straw particles and geometric model were set as shown in Table 1 [37].

Table 1. Setting of simulation parameters.

Straw particle	Density $\rho/(kg/m^3)$	1440
	Poisson ratio	0.25
	Shear elastic modulus $G/(Pa)$	1.5×10^6
Geometric model	Density $\rho/(kg/m^3)$	7800
	Poisson ratio	0.35
	Shear elastic modulus $G/(Pa)$	8×10^{10}
Particle-Particle	Recovery coefficient	0.65
	Static friction coefficient	0.18
	Rolling friction coefficient	0.01
Particle-Geometric mode	Recovery coefficient	0.65
	Static friction coefficient	0.15
	Rolling friction coefficient	0.01

2.2.3. Simulation Process

To find a suitable combination of spiral casing and tooth groove angle, different twin-screw structure combinations were designed for simulation analysis of pulping performance. The combination form of the spiral casing (P for positive spiral casing, N for negative spiral casing) and the arrangement form of tooth groove angle (W for 45°, S for 30°, R for 15°) are shown in Table 2.

Table 2. Combinations of spiral casing and arrangement of groove angle.

Spiral Sleeve Combination		Groove Angle Arrangement	
PNPN	NPPN	45°−30°−15° (WSR)	15°−45°−30° (RWS)
PPNN	PNNN	45°−15°−30° (WRS)	45°−45°−15° (WWR)
PNNP	NPNN	30°−45°−15° (SWR)	45°−45°−30° (WWS)
NPNP	NNPN	30°−15°−45° (SRW)	30°−30°−45° (SSW)
NNPP	NNNP	15°−30°−45° (RSW)	30°−30°−15° (SSR)

The number of straw particles was set to 2000, the rolling friction coefficient of particle-particle was 0.01, and the rolling friction coefficient of the particle-geometry model was 0.01. The computation step was set to 14 times, and the computation was terminated after 10 consecutive convergences, and the result was obtained after 8556 iterations to reach convergence. The simulation period of straw particles in the twin-screw pulping device was 21 s, intercepting the time points 0 s, 3 s, 6 s, 9 s, 12 s, 15 s, 18 s and 21 s of the combination breakage process for analysis. The straw particles have a certain viscosity that caused them to adhere to the twin-screw pulping device during the simulation process, which affected the accuracy of the simulation process. Therefore, the model was modified to ensure the accuracy of the model, as shown in Figure 9.

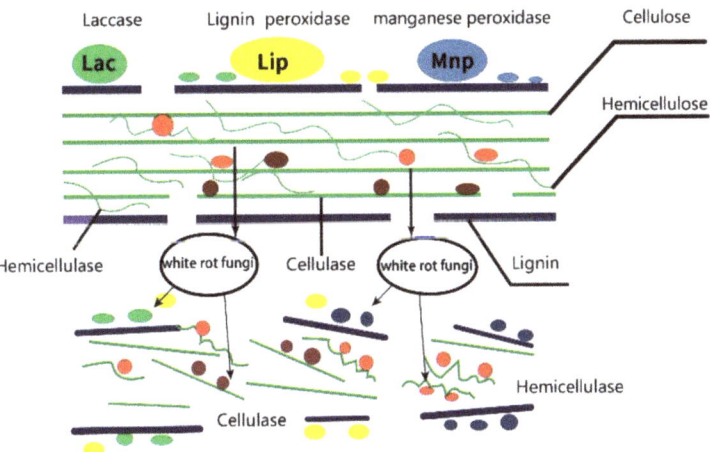

Figure 9. The mechanism of lignin degradation by white rot fungi.

2.3. Experimental Study

2.3.1. Trial Preparation

Rice straw was obtained from the experimental basement of the Shunbang Agricultural Machinery R&D Center in Siping City, Jilin Province, with a plant height of 0.7–1 m. Mature rice straw was selected and pre-treated to 5–10 cm straw by a kneading machine [38].

The CM0085 medium was sterilized at 121 °C for 30 min and cooled to room temperature. White rot fungi (Bio-114,230) were taken into the medium with an inoculation loop and were placed in a constant temperature vibration incubator and incubated continuously at 120 r/min for 7 days at 30 °C. The bacterial liquid was prepared by the expanded cultivation of white rot fungi and water in a ratio of 1:10 and sprayed evenly into the rice straw. The twin-screw pulping machine was operated and the microbiologically treated rice straw was pulped.

The mechanism of lignin degradation by white rot fungi is shown in Figure 9. A large number of extracellular oxidases were secreted during the mycelial growth and spread of white rot fungi, such as laccase (Lac), lignin peroxidase (Lip) and manganese peroxidase (MnP). The formation of H_2O_2 with the participation of molecular oxygen triggered the initiation of a series of free radical chain reactions to break the Cα-Cβ, β-O-4 and α-O-4 between the lignin benzene rings, followed by a series of reactions to degrade spontaneously.

Multi-factor test with tooth groove angle, screw speed and straw moisture content as influencing factors, and with pulp yield as an evaluation index of the performance of straw pulping [39]. Pulping yield is the percentage of pulp mass after lignin degradation of straw to the mass of straw before pulping, calculated as follows:

$$Q_e = \frac{G_2}{G_1} \times 100\% \qquad (8)$$

where Q_e (%) is pulp yield; G_1 (g) and G_2 (g) are straw mass before pulping and pulp mass of straw after lignin degradation.

2.3.2. Box-Behnken Experimental Design

The ternary quadratic orthogonal center of rotation combination test was used to optimize the combination of tooth groove angle, screw speed and straw moisture content, and the factor codes are shown in Table 3.

Table 3. Factor coding.

Coding	Factors		
	Groove Angle A/(°)	Screw Speed B/(r/min)	Straw Moisture Content C/(%)
−1.682	15	450	55
−1	20	500	60
0	25	550	65
1	30	600	70
1.682	35	650	75

2.4. Validation Experiment

2.4.1. Twin-Screw Structure Verification Experiment

In order to find out the effect of twin-screw structure parameters on the straw pulping performance, the optimized design of the twin-screw pulping device was studied by the platform of Shunbang Agricultural Machinery Manufacturing Limited Company (Siping, China). The twin-screw pulping machine as shown in Figure 10.

The combination of the spiral casing of twin-screw pulping machine has a certain influence on the pulping performance [40]. Fibers passing through the twin screw machine always present a certain level of fibrillation and part of the hemicellulose and extractives would be removed [41]. The force of the positive spiral casing on the straw particles is the squeezing pressure, which is too strong for the bond between the straw particles and affects the degradation of the straw lignocellulose [42]. The force of the negative spiral casing on the straw particles is the shearing pressure, which is too strong for some of the straw particles attached to the side of the tooth groove, hindering the flow of straw materials. The arrangement of the tooth groove angle on negative spiral casing has a certain influence on the pulping performance [43]. If the angle of the tooth groove is too large, the shearing force on the straw is small and not enough to break the internal van der Waals force of the straw. If the angle of the tooth groove is too small, the straw material input is too small, and the straw material clearance is too large that the extrusion pressure on the straw is too small to destroy the internal lignocellulosic component structure of the straw. Therefore, the experimental study of the twin-screw pulping performance was carried out by changing the combination of the spiral casing and the arrangement of the tooth groove angle on negative spiral casing.

Figure 10. Twin-screw pulping machine on location.

2.4.2. Straw Moisture Content Verification Experiment

The initial moisture content of rice straw was 40%, and the 36 groups of samples after different degrees of soaking treatment were put into high-temperature drying ovens and dried at a temperature of 105 °C for more than 12 h to obtain the dry mass of rice straw. The moisture content of the samples in real-time was calculated according to the rice straw moisture content formula [44], which is calculated as follows:

$$M_e = \frac{W_1 - W_2}{W_1} \times 100\% \tag{9}$$

where M_e (%) is moisture content of rice straw; W_1 (g) and W_2 (g) are mass of rice straw samples before drying and dry mass of rice straw samples.

The experimental study on the effect of straw moisture content on pulping performance was carried out by using the twin-screw pulping machine to pulping rice straw with moisture content of 50%, 55%, 60%, 65%, 70% and 75%, keeping the screw speed, tooth groove angle and pulping time consistent.

2.4.3. Screw Speed Verification Experiment

Set the twin-screw speed and use the cardboard press roll forming machine to make the pulping into straw board. Particularly, intercept 100 cm² of the straw board, using image recognition technology to calculate the number of stomata and trichome features of the straw board [45].

3. Results and Discussions

3.1. Simulation Results

From Figure 11a,b, it can be seen that at the beginning of the straw particles entering the twin-screw pulping device, the straw particles were in order and did not start to break. From Figure 11c,d, it can be seen that the straw particles entering the twin-screw conveying stage, the straw particles under the forward thrust and tangential force of the twin-screw, breaking the orderly arrangement of the state to irregular and disordered distribution, the friction between the straw particles made the initial breakage of the straw particles and began to accumulate breakage ratio energy. From Figure 11e,f, it can be seen that the straw particles started to enter the extrusion-shear stage with tighter distribution and were under squeezing pressure from the positive spiral casing and shearing force from the negative spiral casing. The heat generated during the extrusion-shear stage could soften the lignin binding the fibers so that the hydrogen and covalent bonds between the bad cellulose and lignin could be easily broken [46]. The accumulated breakage ratio energy can resist part of the van der Waals force inside the straw particles and the straw particles started to break initially, yet the breaking effect is not obvious. From Figure 11g,h, it can be seen that the straw particles completely entered the pulping stage with a tight distribution and were under the pressure of the positive spiral casing and the shearing force of the negative spiral casing. The accumulated breakage ratio energy completely eliminated the van der Waals force inside the straw particles, and the straw particles broke rapidly in large quantities. Comparing and analyzing the number of broken straw particles, when the spiral casing combination form was NPNP and the angle arrangement of the negative spiral casing was WSR, the number of straw particles crushed was highest.

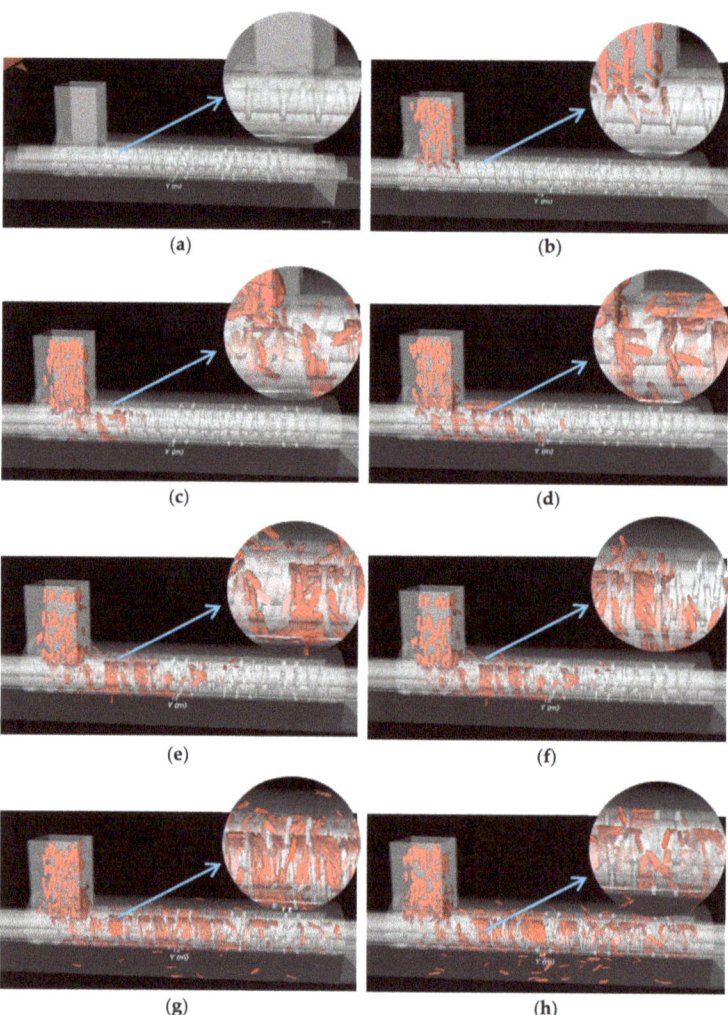

Figure 11. Simulation of breaking of straw particles in a twin screw machine for cellulose in time for 0 s (**a**), 3 s (**b**), 6 s (**c**), 9 s (**d**), 12 s (**e**), 15 s (**f**), 18 s (**g**) and 21 s (**h**).

3.2. Multi-Factor Experiment Results

3.2.1. Multi-Factor Experiment Results and Analysis

The experimental design was carried out with Design Expert 12.0 to analyze the factors to rationalize the data obtained from the experiment, as shown in Table 4.

Table 4. Analyzed factors of the experiment.

Number	Factors			
	Groove Angle A/(°)	Screw Speed B/(r/min)	Straw Moisture Content C/(%)	Pulping Yield/%
1	1	1	1	91.4
2	1	1	−1	82.8
3	1	−1	1	88.9
4	1	−1	−1	90.5
5	−1	1	1	89.3
6	−1	1	−1	86.7
7	−1	−1	1	88.3
8	−1	−1	−1	85.4
9	1.682	0	0	92.5
10	−1.682	0	0	84.2
11	0	1.682	0	89.3
12	0	−1.682	0	91.3
13	0	0	1.682	87.7
14	0	0	−1.682	92.1
15	0	0	0	83.4
16	0	0	0	85.5
17	0	0	0	87.4
18	0	0	0	90.9
19	0	0	0	89.6

A second-order regression model of the pulping performance with the three independent variables of tooth groove angle, screw speed, and straw moisture content was developed, its quadratic polynomial equation is:

$$\theta = 90.2 + 0.93\,A + 3.20\,B + 1.94\,C - 0.10\,AB - 0.03\,AC + 1.8\,BC - 7.86\,A^2 - 16.39\,B^2 - 7.21\,C^2 \quad (10)$$

The results of the ANOVA are shown in Table 5 and the results showed that the equation model was highly significant with $p < 0.001$. The Lack of fit $p = 0.1239$ indicated that the equation fit was credible, reliable, and accurate. Tooth groove angle, screw speed, and straw moisture content had a highly significant effect on the pulping performance ($p < 0.01$). The interaction term BC had a significant effect on the pulping performance ($p = 0.0305$) and the secondary terms A^2, B^2 and C^2 had a significant effect on the pulping performance ($p < 0.05$). The order of influence of each factor on pulping performance is $B > C > A$.

3.2.2. Box-Behnken Experiment Results

The Box-Behnken experiment was designed to obtain the response surface graph of the interaction of tooth groove angle, screw speed, and straw moisture content on the pulping yield, as shown in Figure 12. The center of the response surface plane projection graph was located within the ellipse, indicating that the interactions AB, BC, and AC had a significant effect on the pulping performance.

Table 5. Analysis of variance of the Box-Behnken experiment.

Source	Sum of Squares	Freedom	Mean Square	p
Model	1894.46	9	210.5	<0.0001
A	7.03	1	7.03	0.0076
B	81.92	1	81.92	0.0035
C	30.03	1	30.03	0.0051
AB	0.04	1	0.04	0.9269
AC	0.0025	1	0.0025	0.9817
BC	12.96	1	12.96	0.0305
A^2	260.29	1	260.29	0.0001
B^2	1130.74	1	1130.74	<0.0001
C^2	219.03	1	219.03	0.0002
Residual	30.93	7	4.42	/
Lack of fit	22.57	3	7.52	0.1239
Error	8.36	4	2.09	/
Total value	1925.39	16	/	/

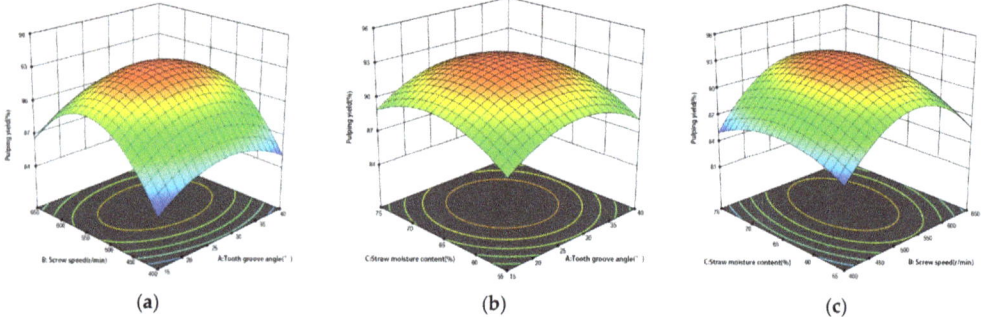

Figure 12. Response surface graph of the effect of interactions on straw the pulping rate. (a) The effects of screw speed (B) vs. tooth groove angle (A) on pulping yield; (b) the effects of straw moisture content (C) vs. tooth groove angle (A) on pulping yield; (c) the effects of straw moisture content (C) vs. screw speed (B) on pulping yield.

From Figure 12a, it was found that the pulp yield was 83.4% when the screw speed was 400 r/min and the tooth groove angle was 15°. Higher screw speed could increase the extrusion force and shear force on the straw and promote the production of extracellular oxidase by white rot fungi in straw and enhance the degree of lignin degradation. Lignin derivatives produced during lignin degradation could inhibit the hydrolysis of cellulase and xylanase, resulting in a relatively higher pulp yield [47]. The pulp yield was 85.5% when the screw speed was 650 r/min and the tooth groove angle was 15°. The larger tooth groove angle will lead to a larger opening area and smaller cross-sectional area, which will increase the feeding amount of the straw and improve the pulp yield.

From Figure 12b, it was found that the pulp yield was 84.2% when the straw moisture content was 55% and the tooth groove angle was 15°. With the increasing moisture content of straw, the internal van der Waals forces and chemical bonding of straw were gradually broken and the pulp yield improved. The pulp yield was 87.7% when the straw moisture content was 75% and the tooth groove angle was 15°. With the increasing angle of the tooth groove, the opening area increased and the cross-sectional area decreased, which improved the feeding amount and pulp yield.

From Figure 12c, it was found that the pulp yield was 82.8% when the screw speed was 400 r/min and the straw moisture content was 55%. With the increase of straw moisture content, free water was generated within the straw to fill the cell cavities, and the internal forces within the straw particles were easily broken leading to the improvement of pulp yield. The pulp yield was 85.4% when the screw speed was 400 r/min and the straw moisture content was 75%. With the increase of screw speed, the twin-screw extrusion pressure and shear force on the straw gradually eliminated the internal van der Waals force of the straw. The lignin degradation rate was increased while the hydrolysis of cellulose and hemicellulose was inhibited and the pulp yield improved.

Straw pulping performance and pulp yield are positively correlated, that is, the straw pulping performance is the macro performance of the pulp yield. Higher pulping yield leads to better straw pulping performance [48].

3.3. Results of the Validation Experiment

3.3.1. Effect of Twin-Screw Structure Parameters on the Straw Pulping Performance

The experiment results are shown in Figure 13, the straw pulping performance was poor when the combination of the spiral casing was PNNN and NNNP, and the adjustment of the tooth groove angle failed to improve the straw pulping performance. The reason is that the number of positive spiral casings was large and the number of negative spiral casings was small, resulting in a larger squeezing force and smaller shearing force on the straw particles. Excessive squeezing pressure caused stronger bonding between straw particles and difficulty in breaking the hydrogen bond between cellulose and lignin, which hindered the degradation of straw lignocellulose and poor straw pulping performance. The straw pulping performance was poor when the spiral casing combination was NPNN and NNPN, while the straw pulping performance get improved when the tooth groove angle arrangement was SSR. The reason is that the small tooth groove angle and large cross-sectional area increased the shear force on the straw particles, making up for the small shear force of the spiral casing combination. It caused an easy disruption of the covalent bond between cellulose and lignin to improve the degradation of straw lignocellulose. However, the shear force was still not enough to break the internal van der Waals force of straw, and the straw pulping performance was also poor [49]. The straw pulping performance was poor when the spiral casing combination was PNNP and NPPN, while the straw pulping performance get improved when the tooth groove angle arrangement was SWR and RWS. The reason is that the angles of tooth groove were arranged as small-large-small, so that the straw material was first ground by large shear force, then flowed with higher flow, and finally ground by large shear force for the second time. The disadvantages of continuous shearing and continuous extrusion of the spiral casing combination were compensated, and the $C\alpha$-$C\beta$, β-O-4 and α-O-4 between the lignin benzene rings were broken to improve the straw pulping performance. The straw pulping performance was better when the combination of spiral casing was PPNN and NNPP, and the degradation of straw lignin was hindered with the combination of SSR and WWR in the tooth groove angle. The reason is that the angle of the tooth groove was continuously too large to small leading to a rapid decrement in flow, resulting in straw material blocking. The angle of the tooth groove was continuously small to large, resulting in the shearing force being too large to small, and the material feeding speed was too fast to grind the straw material completely, which affected the performance of straw pulping. The straw pulping performance was the best when the spiral casing combination was NPNP and PNPN, while the straw pulping performance get improved when the tooth groove angle arrangement was SSR. The reason is that the angle of the tooth groove was gradually smaller enabling the flow of straw material to decrease steadily, while the shearing force was increasing to grind straw more completely. It contributed to the cracking of the fiber surface and enhances the separation of the fiber material. The slowly changing material flow solved the problem of material blockage and achieved the optimal straw pulping performance.

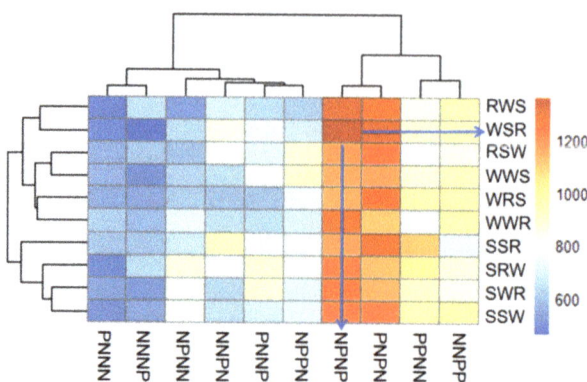

Figure 13. Influence of twin-screw structure parameters on the performance of straw pulping.

In summary, the straw pulping performance was the worst when the twin-screw pulping device spiral casing combination was PNNN and NNNP. The straw pulping performance was poor when the twin-screw pulping device spiral casing combination was NPNN and NNPN. The straw pulping performance was improved when the spiral casing combination was PNNP and NPPN and the tooth groove angle arrangement was SWR and RWS. The straw pulping performance was good when the spiral casing combination was PPNN and NNPP and the tooth groove angle arrangement was SSR and WWR. The straw pulping performance was the best when the spiral casing combination was NPNP and the tooth groove angle arrangement was WSR.

3.3.2. Effect of Straw Moisture Content on Pulping Performance

From Figure 14a, it was found that the straw slurry presented an inhomogeneous multi-angular agglomerate morphology, where the percentage of large multi-angular agglomerates was 65.4%. The reason is that when the straw moisture content was 50%, the internal force of the straw and the lignin benzene rings had not been destroyed and the straw fiber components were not separated so straw pulping performance was the worst [50]. From Figure 14b, it was found that the straw slurry presented inhomogeneous flat-ribbon agglomerate morphology, where the percentage of large inhomogeneous flat-ribbon agglomerates was 45.3%. The reason is that when the straw moisture content was 55%, part of the straw internal force was broken and the straw fiber component separation degree was low. The hydrogen bond between lignin and cellulose was only partially broken so the straw pulping performance was poor [51]. From Figure 14c, it was found that the straw slurry presented an inhomogeneous spherical agglomerate morphology, where the percentage of large spherical agglomerates was 25.7%. The reason is that when the straw moisture content was 60%, The activity of extracellular oxidase produced by white rot fungi that could degrade lignin was increased and the straw fiber components were separated to a higher degree so that the straw pulping performance was good. From Figure 14d, it was found that the straw slurry presented a homogeneous fine-grained agglomerate morphology, where the percentage of large fine-grained agglomerates was 6.5%. The reason is that when the straw moisture content was 65%, the straw internal forces were destroyed extremely high, and the more active oxidase triggered a series of free radical chain reactions that caused spontaneous degradation of the lignin benzene rings in a range of reactions. The straw fiber components were separated highly so that the straw pulping performance was the best. From Figure 14e, it was found that the straw slurry presented an inhomogeneous cottony agglomerate morphology, where the percentage of large cottony agglomerates was 16.8%. The reason is that when the water content of straw was 70%, the internal force of straw and the hydrogen bonds between lignin and cellulose were basically destroyed and the lignin in the straw fiber component was basically degraded so that the straw pulping performance was good. From Figure 14e, it was found that the straw slurry presented an inhomogeneous thin-strip agglomerate morphology, where the percentage of large thin-strip agglomerates was 36.9%. The reason is that when the straw moisture content was 75%, the oxidase activity and the degradation of lignin benzene ring were decreased. Part of the straw internal forces had not been destroyed and the lignin degradation rate in the straw fiber component was reduced so that the straw pulping performance was poor.

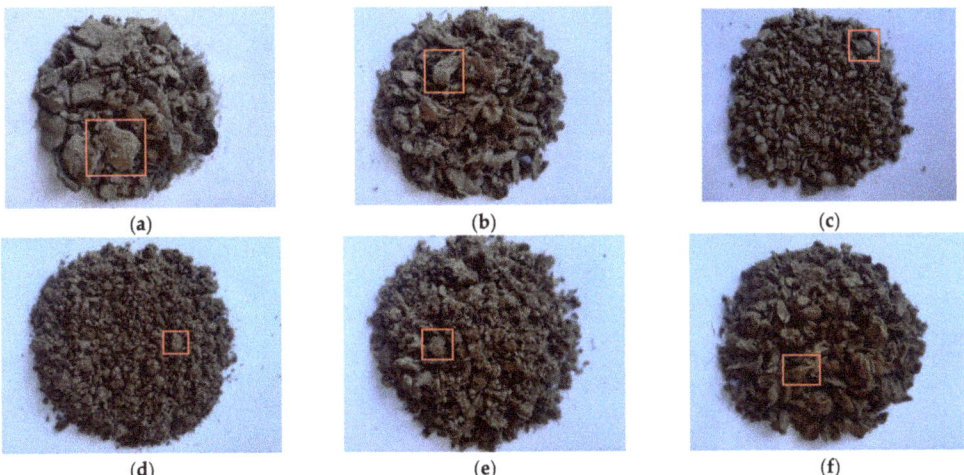

Figure 14. Illustration of straw pulp with different moisture content. Straw moisture content is 50% (a), 55% (b), 60% (c), 65% (d), 70% (e), and 75% (f).

In summary, when the straw moisture content was lower than 60%, only bound water existed inside the straw and the lack of free water to fill the cell cavity. The straw internal van der Waals force and chemical bonding were strong, and the lignin benzene rings were hardly broken leading to difficulty in breaking the straw. When the moisture content of straw was between 60 and 70%, the force within the straw was easily destroyed and the lignin benzene rings were easily disrupted by the effect of highly active lignin-degrading enzymes. The straw was broken to a strong degree resulting in good straw pulping performance. When the water content of straw was higher than 70%, the cell cavity of straw was filled with free water, the squeezing pressure and shear force on straw particles were buffered, and some of the van der Waals forces inside the straw were not destroyed. Part of the hydrogen bond between lignin and cellulose was not broken so some of the lignin was not degraded and the straw pulping performance turned poor. Therefore, the straw moisture content had a significant effect on the pulping performance, which was consistent with the simulation results.

3.3.3. The Effect of Screw Speed on Pulping Performance

From Figure 15a, it was found that part of the straw showed detachment from the main body with a high number of trichomes and rough straw fiber, including a total of 413 trichomes larger than 3 mm and 308 large blowholes. The reason is that when the screw speed was 450 r/min, the extrusion pressure and shear force on the straw failed to resist the van der Waals force inside the straw, and the lignin was not degraded resulting in poor pulping performance [52]. From Figure 15b, it was found that part of the straw fiber was attached to the surface of the slurry, the number of trichomes was relatively small and although the generally close but part of the straw fiber was rough, including 171 trichomes larger than 3 mm and 18 small blowholes. The reason is that when the screw speed was 500 r/min, the extrusion pressure and shear force on the straw eliminated most of the internal van der Waals force of the straw. Cross-linking between fibers was broken and most of the lignin degradation led to better pulping performance. From Figure 15c, it was found that part of the straw fiber was slightly longer but the general straw fiber was close and evenly distributed, with a total of 84 trichomes larger than 3 mm and 10 blowholes. The reason is that when the screw speed was 550 r/min, the extrusion pressure and shear force on the straw basically eliminated the internal van der Waals force of the straw. The production of hair-like fibers on the fiber surface increased the surface area and the lignin basically degraded to allow the best pulp-making performance. From Figure 15d, it was found that part of the straw fiber stuck together into a block, and the straw fiber was rough and unevenly distributed, including 260 trichomes larger than 3 mm and 57 larger blowholes. The reason is that when the screw speed was 600 r/min, although the extrusion pressure and shear force on the straw eliminated part of the straw internal van der Waals force but the elimination effect was uneven. High-speed rotation inhibited the activity of lignin-degrading enzymes resulting in poor pulping performance.

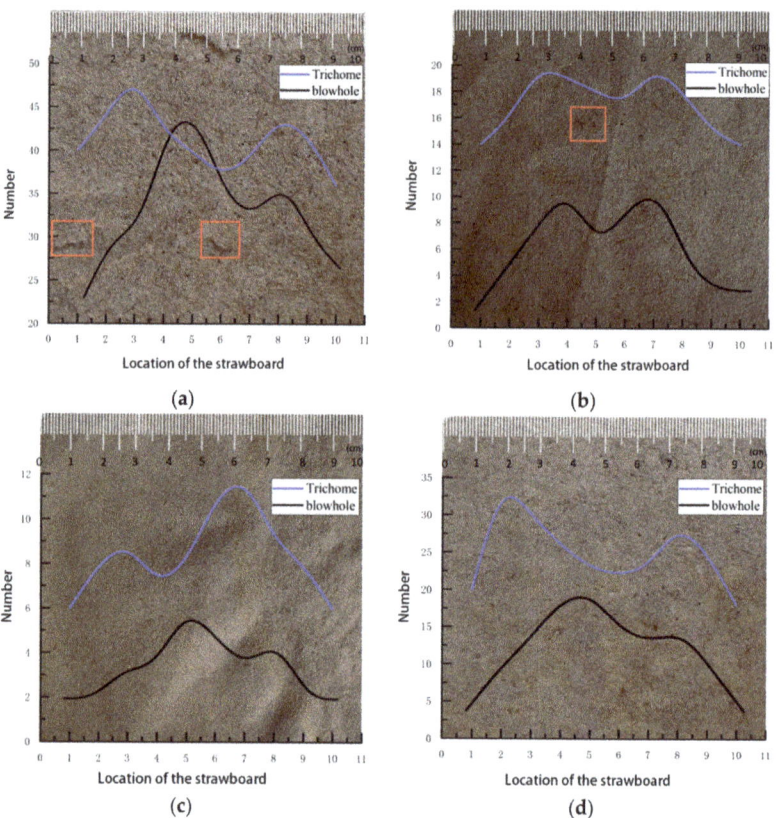

Figure 15. The pattern of trichomes and blowholes in the sheets achieved by different speeds of the screw. The speeds of the screw in (**a**–**d**) are 450 r/min, 500 r/min, 550 r/min and 600 r/min.

In summary, when the screw speed was lower than 500 r/min, the energy generated by the friction between the screw and the straw was lower than the straw breakage ratio energy, resulting in a large amount of unbroken straw and poor pulping performance. When the screw speed is higher than 550 r/min, the straw slurry blocked the twin-screw pulping device, which affected its lifetime. High-speed rotation inhibited the activity of lignin-degrading enzymes and led to a poor straw pulping performance. When the screw speed was between 500 r/min and 550 r/min, the extrusion pressure and shear force on the straw were large and the broken straw fiber was small resulting in good straw pulping performance. Therefore, the screw speed had a significant effect on the pulping performance, which was consistent with the simulation results.

4. Conclusions

This work investigated the pulping performance of rice straw in a twin-screw machine for the rice straw treated with white rot fungi liquid by combining macroscopic and microscopic perspectives and ranking the role of different factors on the pulping performance. A simulation of straw particle breakage processes in twin-screw pulping machines based on the Tavares breakage model and straw particle fragments was produced by the Voronoi fracture equation; according to the distribution law of the Gaudin-Schumann function to calculate the fragmentation probability of straw particles, comparing and analyzing the number of broken straw particles with different combinations of twin-screw structure parameters. When the spiral casing combination is negative-positive-negative-positive (NPNP), the tooth groove angle arrangement of the negative spiral casing is $45°-30°-15°$ (WSR) and the moisture content of straw was 65%, the number of straw particle breakage is highest.

A ternary quadratic orthogonal center-of-rotation combination test was designed to establish a model between the pulp yield and the three independent factors of tooth groove angle, screw speed and straw moisture content. Multi-factor experiments showed that the order of the influencing factors on the pulp yield was screw speed > straw moisture content > tooth groove angle. When the screw speed is 550 r/min, the tooth groove angle is 30°, and the straw moisture content is 65%, the pulp yield reaches the highest value of 92.5%.

The straw pulping performance of the twin-screw pulping machine was verified, and the best pulping performance is achieved when the combination of the spiral casing of the twin-screw pulping machine is NPNP and the angle arrangement of the tooth groove is WSR. When the straw moisture content is 65%, the straw internal force is easily destroyed, and the straw pulping effect is good with a good straw broken degree. When the screw speed is 550 r/min, the extrusion pressure and shearing force of the twin screws on the straw is large, and the straw is finely broken leading to a good pulping performance. It shows that the experimental results are consistent with the simulation results and DEM can be used to study the analysis of the breakage process of straw and other materials.

Author Contributions: Conceptualization: D.R. and Y.G.; software, N.Z. and L.Z.; validation: H.C. and D.L.; supervisory role: D.R.; writing—review and editing: H.C.; funding acquisition and supervision: D.R. and Y.G. All authors have read and agreed to the published version of the manuscript.

Funding: This research was funded by the Educational Department of Liaoning Province (LSNQN202027).

Institutional Review Board Statement: Not applicable.

Informed Consent Statement: Not applicable.

Data Availability Statement: Not applicable.

Conflicts of Interest: The authors declare no conflict of interest.

References

1. Wang, W.; Lai, D.Y.; Sardans, J.; Wang, C.; Datta, A.; Pan, T.; Zeng, C.; Bartrons, M.; Peñuelas, J. Rice straw incorporation affects global warming potential differently in early vs. late cropping seasons in Southeastern China. *Field Crop. Res.* **2015**, *181*, 42–51. [CrossRef]
2. Liang, J.J. Analysis on Comprehensive Utilization of Straw in Agricultural Area. In Proceedings of the 7th ICASS International Conference on Social Science and Information (SSI 2018), Lima, Peru, 27–29 December 2018; pp. 30–34.
3. Hirani, A.H.; Javed, N.; Asif, M.; Basu, S.K.; Kumar, A. A review on first-and second-generation biofuel productions. In *Biofuels: Greenhouse Gas Mitigation and Global Warming*; Springer: New Delhi, India, 2018; pp. 141–154.
4. Sun, M.; Wang, Y.; Shi, L. Environmental performance of straw-based pulp making: A life cycle perspective. *Sci. Total Environ.* **2017**, *616–617*, 753–762. [CrossRef] [PubMed]

5. Weerasekara, N.S.; Powell, M.S.; Cleary, P.W.; Tavares, L.; Evertsson, M.; Morrison, R.; Quist, J.; de Carvalho, R. The contribution of DEM to the science of comminution. *Powder Technol.* **2013**, *248*, 3–24. [CrossRef]
6. Tavares, L.M.; André, F.P.; Potapov, A.; Maliska, C. Adapting a breakage model to discrete elements using polyhedral particles. *Powder Technol.* **2019**, *362*, 208–220. [CrossRef]
7. Tavares, L.M. Review and Further Validation of a Practical Single-Particle Breakage Model. *KONA Powder Part. J.* **2022**, *39*, 62–83. [CrossRef]
8. Barrios, G.K.; Jiménez-Herrera, N.; Tavares, L. Simulation of particle bed breakage by slow compression and impact using a DEM particle replacement model. *Adv. Powder Technol.* **2020**, *31*, 2749–2758. [CrossRef]
9. Tavares, L.M.; de Carvalho, R.M. Modeling breakage rates of coarse particles in ball mills. *Miner. Eng.* **2009**, *22*, 650–659. [CrossRef]
10. Oliveira, A.; Rodriguez, V.; de Carvalho, R.; Powell, M.; Tavares, L. Mechanistic modeling and simulation of a batch vertical stirred mill. *Miner. Eng.* **2020**, *156*, 106487. [CrossRef]
11. Tavares, L.M.; Rodriguez, V.A.; Sousani, M.; Padros, C.B.; Ooi, J.Y. An effective sphere-based model for breakage simulation in DEM. *Powder Technol.* **2021**, *392*, 473–488. [CrossRef]
12. Zeng, Y.; Mao, B.; Jia, F.; Han, Y.; Li, G. Modelling of grain breakage of in a vertical rice mill based on DEM simulation combining particle replacement model. *Biosyst. Eng.* **2022**, *215*, 32–48. [CrossRef]
13. Xu, T.; Zhang, R.; Wang, Y.; Jiang, X.; Feng, W.; Wang, J. Simulation and Analysis of the Working Process of Soil Covering and Compacting of Precision Seeding Units Based on the Coupling Model of DEM with MBD. *Processes* **2022**, *10*, 1103. [CrossRef]
14. Rahman, M.; Avelin, A.; Kyprianidis, K. An Approach for Feedforward Model Predictive Control of Continuous Pulp Digesters. *Processes* **2019**, *7*, 602. [CrossRef]
15. Garzon, L.; Fajardo, J.I.; Rodriguez-Maecker, R.; Fernandez, E.D.; Cruz, D. Thermo-Mechanical and Fungi Treatment as an Alternative Lignin Degradation Method for Bambusa oldhamii and Guadua angustifolia Fibers. *J. Fungi* **2022**, *8*, 399. [CrossRef] [PubMed]
16. Liu, T.Z.; Kittikunakorn, N.; Zhang, Y.; Zhang, F. Mechanisms of twin screw melt granulation. *J. Drug Deliv. Sci. Technol.* **2021**, *61*, 102150. [CrossRef]
17. Pradhan, S.U.; Sen, M.; Li, J.; Litster, J.D.; Wassgren, C.R. Granule breakage in twin screw granulation: Effect of material properties and screw element geometry. *Powder Technol.* **2017**, *315*, 290–299. [CrossRef]
18. Shirazian, S.; Darwish, S.; Kuhs, M.; Croker, D.M.; Walker, G. Regime-separated approach for population balance modelling of continuous wet granulation of pharmaceutical formulations. *Powder Technol.* **2018**, *325*, 420–428. [CrossRef]
19. Bumm, S.H.; White, J.L.; Isayev, A.I. Glass fiber breakup in corotating twin screw extruder: Simulation and experiment. *Polym. Compos.* **2012**, *33*, 2147–2158. [CrossRef]
20. Rol, F.; Belgacem, N.; Meyer, V.; Petit-Conil, M.; Bras, J. Production of fire-retardant phosphorylated cellulose fibrils by twin-screw extrusion with low energy consumption. *Cellulose* **2019**, *26*, 5635–5651. [CrossRef]
21. Theng, D.; Arbat, G.; Delgado-Aguilar, M.; Ngo, B.; Labonne, L.; Evon, P.; Mutjé, P. Comparison between two different pretreatment technologies of rice straw fibers prior to fiberboard manufacturing: Twin-screw extrusion and digestion plus defibration. *Ind. Crop. Prod.* **2017**, *107*, 184–197. [CrossRef]
22. Ho, T.T.T.; Abe, K.; Zimmermann, T.; Yano, H. Nanofibrillation of pulp fibers by twin-screw extrusion. *Cellulose* **2014**, *22*, 421–433. [CrossRef]
23. Kępa, K.; Chaléat, C.M.; Amiralian, N.; Batchelor, W.; Grøndahl, L.; Martin, D.J. Evaluation of properties and specific energy consumption of spinifex-derived lignocellulose fibers produced using different mechanical processes. *Cellulose* **2019**, *26*, 6555–6569. [CrossRef]
24. Espinosa, E.; Rol, F.; Bras, J.; Rodríguez, A. Production of lignocellulose nanofibers from wheat straw by different fibrillation methods. Comparison of its viability in cardboard recycling process. *J. Clean. Prod.* **2019**, *239*, 118083. [CrossRef]
25. Liang, F.; Fang, G.; Jiao, J.; Deng, Y.; Han, S.; Shen, K.; Shi, Y.; Li, H.; Zhu, B.; Pan, A.; et al. The Use of Twin Screw Extruder Instead of Model Screw Device During Bamboo Chemo-mechanical Pulping. *BioResources* **2018**, *13*, 2487–2498. [CrossRef]
26. Yang, X.; Wang, G.; Miao, M.; Yue, J.; Hao, J.; Wang, W. The Dispersion of Pulp-Fiber in High-Density Polyethylene via Different Fabrication Processes. *Polymers* **2018**, *10*, 122. [CrossRef]
27. Kratky, L.; Jirout, T. Modelling of particle size characteristics and specific energy demand for mechanical size reduction of wheat straw by knife mill. *Biosyst. Eng.* **2020**, *197*, 32–44. [CrossRef]
28. Yu, W.; Kai, W.; Yu, S. Effects of raw material particle size on the briquetting process of rice straw. *J. Energy Inst.* **2016**, *91*, 153–162.
29. Lenaerts, B.; Aertsen, T.; Tijskens, E.; De Ketelaere, B.; Ramon, H.; De Baerdemaeker, J.; Saeys, W. Simulation of grain–straw separation by Discrete Element Modeling with bendable straw particles. *Comput. Electron. Agric.* **2014**, *101*, 24–33. [CrossRef]
30. Molari, L.; Maraldi, M.; Molari, G. Non-linear rheological model of straw bales behavior under compressive loads. *Mech. Res. Commun.* **2017**, *81*, 32–37. [CrossRef]
31. Tavares, L.M. Analysis of particle fracture by repeated stressing as damage accumulation. *Powder Technol.* **2009**, *190*, 327–339. [CrossRef]
32. Liu, Y.; Li, Y.; Zhang, T.; Huang, M. Effect of concentric and non-concentric threshing gaps on damage of rice straw during threshing for combine harvester. *Biosyst. Eng.* **2022**, *219*, 1–10. [CrossRef]

33. Zhang, T.; Zhao, M.; Liu, F.; Tian, H.; Wulan, T.; Yue, Y.; Li, D. A discrete element method model of corn stalk and its mechanical characteristic parameters. *BioResources* **2020**, *15*, 9337–9350. [CrossRef]
34. Meng, X.; Han, Y.; Jia, F.; Chen, P.; Xiao, Y.; Bai, S.; Zhao, H. Numerical simulation approach to predict the abrasion rate of rice during milling. *Biosyst. Eng.* **2021**, *206*, 175–187. [CrossRef]
35. Zini, N.H.M.; de Rooij, M.B.; Bazr Afshan Fadafan, M.; Ismail, N.; Schipper, D.J. Extending the Double-Hertz Model to Allow Modeling of an Adhesive Elliptical Contact. *Tribol. Lett.* **2018**, *66*, 30. [CrossRef]
36. Zhang, T.; Peng, W.; Shen, K.; Yu, S. AFM measurements of adhesive forces between carbonaceous particles and the substrates. *Nucl. Eng. Des.* **2015**, *293*, 87–96. [CrossRef]
37. Zhang, S.; Ma, L.; Gao, S.; Zhu, C.; Yan, Y.; Liu, X.; Li, L.; Chen, H. A Value-Added Utilization Method of Sugar Production By-Products from Rice Straw: Extraction of Lignin and Evaluation of Its Antioxidant Activity. *Processes* **2022**, *10*, 1210. [CrossRef]
38. Jia, H.; Deng, J.; Deng, Y.; Chen, T.; Wang, G.; Sun, Z.; Guo, H. Contact parameter analysis and calibration in discrete element simulation of rice straw. *Int. J. Agric. Biol. Eng.* **2021**, *14*, 72–81. [CrossRef]
39. Lai, Y.-H.; Sun, H.-C.; Chang, M.-H.; Li, C.-C.; Shyu, J.-G.; Perng, Y.-S. Feasibility of substituting old corrugated carton pulp with thermal alkali and enzyme pretreated semichemical mechanical rice straw pulp. *Sci. Rep.* **2022**, *12*, 3493. [CrossRef]
40. Razzak, A.; Khiari, R.; Moussaoui, Y.; Belgacem, M.N. Cellulose Nanofibers from Schinus molle: Preparation and Characterization. *Molecules* **2022**, *27*, 6738. [CrossRef]
41. Rol, F.; Karakashov, B.; Nechyporchuk, O.; Terrien, M.; Meyer, V.; Dufresne, A.; Belgacem, M.N.; Bras, J. Pilot-Scale Twin Screw Extrusion and Chemical Pretreatment as an Energy-Efficient Method for the Production of Nanofibrillated Cellulose at High Solid Content. *ACS Sustain. Chem. Eng.* **2017**, *5*, 6524–6531. [CrossRef]
42. Guo, H.; Chang, J.; Yin, Q.; Wang, P.; Lu, M.; Wang, X.; Dang, X. Effect of the combined physical and chemical treatments with microbial fermentation on corn straw degradation. *Bioresour. Technol.* **2013**, *148*, 361–365. [CrossRef]
43. Taheri, H.; Hietala, M.; Oksman, K. One-step twin-screw extrusion process of cellulose fibers and hydroxyethyl cellulose to produce fibrillated cellulose biocomposite. *Cellulose* **2020**, *27*, 8105–8119. [CrossRef]
44. Robinson, J.; Aoun, H.K.; Davison, M. Determining moisture levels in straw bale construction. In Proceedings of the 3rd International Conference on Sustainable Civil Engineering Structures and Construction Materials—Sustainable Structures for Future Generations (SCESCM), Bali, Indonesia, 5–7 September 2016; pp. 1526–1534.
45. Lu, Z.; Hu, X.; Lu, Y. Particle Morphology Analysis of Biomass Material Based on Improved Image Processing Method. *Int. J. Anal. Chem.* **2017**, *2017*, 5840690. [CrossRef] [PubMed]
46. Hänninen, T.; Thygesen, A.; Mehmood, S.; Madsen, B.; Hughes, M. Mechanical processing of bast fibres: The occurrence of damage and its effect on fibre structure. *Ind. Crop. Prod.* **2012**, *39*, 7–11. [CrossRef]
47. Jing, X.; Zhang, X.; Bao, J. Inhibition Performance of Lignocellulose Degradation Products on Industrial Cellulase Enzymes During Cellulose Hydrolysis. *Appl. Biochem. Biotechnol.* **2009**, *159*, 696–707. [CrossRef]
48. Xing, L.; Xu, M.; Pu, J. The Properties and Application of an Ultrasonic-Assisted Wheat Straw Pulp having Enhanced Tendency for Ash Formation. *BioResources* **2016**, *12*, 871–881. [CrossRef]
49. Jiang, H.; Liu, J.; Wang, H.; Yang, R.; Zhao, Y.; Yin, S.; Shen, L. Study on Combined Vacuum–Mechanical Defoaming Technology for Flotation Froth and Its Mechanism. *Processes* **2022**, *10*, 1183. [CrossRef]
50. Nuengwang, W.; Srinophakun, T.R.; Realff, M.J. Real-Time Optimization of Pulp Mill Operations with Wood Moisture Content Variation. *Processes* **2020**, *8*, 651. [CrossRef]
51. Quispe-Quispe, L.G.; Limpe-Ramos, P.; Arenas-Chávez, C.A.; Gomez, M.M.; Mejia, C.R.; Alvarez-Risco, A.; Del-Aguila-Arcentales, S.; Yáñez, J.A.; Vera-Gonzales, C. Physical and Mechanical Characterization of a Functionalized Cotton Fabric with Nanocomposite Based on Silver Nanoparticles and Carboxymethyl Chitosan Using Green Chemistry. *Processes* **2022**, *10*, 1207. [CrossRef]
52. Gardea, F.; Glaz, B.; Riddick, J.; Lagoudas, D.C.; Naraghi, M. Energy Dissipation Due to Interfacial Slip in Nanocomposites Reinforced with Aligned Carbon Nanotubes. *ACS Appl. Mater. Interfaces* **2015**, *7*, 9725–9735. [CrossRef]

Article

Analysis of Hydrothermal Solid Fuel Characteristics Using Waste Wood and Verification of Scalability through a Pilot Plant

Tae-Sung Shin [1], Seong-Yeun Yoo [2], In-Kook Kang [2], Namhyun Kim [3], Sanggyu Kim [4], Hun-Bong Lim [5], Kangil Choe [2,*], Jae-Chul Lee [6,*] and Hyun-Ik Yang [1,*]

[1] Department of Mechanical Design Engineering, Hanyang University, Wangsibri-ro 222, Seongdong-gu, Seoul 04763, Republic of Korea
[2] Bioenergy Center, Kinava Co., Ltd., #701-704 7 Heolleung-ro, Seocho-gu, Seoul 06792, Republic of Korea
[3] Carbon Neutral Division, Korea East-West Power Co., Ltd., #395 Jongga-ro, Jung-gu, Ulsan 44543, Republic of Korea
[4] Construction Division, Korea East-West Power Co., Ltd., #395 Jongga-ro, Jung-gu, Ulsan 44543, Republic of Korea
[5] Department of Mechanical Design Engineering, Myongji College, 134, Gajwa-ro, Seodaemun-gu, Seoul 03656, Republic of Korea
[6] Material & Component Convergence R&D Department, Korea Institute of Industrial Technology (KITECH), Hanggaul-ro 143, Sangnok-gu, Ansan-si 15588, Republic of Korea
* Correspondence: kc15@caa.columbia.edu(K.C.); jc2@kitech.re.kr (J.-C.L.); skynet@hanyang.ac.kr (H.-I.Y.); Tel.: +82-10-9257-7851(K.C.); +82-31-8040-6244 (J.-C.L.); +82-31-400-5285 (H.-I.Y.)

Abstract: Increases in energy demand and waste are a major cause of natural resource depletion and environmental pollution, and technology capable of processing waste to convert it into energy is required to mitigate this issue. Hydrothermal carbonization (HTC) is an example of this technology that can convert waste into energy, and various studies have been conducted using it for fuel conversion. This study focused on the production of a solid fuel equivalent to coal for power generation through HTC processes using waste wood. Unlike previous work, which consists only of laboratory-scale HTC experiments, we confirmed scalability through pilot-scale HTC experiments. Overall, it was possible to convert waste wood into HTC solid fuel with a calorific value of over 27,000 kJ/kg through the pilot plant HTC process. Additionally, heavy metal and hazardous substance analyses proved that it can be used as a biosolid fuel.

Keywords: hydrothermal carbonization; waste wood; solid fuel; biosolid fuel; laboratory-scale; pilot-scale

1. Introduction

Increases in energy demand and waste are causing natural resource depletion and environmental pollution, making technologies capable of solving these problems crucial worldwide [1]. Many Asian, North American, and European governments are focusing on developing technologies that convert various organic wastes into energy as a solution to the zero organic waste policy. Among organic waste-to-energy technologies, hydrothermal carbonization (HTC) is used in various fields, such as energy conversion, environmental improvement, and nutrient recovery, and can convert waste into high-energy-density carbon materials [2].

HTC is a reaction that combines dehydration and decarboxylation processes, and its solid fuels have high calorific values with increasing carbon contents [3]. To convert organic waste into solid fuel using HTC, a temperature of approximately 180–250 °C is required to be maintained for 1–12 h [4,5]. It is relatively low compared with other energization technologies, such as pyrolysis requiring 400 °C and gasification requiring 800 °C. Thus, there is an advantage in that less input energy is required. Additionally, HTC has the advantage of producing solid fuel without dehydration and drying in the pretreatment process [6].

Recently, many studies have been conducted to confirm the possibility of biofuel conversion from various wastes using HTC technology. In these studies, various organic wastes have been used including waste wood, paper, MSW, food waste, and animal waste [7–10]. Additionally, the calorific value of hydrochar after HTC is relatively high in lignocellulosic biomass, such as wood or wood chips, which contain a substantial amount of carbon [11]. However, considering commercial facility design, if the HTC reaction time has to be increased from 4 to 72 h when producing biofuels, it is not an efficient fuel production per hour [7,8,12,13].

Research on HTC biofuel generation using catalysts is also being conducted. The catalysts are made by dissolving in distilled water by mixing with one or more chloride-based metal salts and acids. The catalyst concentration in the aqueous solution is in the range of 2–20 g/L. Hence, the catalyst is used to lower the pH concentration, thereby lowering the activation energy and reducing the hydrothermal carbonization reaction time. When lignocellulosic biomass feedstock HTC was compared according to the presence or absence of catalysts, the reaction time was reduced from 4 to 1 h when catalysts were absent and present, respectively [14,15]. The reaction temperature can be lowered from 10 to 40 °C, and the reaction pressure can be lowered by 1–2 MPa, which reduces the cost when manufacturing the reactor. Additionally, after HTC, the lignin component in the organic material decomposes in the hydrochar, but catalytic HTC showed less lignin degradation by FTIR spectroscopy [15]. Furthermore, when the lignin components are maintained, they have an adsorption effect on each other, making it possible to produce advantageous pellets without the additional attachment of materials during fuel transport. It was confirmed that there was no breakdown through the electrochemical water ingress (EWI) test [16].

Some researchers have expanded their facility capacity for commercialization. For example, Hoekman et al. [17] fabricated an expanded hydrothermal carbonization reactor (40 L) to process 3 kg of lignocellulosic biomass and showed that the experimental results were similar to those of a laboratory-scale 2 L reactor. They compared the experimental results when the HTC reaction temperature was 235 and 275 °C. At 275 °C, the calorific value increased by 40% compared to the supplied raw material. However, considering the effect of increasing the calorific value, when the reaction temperature condition is 275 °C, energy cost and cooling time increase. Furthermore, Ismail et al. [18] numerically analyzed the expected energy and production volume when converting municipal solid waste (MSW) into solid fuel in a pilot-scale facility. They proposed a numerical model for a pilot plant facility that converts MSW to coal through numerical analysis. These studies are meaningful from the perspective of commercial design, but verification has not been conducted through commercialization facilities. Mackintosh et al. [19] designed a pilot plant-scale binary reactor, performed proximate analyses, and determined the calorific value of hydrochar according to the HTC reaction of the catalyst [15] they applied to their lab-scale HTC experiments. They further analyzed the fuel characteristics by expanding the pilot scale, but a comparative study with and without a catalyst is required to verify the effect of the catalyst.

In this study, biofuel with a calorific value equivalent to that of coal for power generation was produced in a pilot plant through waste wood HTC. To verify the scalability, HTC reaction experiments were conducted using lab-scale (0.2 L—3 sphere) and pilot-scale (500 L) reactors, and the pilot-scale reactor was composed of the equipment for the entire process. First, the calorific values and mass yields of solid fuels generated through laboratory-scale HTC experiments were compared according to the presence or absence of catalysts and catalyst concentration. The added catalyst is a mixture of a chloride-based metal salt and an acid, which is a different combination from the catalysts of previous studies. Next, heavy metal and hazardous substance analyses were used to confirm whether the solid fuel that satisfied the desired calorific value and yield condition met the fuel use standard. Based on the determined catalyst conditions, a pilot-scale HTC experiment was performed according to the catalyst concentrations. The effect of catalyst addition was confirmed via comparative analyses of the generated solid fuel according to the presence or absence of catalysts. Overall, the trends of the pilot- and laboratory-scale experimental results were consistent, and their scalability was verified. The biosolid fuel produced through HTC with catalyst additions in the pilot plant proved that it can be used as fuel through heavy metal and hazardous substance analyses.

2. Experimental Setup and Procedure

2.1. Fuel Characteristics of the Waste Wood Raw Material

The waste wood used in the experiment was selected as a product that can be supplied in large quantities because it is required for both laboratory- and pilot-scale HTC experiments. The waste wood is acquired through a company that specializes in recycling. In the test site, only waste wood that has not been stained with adhesives, paints, oil, or preservatives was permitted to be brought in. The same waste wood was used for both the laboratory- and pilot-scale HTC experiments. To compare the fuel properties of the hydrochar after HTC, we analyzed the fuel properties of the raw materials (Table 1). Table 1 shows the results of the fuel characteristic analyses of the waste wood. Detailed fuel characteristic analyses can be divided into proximate, elemental, and calorific value analyses. The collected samples were subjected to proximate analysis before drying, while elemental and calorific value analyses were performed after drying. The analysis method used followed the American Society for Testing and Materials (ASTM), and different equipment were used according to each analysis method. Proximate analysis was performed using the TGA-701 Proximate analyzer, and the ratio of volatile substances, ash, water, and fixed carbon can be known. Elemental analysis was performed using TruSpec Elemental Analyzer. To measure carbon and hydrogen, the sample was burned in a tube furnace and the water and CO_2 produced were absorbed and analyzed. The nitrogen concentration was measured by the Kjeldahl–Gunning method [20]. Sulfur was analyzed using an SC-832DR Sulfur Analyzer. Sulfur was measured by conversion to sulfur dioxide, followed by absorption in hydrogen peroxide solution, and titration with barium acetate solution. The calorific value is divided into a Higher Heating Value (HHV) and a Lower Heating Value (LHV). In this study, the calorific value was confirmed to be measurable HHV. The calorific value was analyzed using an AC600 Semi-Auto Calorimeter. The calorific value is confirmed by analyzing the temperature change after the combustion of the sample based on the initial temperature.

Table 1. Waste wood chip fuel characteristics.

Element Analysis (wt%)					Proximate Analysis (wt%)				Calorific Value (kJ/kg)	
C	H	O	N	S	M	VM	Ash	F.C	HHV	LHV
43.97	6.05	30.19	0.1	0.016	7.98	73.56	0.77	17.69	19,929	18,338

2.2. Specification of a Laboratory-Scale Reactor for HTC Experiments

Three 200 mL reactors were installed in the laboratory-scale HTC experiment equipment (Figure 1), and a heating jacket was used as the heat source of the reactor. The maximum allowable temperature was 250 °C, and the reactor temperature was capable of reaching the reaction temperature using the heating jacket and temperature sensor. The time to reach the reaction temperature is about 50 min. Therefore, the average heating rate of the reactor is 4.5 °C/min. The electric capacity supplied to one unit of the heating jacket was 1.5 kW, and the maximum electric power of the HTC test equipment for the experiment was 5 kW. A temperature sensor that could measure the temperature and a control device that could maintain the reaction temperature were installed. Additionally, a pressure sensor was installed to check the pressure according to the temperature inside the reactor, which was measured to be approximately 2.32 MPa when the temperature inside the reactor was 220 °C. SUS316L, which has strong corrosion resistance, was used as the reactor material in consideration of corrosion and the catalyst, and the maximum allowable pressure of the reactor was designed to be 5.39 MPa considering safety.

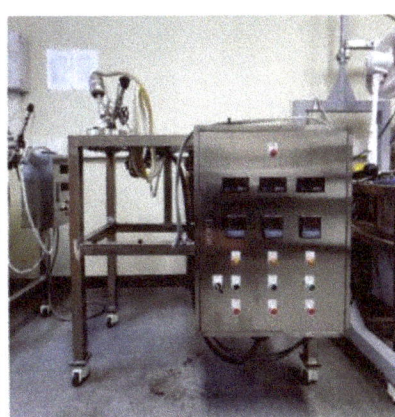

Figure 1. Laboratory-scale hydrothermal carbonization reactor (0.2 L).

2.3. Laboratory-Scale Reactor HTC Experimental Conditions and Process

The HTC process can be divided into pretreatment, reaction, and post-treatment steps. In the pretreatment step, the raw material was pulverized into particles of 2 mm or less to produce uniform particles. The moisture content of the pulverized raw material was then measured and placed into an aqueous solution inside the reactor to obtain the appropriate moisture content. The total volume of the raw material and the aqueous solution was 60–70% of the total volume of the reactor. The amount of catalyst added to the input aqueous solution varied depending on the catalyst concentration, including the case with or without the catalyst. The added catalyst contains inorganic metals and acids. At this time, the case where a catalyst is added to HTC is called catalytic HTC. Additionally, the amount of catalyst added is determined by catalyst conditions. Catalytic conditions according to the experiment were similarly performed according to previous studies [14,15]. The catalyst is a combination of specific inorganic metals and acids. Two combinations designated by the KINAVA Company were used in the above experiment. The first case (Catalyst #1) is a combination of strong acid-based catalysts. The second case (Catalyst #2) is a combination of weak acid-based catalysts. In all cases, they were provided in the form of liquid catalysts prepared by an already specified method. Then, adding the catalyst, the catalyst was diluted in 40% of the total aqueous solution, mixed with the remaining aqueous solution, stirred for 20 min, then added into the reactor. In the reaction step, a heat source was supplied such that the temperature inside the reactor reached the reaction temperature, and the reaction temperature was maintained for the reaction time. In the post-treatment step, it was cooled after the reaction holding time while discharging the vapor inside the reactor. When the temperature reached 80 °C, the HTC product inside the reactor was recovered, and the liquid and solid wastes were separated. Finally, the calorific values and mass yields were measured by drying the solid.

2.4. Pilot Plant Reactor Configuration and Process

The pilot plant facility for the pilot-scale HTC experiment was connected from raw material storage to the pellet production process, as shown in Figure 2. The capacity of the HTC reactor was 500 L, and the heat source was waste heat steam generated at the installation site and electrical energy through the heating jacket. First, waste heat steam with a pressure and temperature of 1.96 MPa and 200 °C raises the reactor temperature to 200 °C. After that, the reactor's internal temperature is increased through the heating jacket. The temperature could be maintained constant after reaching the target reaction temperature because the PID controller was connected to the heating jacket.

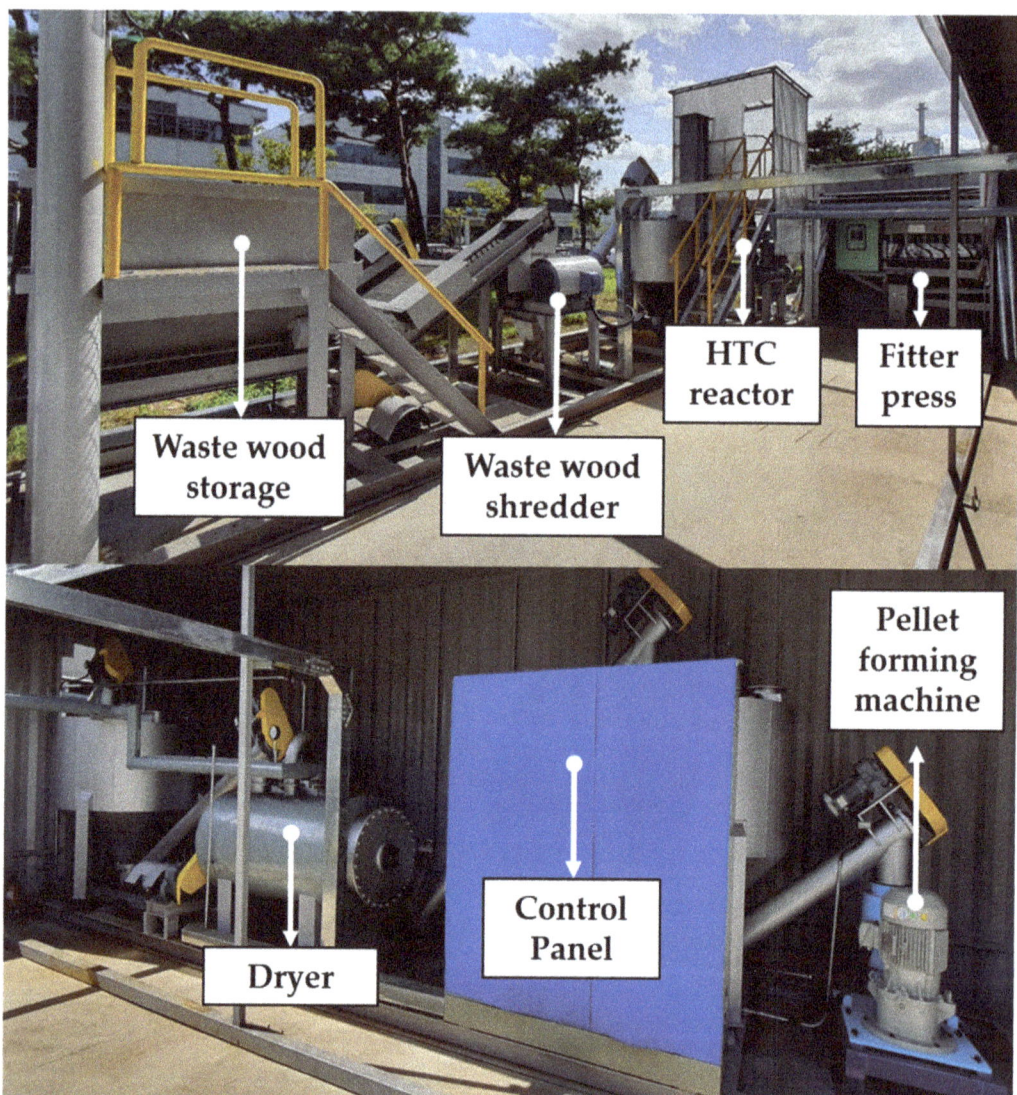

Figure 2. HTC biofuel production pilot plant.

For each process of the pilot plant, the conversion of waste wood to HTC solid fuel through the process is shown in Figure 3. The raw material of waste wood was homogenized into particles within 2 mm using a shredder. Pulverized waste wood produced HTC solid fuel through HTC, and its shape changed depending on its moisture content and molding. The process consisted of the following steps: raw material storage and pretreatment, HTC, solid–liquid separation, drying, and pellet production.

Figure 3. HTC solid fuel according to the pilot plant process.

Step 1. Raw Material Storage and Pretreatment Process

Waste wood, which is a raw material, is transported to a waste wood storage tank where it could be stored and transported. To pulverize into uniform particles, metal materials that are mixed in the waste wood are first removed using a magnetic separator. The inclusion of non-ferrous metals is minimized in sorting waste wood before transport. However, the reason for installing the magnetic separator is that the residual metal mixed in the waste wood may cause a malfunction of the conveying machine. The waste wood is then pulverized into uniform particles within 2 mm using a grinder. The pulverized waste wood is transported to the pulverized wood storage tank through a screw conveyor and stored for use in the next process.

Step 2. HTC Process

The pulverized waste wood is moved to the reactor inlet via a bucket conveyor. The feeding speed of the bucket conveyor can be adjusted through the control panel. At this time, the input amount can be known through the mass sensor in the pulverized waste wood storage tank. The reactor is a batch type, and the inlet and outlet can also be opened and closed through the control panel. The water for processing is placed in the reactor, and the stirrer is operated simultaneously so that the pulverized waste wood and water content is appropriate. At this time, approximately 60–70% of the total volume of the HTC reactor is filled. When the input is completed, the input valve of the HTC reactor is shut off, and a heat source is supplied to the HTC tank to increase the target reaction temperature. The reaction temperature is raised by introducing steam up to 200 °C, at which point the waste heat steam is cut off. At 200 °C or higher, the heat source of the heating jacket is used to increase the temperature to the target reaction temperature. After maintaining the reaction temperature for the target reaction time, the HTC reactor is cooled through steam discharge. When the internal temperature of the HTC reactor is 120 °C, it is discharged into the carbide storage tank using the internal pressure of the HTC reactor.

Step 3. Solid–Liquid Separation Process

The water contained in the hydrothermally carbonized solid is removed by transporting it to a high-pressure filter press using a pump. The solid–liquid separation liquid removed from the solid–liquid separation process is stored in a separate wastewater storage tank. Furthermore, the water content of the hydrothermally carbonized solid is approximately 50%. Subsequently, it is moved to a solid storage tank.

Step 4. Drying Process and Pellet Production Process

The hydrothermally carbonized solid is dried to a moisture content of 20% after being transferred to a dryer. The moisture that evaporated during the drying process is condensed using a condenser, and the condensed water is moved to the wastewater storage tank. It is transported to the condensed gas treatment facility to remove pollutants and then discharged to the outside. After the drying process, the HTC solid fuel is moved to a pellet-forming machine. The pellet-forming machine produces HTC solid fuel in the form of pellets with a length of 2–3 mm.

3. Results

3.1. Setting of HTC Reaction Conditions of the Pilot Plant through Lab-Scale HTC Effect Analyses

The HTC experiment was conducted (Table 2) using a laboratory-scale reactor. The change in calorific value and mass yield of the HTC solid fuel was confirmed according to the reaction temperature and time. Sawdust has properties similar to those of waste wood, and because the particles are uniform, uniform results can be confirmed when analyzing carbides after HTC. The experimental conditions were the same as 80% moisture content according to the waste wood raw material in a 0.2 L laboratory-scale reactor, and the HTC reaction was performed by changing the reaction temperature and time.

Table 2. Comparison of calorific value and mass yield after HTC of sawdust.

Reaction Condition		Lab-Scale Results	
Time (h)	Temperature (°C)	HHV (kJ/kg)	Mass Yield (%)
1	180	19,762	73.8
	190	20,100	71.7
	200	20,372	70.6
	210	22,839	70.3
	220	23,777	65.4
	230	25,158	59.7
2	180	20,750	72.3
	190	21,060	70.2
	200	22,052	66.5
	210	22,948	65.3
	220	24,095	63.4
	230	27,001	50.1
3	180	20,804	71.6
	190	21,034	69.5
	200	22,491	65.5
	210	23,777	65.1
	220	25,309	57.3
	230	29,898	35.1

Overall, as the reaction temperature and time increased, the calorific value of the hydrothermal carbides increased, and the mass yield decreased. The reaction temperature did not show a sharp change at 180–200 °C but a rapid change at 200–230 °C. The experimental results confirmed that the effective reaction temperature for converting waste wood to hydrochar should be at least 200 °C. In this experiment, as the reaction temperature and time increased, the mass yield of the reactants decreased to 35.1% when the calorific value increased to a maximum of 29,898 kJ/kg. The conditions for producing 23,027 kJ/kg or more, equivalent to coal for power generation with a mass yield of 60% or more, were analyzed. As a result, when the reaction time was 1–2 h, the reaction temperature was 220 °C. Furthermore, When the reaction time is 3 h, the condition is satisfied if the reaction temperature is 210 °C. However, in terms of production per hour, the reaction time is preferably within 1 h. Since the pilot plant may require more reaction time, the reaction time was set in the range of 1–1.5 h.

Under the HTC test conditions using waste wood, the moisture content was 80%, and the reaction temperature was in the range of 200–220 °C. The reaction time was tested in the range of 1–1.5 h to be applied to the pilot plant. As with sawdust, as the reaction temperature and time increased, the calorific value increased and the mass yield decreased. It was confirmed that the change in the reaction temperature had a greater effect on the HTC reaction than the change in the reaction time. Unlike sawdust, in waste wood, the calorific value was low at reaction temperatures of 210 °C or higher. In this experiment (Table 3), a calorific value of 23,000 kJ/kg or more and the conditions for production with a mass yield of 60% or more was when the reaction temperature was 220 °C. Therefore, in the pilot-scale HTC experiment, the equipment were configured such that the heat source of the heating jacket could be raised to at least 220 °C.

Table 3. Comparison of calorific value and mass yield after HTC of waste wood.

Reaction Condition		Lab-Scale Results	
Time (h)	Temperature (°C)	HHV (kJ/kg)	Mass Yield (%)
1	200	20,373	78.0
	210	21,629	74.9
	220	23,082	67.5
1.5	200	21,265	76.0
	210	21,914	71.2
	220	23,266	66.4

3.2. Laboratory-Scale Catalytic HTC Effect Analysis

A HTC experiment was conducted (Table 4) with reference to the results of a previous study [14,15] for catalyst types and concentrations. When the catalyst concentration ratio was 2-fold, the calorific value and mass yield after HTC were compared. If the calorific value increases after catalytic HTC, it is a combination of Catalyst #1 (a strong acid-based catalyst). The higher the concentration of Catalyst #1, the better the catalytic HTC reaction. Additionally, as in the previous experiment, the yield tends to decrease as the reaction progresses well. The catalyst concentration of Catalyst #1 was determined to have a calorific value of ≥25,120 kJ/kg that was a minimum of 1.5-fold more than the catalytic density ratio. When the catalyst concentration was reduced to the initial catalytic density ratio, the calorific value decreased to 25,120 kJ/kg or less, which did not reach the target calorific value. Therefore, in the pilot plant HTC experiment, we decided to add the catalyst at a concentration equal to or greater than 2-fold the catalyst density ratio.

Table 4. Comparison of calorific value and mass yield after catalytic HTC of waste wood at the laboratory scales.

Input Condition	Reaction Condition		Lab-Scale Results	
Catalyst	Time (h)	Temperature (°C)	HHV (kJ/kg)	Mass Yield (%)
None	1	220	23,082	67.5
Catalyst #1	1	220	26,687	61.3
	1.5	220	27,369	61.0
Catalyst #2	1	220	25,263	67.4
	1.5	220	26,038	65.3

Furthermore, a HTC experiment was conducted with a catalyst with a reaction time of 1.5 h, and as a result, the calorific value increased slightly. This was consistent with the tendency of HTC to occur more easily as the reaction time increased under the same reaction temperature conditions. Although the catalyst concentration ratios were the same, the strong acid was added in a larger amount than the weak acid, considering the purity of the catalyst. When the strong acid-based catalyst (Catalyst #1) was added, the calorific value was high, but the yield was lower and the amount of catalyst added was increased. When the weak acid-based catalyst (Catalyst #2) was added, the calorific value was lower than when Catalyst #1 was used, but a stable yield was obtained.

3.3. Analysis of Heavy Metals and Hazardous Substances on Laboratory-Scale HTC Solid Fuel

Prior to the production of HTC solid fuel in the pilot plant, heavy metals and hazardous substances were analyzed using specific samples in laboratory-scale HTC experiments. Table 5 is the quality standard applied when producing biosolid fuel at the test site where the pilot plant is installed. Therefore, it was a rule that had to be followed when conducting this study. The analysis method was carried out according to ISO 17022-1 biosolid fuel. Chlorine and sulfur were analyzed by Inductively Coupled Plasma-Optical Emission Spectroscopy (ICP-OES), and metals were analyzed by Inductively Coupled Plasma-Mass Spectrometer (ICP-MS). When the HTC reaction proceeded, the mass yield decreased, but heavy metals and harmful substances did not decrease; therefore, the content per unit mass or volume increased. The heavy metals and toxic substances contained in the non-catalyst HTC solid fuel slightly increased from the raw material waste wood standard; however, in the catalyst HTC solid fuel, among the heavy metals, chromium increased by more than 7 fold and lead increased by more than 2 fold. Among the harmful substances, sulfur has a synergistic effect because the amount of raw material is small, but chlorine increased by 8–37 fold. It was confirmed that Catalyst #1 had the highest calorific value, but it was not a catalyst combination that could be used in the pilot plant because the chlorine content exceeded the standard for hazardous substances. Therefore, we decided to use Catalyst #2, which satisfied the standard of biosolid fuel as being suitable for the pilot plant-scale experiment.

Table 5. Laboratory-scale HTC solid fuel analysis according to heavy metal and hazardous substance standards of biosolid fuel.

Biosolid Fuel Production Condition				Lab-Scale Results		
List	Unit	On-Site Standard	Waste Wood (Raw Material)	HTC (Non)	Catalytic HTC (Catalyst #1)	Catalytic HTC (Catalyst #2)
Cl	wt%	0.5	0.02	0.06	0.75 (excess)	0.17
S	wt%	0.6	0.0161	0.04	0.06	0.03
Hg	ppm	0.6	0.00289	≤0.001	0.002	≤0.001
Cd	ppm	5	≤0.1	≤0.1	≤0.1	≤0.1
Pb	ppm	100	0.52	0.96	1.15	1.45
As	ppm	5	≤0.1	≤0.1	0.15	0.11
Cr	ppm	70	3.6	4.34	30	28.1

3.4. Pilot-Scale HTC Effect Analysis and Scalability Verification

In the pilot-scale HTC experiment, the analysis was performed based on the HTC solid fuel discharged from the filler press during the process. The analyzed results were compared with the laboratory-scale experimental results, as shown in Table 6. The reaction temperature was fixed at 220 °C, and the reaction times were 1 and 1.5 h, and it was confirmed that there was no significant difference with the laboratory-scale test result. The calorific value was 23,027 kJ/kg and the mass yield was 60% or more.

Table 6. Comparison of calorific value and mass yield after HTC of waste wood at the laboratory and pilot scales.

Reaction Condition		Laboratory-Scale Results		Pilot-Scale Results	
Time (h)	Temperature (°C)	HHV (kJ/kg)	Mass Yield (%)	HHV (kJ/kg)	Mass Yield (%)
1	220	23,082	67.5	22,960	67
1.5	220	23,266	66.4	23,236	65

The catalytic HTC experiment was performed according to the reaction conditions listed in Table 7. Unlike the laboratory-scale HTC experiment, the calorific value did not exceed 25,120 kJ/kg at a catalyst concentration of the same condition. The reduction in calorific value and mass yield is an effect of catalyst addition, thus additional experiments were conducted by increasing the catalyst concentration. It was increased by the initial density ratio, and the calorific value was measured to be higher than 25,120 kJ/kg at a 3-fold density ratio. It was also confirmed that the mass yield was more than 60% up to a 4-fold density ratio.

Table 7. Comparison of calorific value and mass yield after catalytic HTC of waste wood according to catalytic density ratio at the pilot scale.

	Reaction Condition			Pilot-Scale Results	
Catalyst	Catalytic Density Ratio	Time (h)	Temperature (°C)	HHV (kJ/kg)	Mass Yield (%)
Non	0	1.5	220	23,027	65
Catalyst #2	2	1.5	220	23,697	64
	3	1.5	220	25,787	61.2
	4	1.5	220	27,189	60

The chemical positions of the waste wood used as a reactant in these experiments and the biosolid fuels produced from the pilot-scale HTC processes were compared with a Van Krevelen diagram (Figure 4). The atomic H/C and O/C ratios of waste wood were 1.65 and 0.51, which are similar to that of general biomass. After HTC of waste wood at 220 °C for 1.5 h without a catalyst, the atomic H/C ratios of the biosolid fuel decreased from 1.65 to 1.13, and the atomic O/C ratios decreased from 0.51 to 0.39. This reduced the atomic H/C and O/C ratios by 31.5% and 23.5%, respectively, compared to those of the raw material, and showed intermediate levels of peat and lignite. On the other hand, the atomic H/C and O/C ratios of the biosolid fuel produced from the catalytic HTC (Catalyst #2) under the same conditions were reduced to 0.83 and 0.24, respectively. These figures showed reduction rates of 49.7% and 52.9%, respectively, compared to those of the raw material, and showed a degree of carbonization similar to that of general coal. In addition, the atomic H/C and O/C ratios decreased by 26.5% and 38.5%, respectively, compared to the biosolid fuel produced from the HTC without a catalyst. From these results, it was confirmed that Catalyst #2 provided by KINAVA greatly increased the selectivity for dehydration even in the HTC reaction under the same conditions, enabling the production of biosolid fuel with a high calorific value due to a high degree of carbonization.

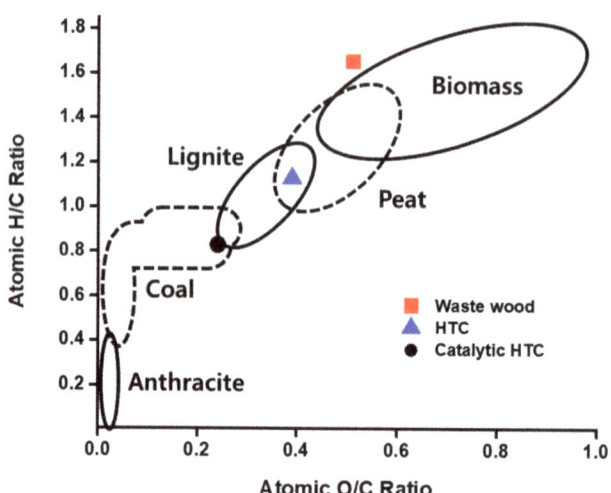

Figure 4. Van Krevelen diagram of waste wood and biosolid fuels produced by different HTC processes at the pilot scales.

In the next step, heavy metals and hazardous substances were analyzed to determine whether the HTC solid fuel produced by the pilot plant satisfied the biosolid fuel standard. As shown in Table 8, it was confirmed that the HTC solid fuel produced in the laboratory satisfied the prescribed conditions. As a result, it did not exceed the standard for hazardous substances in the test site, and it was analyzed with a result similar to that of the laboratory-scale HTC solid fuel.

Table 8. Laboratory- and pilot-scale hydrothermal carbonization solid fuel comparison according to heavy metal and hazardous substance standards of biosolid fuel.

Biosolid Fuel Production Condition			Laboratory-Scale Results	Pilot-Scale Results
List	Unit	On-Site Standard	Catalytic HTC (Catalyst #2)	Catalytic HTC (Catalyst #2)
Cl	wt%	0.5	0.17	0.2
S	wt%	0.6	0.03	0.02
Hg	ppm	0.6	≤0.001	0.0022
Cd	ppm	5	≤0.1	≤0.1
Pb	ppm	100	1.45	1.44
As	ppm	5	0.11	≤0.1
Cr	ppm	70	28.1	29.8

4. Conclusions

The purpose of this study was to convert waste wood into solid fuel with a calorific value equivalent to coal for power generation using HTC. First, a laboratory-scale HTC experiment confirmed that the mass yield was lowered with a higher calorific value than when no catalyst was added under the same reaction conditions. Then, to verify the scalability of the HTC effect, pilot-scale HTC test conditions were determined based on laboratory-scale HTC test results. The pilot-scale HTC experiment confirmed that the reaction temperature was 220 °C, and the reaction time was increased to 1.5 h, similar to the laboratory-scale HTC experiment result. However, in the catalytic HTC experiment, the laboratory-scale experimental results were similar when the catalyst concentration was increased by 1.5 fold or more. Therefore, it was confirmed that optimization of the catalyst concentration is necessary for a commercial HTC plant to which a catalyst is added. In conclusion, the optimal catalyst combination and concentration through the pilot plant produced solid fuel with a calorific value 18% higher than coal for power generation. In addition, it was verified through scalability that it could be used as a biosolid fuel.

Author Contributions: Conceptualization, T.-S.S., K.C. and H.-I.Y.; methodology, T.-S.S., K.C. and H.-I.Y.; validation, T.-S.S.; formal analysis, T.-S.S. and J.-C.L.; investigation, J.-C.L.; Resources, I.-K.K.; data curation, S.-Y.Y. and I.-K.K.; Pilot Plant design verification, H.-B.L.; Pilot Plant test results analysis and verification, N.K. and S.K.; writing—original draft preparation T.-S.S.; writing—review and editing, K.C., J.-C.L. and H.-I.Y.; supervision, H.-I.Y.; Project administration, K.C.; funding acquisition, N.K. and S.K. All authors have read and agreed to the published version of the manuscript.

Funding: This study was supported by the Korea East-West Power Company of the Republic of Korea (Pilot Plant Development for Green Pellet Production from Woodwaste Using Hydrothermal Polymerization Technology (2019)).

Institutional Review Board Statement: Not applicable.

Informed Consent Statement: Not applicable.

Data Availability Statement: This experimental data is a rough representation of what was derived from the joint experiment between the lead author and KINAVA Company. The data presented in this study are available on request from the corresponding authors and KINAVA Company.

Acknowledgments: We appreciate Korea East-West Power Company and KINAVA Company for their assistance in this study.

Conflicts of Interest: S.-Y.Y., I.-K.K. and K.C. are employees of the KINAVA Company. N.K. and S.K. are employees of Republic of Korea East-West Power Company.

References

1. Gil, A. Challenges on Waste-to-Energy for the Valorization of Industrial Wastes: Electricity, Heat and Cold, Bioliquids and Biofuels. *Environ. Nanotechnol. Monit. Manag.* **2022**, *17*, 100615. [CrossRef]
2. Maniscalco, M.P.; Volpe, M.; Messineo, A. Hydrothermal Carbonization as a Valuable Tool for Energy and Environmental Applications: A Review. *Energies* **2020**, *13*, 4098. [CrossRef]
3. Funke, A.; Ziegler, F. Hydrothermal Carbonization of Biomass: A Summary and Discussion of Chemical Mechanisms for Process Engineering. *Biofuels Bioprod. Biorefin.* **2010**, *4*, 160–177. [CrossRef]
4. Wu, L.M.; Tong, D.S.; Li, C.S.; Ji, S.F.; Lin, C.X.; Yang, H.M.; Zhong, Z.K.; Xu, C.Y.; Yu, W.H.; Zhou, C.H. Insight into Formation of Montmorillonite-Hydrochar Nanocomposite under Hydrothermal Conditions. *Appl. Clay Sci.* **2016**, *119*, 116–125. [CrossRef]
5. Libra, J.A.; Ro, K.S.; Kammann, C.; Funke, A.; Berge, N.D.; Neubauer, Y.; Titirici, M.M.; Führer, C.; Bens, O.; Kern, J.; et al. Hydrothermal Carbonization of Biomass Residuals: A Comparative Review of the Chemistry, Processes and Applications of Wet and Dry Pyrolysis. *Biofuels* **2011**, *2*, 71–106. [CrossRef]
6. Fakkaew, K.; Koottatep, T.; Pussayanavin, T.; Polprasert, C. Hydrochar Production by Hydrothermal Carbonization of Faecal Sludge. *J. Water Sanit. Hyg. Dev.* **2015**, *5*, 439–447. [CrossRef]
7. Berge, N.D.; Ro, K.S.; Mao, J.; Flora, J.R.V.; Chappell, M.A.; Bae, S. Hydrothermal Carbonization of Municipal Waste Streams. *Environ. Sci. Technol.* **2011**, *45*, 5696–5703. [CrossRef] [PubMed]
8. Hwang, I.-H.; Aoyama, H.; Matsuto, T.; Nakagishi, T.; Matsuo, T. Recovery of Solid Fuel from Municipal Solid Waste by Hydrothermal Treatment Using Subcritical Water. *Waste Manag.* **2012**, *32*, 410–416. [CrossRef] [PubMed]
9. Cao, X.; Ro, K.S.; Chappell, M.; Li, Y.; Mao, J. Chemical Structures of Swine-Manure Chars Produced under Different Carbonization Conditions Investigated by Advanced Solid-State ^{13}C Nuclear Magnetic Resonance (NMR) Spectroscopy. *Energy Fuels* **2011**, *25*, 388–397. [CrossRef]
10. Goto, M.; Obuchi, R.; Hirose, T.; Sakaki, T.; Shibata, M. Hydrothermal Conversion of Municipal Organic Waste into Resources. *Bioresour. Technol.* **2004**, *93*, 279–284. [CrossRef] [PubMed]
11. Simsir, H.; Eltugral, N.; Karagoz, S. Hydrothermal Carbonization for The Preparation of Hydrochars from Glucose, Cellulose, Chitin, Chitosan and Wood Chips via Low-temperature and Their Characterization. *Bioresour. Technol.* **2017**, *246*, 82–87. [CrossRef] [PubMed]
12. Lu, X.; Pellechia, P.J.; Flora, J.R.V.; Berge, N.D. Influence of Reaction Time and Temperature on Product Formation and Characteristics Associated with the Hydrothermal Carbonization of Cellulose. *Bioresour. Technol.* **2013**, *138*, 180–190. [CrossRef] [PubMed]
13. Qi, X.; Lian, Y.; Yan, L.; Smith, R.L. One-step Preparation of Carbonaceous Solid Acid Catalysts by Hydrothermal Carbonization of Glucose for Cellulose Hydrolysis. *Catal. Commun.* **2014**, *57*, 50–54. [CrossRef]

14. Joo, B.; Yeon, H.; Lee, S.; Ahn, S.; Lee, K.; Jang, E.; Won, J. Conversion of Wood Waste into Solid Biofuel Using Catalytic HTC Process. *New Renew. Energy* **2014**, *10*, 12–18. (In Korean) [CrossRef]
15. Mackintosh, A.F.; Shin, T.; Yang, H.; Choe, K. Hydrothermal Polymerization Catalytic Process Effect of Various Organic Wastes on Reaction Time, Yield, and Temperature. *Processes* **2020**, *8*, 303. [CrossRef]
16. Ghaziaskar, A.; McRae, G.A.; Mackintosh, A.; Basu, O.D. Catalyzed Hydrothermal Carbonization with Process Liquid Recycling. *Energy Fuels* **2019**, *33*, 1167–1174. (In English) [CrossRef]
17. Hoekman, S.K.; Broch, A.; Robbins, C.; Purcell, R.; Zielinska, B.; Felix, L.; Irvin, J. Process Development Unit (PDU) for Hydrothermal Carbonization (HTC) of Lignocellulosic Biomass. *Waste Biomass Valoriz.* **2013**, *5*, 669–678. [CrossRef]
18. Ismail, T.M.; Yoshikawa, K.; Sherif, H.; El-Salam, M.A. Hydrothermal treatment of municipal solid waste into coal in a commercial Plant: Numerical assessment of process parameters. *Appl. Energy* **2019**, *250*, 653–664. [CrossRef]
19. Mackintosh, A.F.; Jung, H.; Kang, I.-K.; Yoo, S.; Kim, S.; Choe, K. Experimental Study on Hydrothermal Polymerization Catalytic Process Effect of Various Biomass through a Pilot Plant. *Processes* **2021**, *9*, 758. [CrossRef]
20. Liu, J.I.; Paode, R.D.; Holsen, T.M. Modeling the Energy Content of Municipal Solid Waste Using Multiple Regression Analysis. *J. Air Waste Manag. Assoc.* **1996**, *46*, 650–656. [CrossRef]

Communication

Techno-Economic Analysis of an Integrated Bio-Refinery for the Production of Biofuels and Value-Added Chemicals from Oil Palm Empty Fruit Bunches

Kean Long Lim [1,*], Wai Yin Wong [1], Nowilin James Rubinsin [1], Soh Kheang Loh [2] and Mook Tzeng Lim [3,4]

1. Fuel Cell Institute, Universiti Kebangsaan Malaysia, Bangi 43600, Selangor, Malaysia
2. Energy and Environment Unit, Engineering and Processing Division, Malaysian Palm Oil Board, 6, Persiaran Institusi, Bandar Baru Bangi, Kajang 43000, Selangor, Malaysia
3. TNB Research Sdn. Bhd., Kajang 43000, Selangor, Malaysia
4. Department of Mechanical Engineering, Faculty of Engineering and Quantity Surveying, INTI International University, Nilai 71800, Negri Sembilan, Malaysia
* Correspondence: kllim@ukm.edu.my; Tel.: +60-3-8911-8494; Fax: +60-3-8911-8530

Abstract: Lignocellulose-rich empty fruit bunches (EFBs) have high potential as feedstock for second-generation biofuel and biochemical production without compromising food security. Nevertheless, the major challenge of valorizing lignocellulose-rich EFB is its high pretreatment cost. In this study, the preliminary techno-economic feasibility of expanding an existing pellet production plant into an integrated bio-refinery plant to produce xylitol and bioethanol was investigated as a strategy to diversify the high production cost and leverage the high selling price of biofuel and biochemicals. The EFB feedstock was split into a pellet production stream and a xylitol and bioethanol production stream. Different economic performance metrics were used to compare the profitability at different splitting ratios of xylitol and bioethanol to pellet production. The analysis showed that an EFB splitting ratio below 40% for pellet production was economically feasible. A sensitivity analysis showed that xylitol price had the most significant impact on the economic performance metrics. Another case study on the coproduction of pellet and xylitol versus that of pellet and bioethanol concluded that cellulosic bioethanol production is yet to be market-ready, requiring a minimum selling price above the current market price to be feasible at 16% of the minimum acceptable return rate.

Keywords: empty fruit bunches; pellet; xylitol; bioethanol; integrated process

1. Introduction

Malaysia is currently the world's second-largest palm oil producer after Indonesia. As in 2019, 5.90 Mha of land in Malaysia is covered with oil palms, 46.9% of which is in peninsular Malaysia, and the remaining 26.1% and 26.9% are in Sabah and Sarawak, respectively [1]. With the rapid expansion of the palm oil industry in Malaysia, sustainability issues related to oil palm have accelerated in recent years [2], especially environmental issues associated with the palm oil industry. These issues have increased the urgency for the industry to find a balance between environmental and economic sustainability. One of the many options is to convert the excessive biomass leftover from the palm oil mills, especially the empty fruit bunches (EFBs), into value-added biomass and generate revenue from the waste [3]. EFBs are the remaining parts after oil palm fruitlets are stripped from fresh fruit bunches (FFBs) [4,5]. According to Hamzah et al. [6], the amount of EFB produced is estimated to be 22% of the FFB (in wet weight), which is the largest proportion of oil palm plantation solid waste. Based on the FFB yield data from MPOB (2019) [7], it is estimated that the average total amount of EFB produced from 2017 to 2019 is 22.42 million tons annually. EFB is considered a waste that needs to be continuously removed to avoid piling at the site because it can lead to methane emissions that contribute to air pollution and negative health impacts [8,9]. Conventionally, EFB is used as a mulch or organic fertilizer in oil palm plantations because of its high alkali content or fuel in the boiler to reduce diesel consumption [10,11]. However, feeding EFB directly into the boiler without removing its alkali content contributes to slagging and fouling, which will eventually reduce the operation efficiency of the boiler [12]. In addition, EFB is a lignocellulosic material that can potentially be utilized to produce high-value-added products such as biofuel, biochemical materials, industrial sugar, and biofertilizer. If the potential is unlocked and fully exploited, the palm oil industry will be one step closer to sustainable development and circular economy [13].

There were several studies related to the valorization of EFB to fuel (pellets, briquettes, bioethanol) and value-added products (charcoal, long fibers, biochemical) as a promising and sustainable alternative to the replacement of fossil fuels and chemical products [14–49]. The lignocellulose composition of EFB consists of 36–43% cellulose, 15–25% hemicellulose, and 22–34% lignin [20], as well as approximately 4% ash or mineral content [21] that contains compounds such as K_2O, P_2O_5 and SiO_2, which can be a suitable source for pellet, biofertilizer, xylitol and bioethanol production. Instead of burning EFB as a fuel directly, processing EFB into pellets by increasing its lignin content and removing its mineral content improves the calorific value [22,23]. Renewable solid fuel for power generation is in high demand in Europe, Japan, Korea, and China. Currently, an industrial establishment in Malaysia has the capacity to produce 1000 to 3000 tons of pellets monthly [24], but the average annual global growth of pellet demand is of 960,000 tons/year [25–27]. This huge supply–demand mismatch offers an opportunity for EFB pellets to fill in the supply gap because of their cost competitiveness, low moisture content, high calorific value, and low smoke and fume generation during combustion [28]. Cellulosic bioethanol is considered a second-generation (2G) biofuel produced from the cellulose component of EFB. 2G biofuel has gained considerable demand because it offers an alternative to minimize possible conflicts between fuel and food security [29]. However, cellulosic bioethanol is inherently more challenging to produce than sugar- or starch-based bioethanol and more costly than fossil-based ethanol. Hence, it is necessary to produce valuable biochemicals such as xylitol simultaneously with 2G cellulosic bioethanol to improve the overall feasibility of the production process [30]. EFB with a substantially high amount of polymeric xylan (hemicellulose) is a suitable xylose source for xylitol production. Xylitol is a highly sought value-added product in the food and pharmaceutical industries. It can only be extracted from plant biomass [31]. The global demand for xylitol is approximately 125,000 tons, with an average market price of 5000 to 20,000 USD/ton, and xylitol is mostly used as a substitute for sugar because of its lower caloric content but with a similar sweetening power [32].

Although the demand for these products is high worldwide, the recalcitrant structure of lignocellulose in EFB is the bottleneck to yield high amounts of bioproducts [33,34]. Therefore, multiple pretreatment steps are required to fractionate the complex structure, increase the EFB porosity, reduce cellulose's crystallinity, solubilize hemicellulose, and modify the lignin structure [35–37]. These pretreatment methods include physical, biological, and chemical processes to condition the EFB before feeding to the pellet and biochemical production. The physical method aims to reduce the particle size and crystallinity of biomass by milling, grinding, or chipping. Further processing of biomass is easier and more effective. Biological methods use microorganisms such as fungi and bacteria to degrade lignin, hemicellulose, and cellulose. Biological methods are usually cost-effective, have low energy requirements, and are environmentally friendly. No chemical waste is generated, but the degradation process is slow. Either acid or alkali is often used to treat biomass in chemical treatment [38]. A combination of physical and chemical methods has also been used to reduce the recalcitrance of lignocellulosic biomass. The most commonly used physicochemical methods are liquid hot water, steam explosion, microwave pretreatment, and ozonolysis treatment [39–42]. Most of these physicochemical methods are conducted at a high temperature and pressure to accelerate biomass degradation. Still, these methods are less efficient because they can cause severe degradation of the EFB components [43].

There were three main processes in producing EFB pellets: moisture removal, composition adjustment (lignin increment and ash reduction), and pellet densification. Removal of high alkali content in EFB, such as K and Na, in the form of ash, is a crucial step to produce premium-grade pellets for boiler applications. The ash deposition of EFB can be removed with washing treatment. This washing process's effluent contains essential nutrients for plant growth and can be a suitable source of N-P-K fertilizers [44]. In addition, EFB contains highly hydrophilic hemicellulose and has approximately 67% moisture. It requires a high intensity of drying and hemicellulose removal to be used as fuel in the boilers. The removal of hemicellulose content increases the lignin percentage, which has a high calorific or heating value [23,45]. Pretreatment processes such as torrefaction, steam explosion, and hydrothermal treatment are commercial thermal treatment techniques to improve the calorific value [46]. Torrefaction is a heat treatment process to carbonize EFB and increase its C/O ratios [47]. Torrefaction also reduces the moisture content of EFB. It reduces the mass by almost half, enhancing the transportability and prolonging the storage duration of pellets [48]. Steam explosion uses high-temperature saturated steam to penetrate through the cell wall structure at high pressure and solubilize hemicellulose [41]. Upon instantaneous controlled pressure drop, the cell wall expands adiabatically and undergoes explosive decomposition, making cellulose more accessible [23]. Hydrothermal treatment (HTT), also known as wet torrefaction, is another pretreatment process suitable for biomass with high moisture content, with a typical treatment temperature of 150 to 350 C [49]. Both Ahda et al. [50] and Novianti et al. [45] have shown that the HTT process upgrades the EFB into more stable, hydrophobic, and more lignin-containing feedstock with lower mineral content. While pretreatment of EFB in pellet production improves the fiber's quality, the loose EFB fiber is usually bulky and low in density. The densification process can increase the energy potential of the biomass. There are two densification methods: screw press and piston press. Pellets produced through this process can be used in direct combustion for energy generation [28].

In a coproduction of bioethanol and xylitol, EFB is first chemically pretreated with dilute acid or alkali to extract cellulose from EFB [51] to produce hydrolyzed hemicellulose or xylose [20]. Acid hydrolysis pretreatment is commonly used because it can hydrolyze hemicellulose, which has a lower degree of polymerization and amorphous structure, much faster than cellulose. The pretreatment process was able to increase the cellulose content from 41% to 72% [52–55]. The most commonly used acid is diluted sulfuric acid at a concentration below 4% (w/w) [56]. Other mineral acids, such as hydrochloric, nitric, and phosphoric acids, can be used, but sulfuric acid results in an efficient process, lower cost, and shorter reaction time [57].

For bioethanol production, the cellulose-rich hydrolysate undergoes subsequent hydrolysis in either biological method through enzymatic hydrolysis or chemical method through acid hydrolysis. The enzymatic hydrolysis method is preferable because it is cost-effective and produces a higher sugar content than acid hydrolysis [58,59]. Enzymatic hydrolysis degrades cellulose into simple sugar by cellulolytic enzymes, which require an optimal temperature between 45 and 55 °C and a pH range of 4–5 [36]. Ghazali and Makhtar [60] used cellulase enzymes produced from the fungus *Trichoderma reesei* to increase glucose yield. The enzymes produced a constant maximum yield of 2.5 g/L at an enzyme to substrate ratio of 0.05 (0.5 g enzyme/10 g EFB) at 50 °C and pH 5. Zhai et al. [51] reported that the sugar yield was improved from 30.5% to 66.9% by increasing the enzyme dose from 10 to 60 FPU/g EFB at 50 °C and pH 4.8. Both studies have indicated that optimal enzymatic hydrolysis conditions, such as temperature, pH, and substrate concentration, are essential in improving sugar yield [61]. Other fungi can be used to produce enzymes, but *Trichoderma reesei* is the most commonly used fungus in industrial enzymatic processes [62]. The sugar from enzymatic hydrolysis is fermented with yeast to produce bioethanol. The most frequently used yeast is *Saccharomyces cerevisiae* because it can provide a high ethanol yield [59,63]. Sugar can be fermented into ethanol by two processes: separate hydrolysis and fermentation (SHF) and simultaneous saccharification and fermentation (SSF). For an SHF system, both operating conditions of hydrolysis and fermentation are operated independently at different optimal conditions. Using this method, the optimization and control of the process conditions, such as the temperature and pH of the hydrolysis and fermentation, can be performed effectively [64]. In contrast, SSF system performs better than SHF, where SSF has a shorter processing time, lower cost, and higher bioethanol yield. Nevertheless, it is challenging to obtain optimum pH and temperature conditions for both saccharification and fermentation in the SSF system. Dahnum et al. [64] showed that SSF has a shorter processing time of 72 h than SHF in producing bioethanol at a temperature of 32 °C and pH of 4.8. Similarly, Sukhang et al. [55] concluded that SSF provided a higher bioethanol yield of 0.281 g/g EFB than SHF, with a yield of 0.258 g/g EFB at a temperature of 36.94 °C and pH of 4.5. Both studies implied that the optimal temperature for both enzyme and yeast activity was in the range of 32 °C to 39.8 °C, while the pH was acidic in the range of 4.0–5.0 [59].

For xylitol production, the hydrolyzed hemicellulose (xylose) from the pretreatment is further converted to xylitol via chemical (hydrogenation) or biological (fermentation) routes [65]. In the chemical route, xylitol production consists of three stages after the pretreatment stage: xylose purification, catalytic hydrogenation, and xylitol purification. Xylose purification is essential to obtain a high concentration of xylose, reduce unwanted side products and deactivate the acid catalyst used [31,57]. The hydrogenation step occurs in the presence of the Raney-nickel catalyst. This catalyst is widely used in industry because it offers a high yield of 80% to 95% xylitol and conversion efficiency of 98% [66–68]. Then, the produced xylitol is purified with filtration or ion exchange to recover xylitol at higher purity. At the industrial level, the chemical route is commonly used [31,68], but the drawbacks are high separation and purification costs, high energy consumption, and environmental impacts, such as toxic catalysts and high-pressure hydrogen gas [65]. The biological route has recently become more attractive because the process is sustainable and has a lower cost than the chemical route. Yeast that belongs to the genus *Candida* sp. [69] can convert xylose into xylitol with a yield of up to 90% [66]. Another report from Tamburini et al. [70] showed that the genus *Candida* sp. can produce xylitol at a maximum yield of 86.84% at 32 °C, 80 g/L xylose and pH 2.5. Nevertheless, the biological route's limitations are the expensive separation process of xylitol from the fermentation broth and the toxicity effects of xylitol to yeast [31,57]. Kresnowati et al. [71] proposed using ultrafiltration membrane technology to obtain high xylitol concentrations from fermentation broths. The proposed method has the potential for energy savings and higher purity, but the fouling problems need to be addressed.

According to the Malaysia National Biomass Strategy 2020 [72], pellet production is identified as a low entry point to generate wealth from biomass because of its technological maturity, relatively low capital investment, and short payback time. Expansion investment to produce 2G biofuels and biochemicals has a higher risk; however, the potential value creation is up to 5 times the revenue of pellets per dry ton of solid biomass input. In the long term, valorizing EFB biomass will minimize waste and recover more valuable products that will increase the profitability of the investment. More effort is required to remove the barriers of unfavorable high processing costs and low profitability of final products [20,73]. Therefore, a techno-economic feasibility study of an integrated bio-refinery to produce pellets, xylitol, and bioethanol was developed and evaluated in this work. The aim is to identify the process that is profitable with the production of these three products. If the higher-value biochemicals global market materializes earlier, EFB can be swiftly diverted to these biochemicals' production. In this work, a pilot scale pretreatment plant in Malaysia was used as a case study to investigate the potential to expand the products to pellets, biofertilizer, xylitol, and bioethanol. The respective products' market demand, technological production, and economic potential are investigated as well.

2. Methodology

The following section describes the methodology required to develop the analysis. Figure 1 shows the flow of techno-economic analysis of the bio-refinery plant starting from developing the process flow diagram until evaluating the economic performance metrics of different scenarios. The first step was to develop the process flow diagram (PFD) of a bio-refinery to produce pellet, xylitol and bioethanol from empty fruit bunches, where feed-in streams, output streams, conversion factor, process flow, unit operations and its corresponding operating conditions were identified. The law of mass conservation is still applicable even though there are chemical reactions and physical transformations of feedstock. Thereafter, materials balance was performed by accounting for the materials entering and leaving the system. With the mass balance information, the total capital investment (TCI), total production cost (TPC) and revenue were estimated. The net present value (NPV), return on investment (ROI), payback period (PBP) and internal rate of return (IRR) were calculated from the cash flow analysis based on the predefined scenarios. It should be noted that both TCI and TPC were only preliminary estimations due to the limited data available. However, the estimations will not affect the overall analysis at different scenarios because the comparisons were made on the same ground.

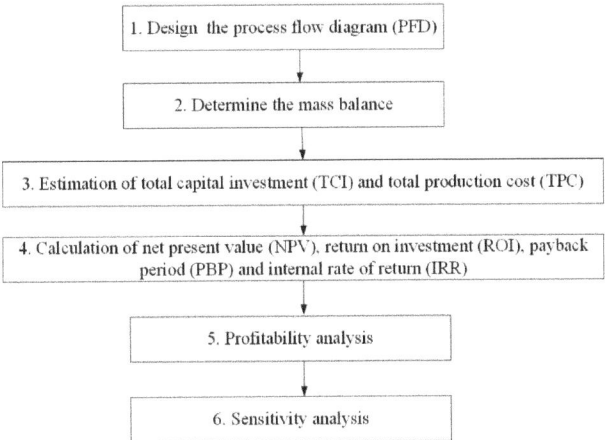

Figure 1. Methodology of techno-economic analysis.

2.1. Process Design Description

The bio-refinery was designed based on an EFB feedstock capacity of 126,720 tons/year with an actual annual operation of 5280 h (330 days, based on the existing pellet production plant) to produce three main products: pellet, xylitol, and bioethanol. The process flow diagram of the bio-refinery is shown in Figure 2. The detail process description is described in Sections 2.1.1 and 2.1.2 below. The EFB feedstock was split into two streams, one for pellet production and the other for xylitol and bioethanol. In this study, the profitability of the coproduction of these three products was investigated based on the splitting ratio between these two streams. Table 1 shows an example of the material balance of the bio-refinery at 30% EFB fed into the pellet production plant.

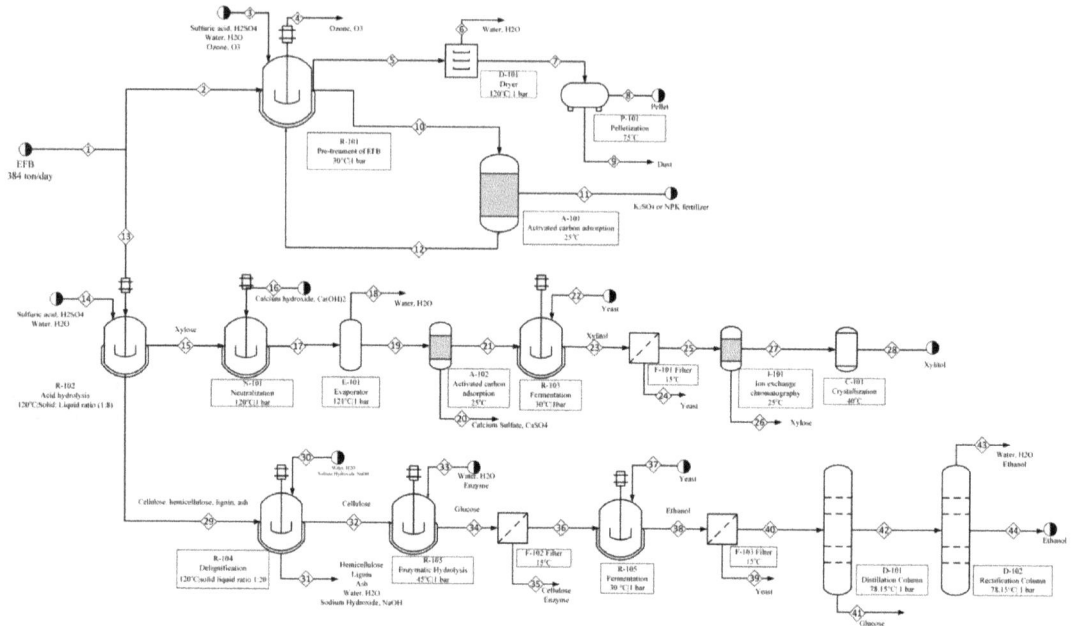

Figure 2. Process flow diagram of the proposed integrated bio-refinery.

Table 1. Overall Mass Balance Summaries at a 30% EFB Splitting Ratio to Pellet Production.

	Materials	Consumption or Yield (ton/day)
In	EFB (dry)	384.00
	Water (H_2O)	5776.40
	Sulfuric acid (H_2SO_4)	27.46
	Ozone (O_3)	8×10^{-3}
	Calcium hydroxide ($Ca(OH)_2$)	20.31
	Sodium hydroxide (NaOH)	31.21
	Enzyme	0.88
	Yeast	12.54
	Total	6252.79
Out	Delignification residue	122.17
	Water (H_2O)	5786.27
	Sulfuric acid (H_2SO_4)	0.58
	Depleted ozone (O_3)	8×10^{-3}
	Calcium sulfate ($CaSO_4$)	37.31
	Sodium hydroxide (NaOH)	31.21
	Spent enzyme	0.88
	Spent yeast	12.54
	Xylose	0.96
	Glucose	29.17
	Xylitol	72.75
	Bioethanol	43.75
	Pellet	100.80
	Dust	10.86
	Fertilizer	3.54
	Total	6252.79

2.1.1. Pellet Production Plant

The EFB feed for pellet production is pretreated with solvents consists of dilute sulfuric acid (H_2SO_4) (0.5% w/w), ozone (O_3) (6.92 × 10^{-4}% w/w) and water (H_2O) (10 ton/ton of EFB) in the pretreatment reactor (R-101) at 180 °C and 1 bar [74]. The pretreatment process is aiming to exposing its cellulose, hemicellulose and lignin fraction and reducing the amount of ash content in the EFB [45]. Table 2 shows the EFB compositions before and after the pretreatment. The pretreated EFB is then dried (D-101) at temperature of 120 °C and pressure of 1 bar, which has a moisture content of approximately 50 wt% [75,76]. The dried EFB is then fed into the pelletization chamber (P-101) to produce pellet at 75 °C and at a maximum compression pressure of 200 MPa [77]. About 87.5% of raw EFB is converted into pellet form and the remaining is dust [78]. Meanwhile, the effluent from the pretreatment reactor (R-101) contains of 3.08% dissolved solids [74] is fed into the activated carbon adsorption (A-101) at 25 °C to separate solubilized N, P and K nutrient [79]. The solvent is recycled back to be used in the pretreatment stage while the remaining solid is removed as fertilizer.

Table 2. Overall Mass Balance Summaries at a 30% EFB Splitting Ratio to Pellet Production [74,80].

Chemical Composition	Percentage (wt%, Dry Basis)	
	Raw EFB	Pretreated EFB
Cellulose	45.2%	80.8%
Hemicellulose	29.5%	6.0%
Lignin	23.6%	13.0%
Ash	1.7%	0.3%

2.1.2. Xylitol and Bioethanol Production Plant

The EFB feed for xylitol and bioethanol production is pretreated with dilute acid in the acid hydrolysis reactor (R-102) with H_2SO_4 (1.25% w/v) and water at 120 °C and at a solid-liquid ratio of 1:8 [75,81]. At these reaction conditions, 93% of hemicellulose is converted into soluble xylose, whereas the remaining insoluble hemicellulose, cellulose and lignin are separated as feedstock for bioethanol production. The xylose fraction thereafter is dosed with calcium hydroxide ($Ca(OH)_2$) in neutralization reactor (N-101) to neutralize the H_2SO_4 at 130 °C [75,82]. The amount of $Ca(OH)_2$ consumed is 0.77 ton/ton of H_2SO_4 used in the acid hydrolysis process [75,83]. The neutralized xylose is sent to the evaporator (E-101) to remove the moisture content at 121 °C and then to the activated carbon adsorption unit (A-102) at 25 °C to remove calcium sulfate ($CaSO_4$) (1.25 ton/ton of $Ca(OH)_2$) formed during the neutralization process [75,79]. Xylose is then fermented with *Candida guiliermondii* yeast (0.17 ton/ton xylose) in fermentation reactor (R-103) to produce xylitol at 30 °C with a yield of 98.7% [75]. The fermented liquid containing xylitol is filtered (F-101) to remove the yeast. The xylitol is then purified using ion-exchange chromatography (I-101) at 25 °C and crystallized with a crystallizer (C-101) and 40 °C to obtain xylitol in the form of solid crystal [75].

The cellulose and lignin-rich solid phase that leaves the acid hydrolysis reactor (R-102) is further processed for bioethanol production. The solid is first treated with 2% (w/v) sodium hydroxide (NaOH) solution in the delignification reactor (R-104) at a solid-liquid ratio of 1:20 and at 120 °C [75]. The treatment aims to remove lignin and other components from cellulose. Delignification process is an important step to liberate cellulose and hemicellulose from their complex with lignin, so that these compounds can undergo hydrolysis to produce fermentable sugars. The effluent that is in the form of black liquor, containing hemicellulose, ash, NaOH and water, is removed from the delignification reactor as waste. The cellulose-containing stream is fed into enzymatic hydrolysis reactor (R-105) to further degrade into glucose at temperature of 45 °C [75]. *Trichoderma reesei* cellulase with a consumption rate of 0.02 ton/ton of ethanol is used as the enzyme [74]. The hydrolysis process produces 60% of glucose-rich hydrolysate [75]. The glucose is then separated from the solid residue containing unreacted cellulose and enzyme in the filter unit (F-101). The glucose is fermented to produce ethanol in fermentation reactor (R-105) at 30 °C, using *Zymomonas mobilis* yeast with a consumption rate of 0.0004 ton/ton of ethanol [84]. This process converts 60% of glucose into bioethanol [75]. The bioethanol-containing stream is separated from the solid residue containing the yeast in the filter unit (F-102) and the unreacted glucose or stillage using distillation column (D-101) at 78.15 °C [85]. The bioethanol is finally dehydrated with rectification column to remove excess water (D-102) at 78.15 °C [74].

2.2. Estimation of Total Capital Investment (TCI) and Total Production Cost (TPC)

The total capital investment (TCI) of the bio-refinery was estimated using the power law or exponential method and Chemical Engineering Plant Cost Index (CEPCI), as shown in Equations (1) and (2), respectively.

$$C_2 = C_1 \times \left(\frac{S_2}{S_1}\right)^n \tag{1}$$

where

C_1 = Cost of the reference plant at capacity of S_1;
C_2 = Cost of the plant at desired capacity of S_2;
n = Scale exponent or cost-capacity factor.

The cost of the plant at the desired capacity (C_2) is the result of multiplication between the reference plant cost (C_1) and the capacity ratio of the new capacity (S_2) to the reported capacity (S_1), to the power of sizing exponent (n). The n value of 0.6 was used in this preliminary study because a typical chemical plant typically follows the six-tenths rule [86].

$$TCI_{p,x,b} = \sum_{p,x,b} C_2 \times \frac{CEPCI_{new}}{CEPCI_{install}} \tag{2}$$

where

$TCI_{p,x,b}$ = Total capital cost of the bio-refinery;
C_2 = Cost of the plant at desired capacity;
$CEPCI_{new}$ = Chemical engineering cost price index at present year;
$CEPCI_{install}$ = Chemical engineering cost price index at reference year.

As shown in Equation (2), the cost of the desired plant capacity (C_2) is then scaled to the desired time value of TCI using the CEPCI, which is a dimensionless number to estimate the capital cost from the past year to the year 2020. The total capital investment of all three products, $TCI_{p,x,b}$, is the summation of all three TCI estimated individually from their respective CEPCI at its respective year. The reference capital, capacity and $CEPCI_{install}$ of pellet, xylitol, and bioethanol production plant used for this study are listed in Table 3. All CEPCI values were obtained from chemengonline [87], where the $CEPCI_{new}$ of year 2020 is 588.06.

Table 3. Total capital investment of individual pellet, xylitol, and bioethanol production plant.

Products	Cost of the Reference Plant, C_1 (Million USD)	Production Capacity, S_1 (Ton/Year)	$CEPCI_{install}$ Value	Ref.
Pellet	14.05	110,880	CEPCI (2018): 603.10	[74]
Xylitol	220.06	30,624	CEPCI (2016): 541.70	[88]
Bioethanol	40.58	9966	CEPCI (2016): 556.80	[82]

It should be noted that the TCI is estimated based on order of magnitude with limited information, hence the accuracy range is rather wide. Nonetheless, such simplified method allows us to estimate the cost quickly at a different EFB splitting ratio to pellet production.

The total production cost (TPC) is the total cost incurred for the production of a particular amount of products. The TPC consists of two components, namely cost of manufacturing, $COM_{p,x,b}$, and general expenses, $GE_{p,x,b}$, as shown in Equation (3). The COM consists of the variable cost of production, $VCOP_{p,x,b}$, fixed cost of production, $FCOP_{p,x,b}$, and plant overhead, $PO_{p,x,b}$, as shown in Equation (4). Sinnott and Towler [89] have suggested the percentage shares of each component in Equations (3) and (4); these percentage shares are listed in Table 4. The cost of raw materials is listed in Table 5.

$$TPC_{p,x,b} = COM_{p,x,b} + GE_{p,x,b} \qquad (3)$$

$$COM_{p,x,b} = VCOP_{p,x,b} + FCOP_{p,x,b} + PO_{p,x,b} \qquad (4)$$

Table 4. Summary of percentage shares of TPC [89,90].

Total Production Cost (TPC)	Percentages Share
1. Variable cost (VCOP)	66% of TPC
(a) Operating labor	10% of VCOP
(b) Utility	10% of VCOP
(c) Patents and royalties	6% of VCOP
(d) Direct supervisory and clerical labor	4% of VCOP
(e) Maintenance and repair	4% of VCOP
(f) Operating supplies	4% of VCOP
(g) Laboratory charges	4% of VCOP
2. Fixed cost (FCOP)	10% of TPC
(a) Local taxes	2% of FCOP
(b) Insurance	2% of FCOP
(c) Financial cost (interest)	3% of FCOP
(d) Rent	3% of FCOP
3. Plant overhead cost	9% of TPC
4. General expenses (GE)	15% of TPC
(a) Administrative expenses	5% of GE
(b) Distribution and marketing expenses	10% of GE
(c) Research and development expenses	5% of GE

Table 5. Cost of raw materials.

Raw Material	Cost (USD/Ton)	Ref.
EFB	6	[14]
Sulfuric acid	41	[74]
Enzyme	6310	[84]
Yeast	5700	[84]
Calcium hydroxide	75	[88]
Sodium hydroxide	98	[75]
Water	0.63	[91]

2.3. Calculation of Economic Performance Metrics

There are a few tools that can be used to evaluate the economic feasibility of an investment; some take the time value of money (TVM) into consideration, and others do not. Example tools for the former are NPV and IRR, whereas the latter are PBP and ROI. The tools that do not take TVM into consideration are relatively less complicated and straightforward to use, thus providing a rapid assessment of the viability of a project. However, if the duration length of investment is long, then the tools with TVM provide a more realistic analysis. Eventually, all these tools analyze the cash flow with or without discounted factors from various perspectives to provide a decision-making value for investors to consider. To begin with, this study considered the project has a life expectancy of 20 years, a 100% TCI was spent in year 0, and the minimum acceptable rate of return (MARR) was set at 16%, which indicates the level of risk of the investment. The level of risk is low because the pellet, xylitol, and bioethanol are considered biorefineries with new capacity with the established corporate market position [92]. The prevailing corporate tax rate was set at 24% in reference to the Inland Revenue Board of Malaysia. Although asset depreciation has no direct impact on cash flow, it changes the tax liabilities. Herewith, the straight-line annual depreciation method was used to estimate the asset depreciation.

NPV, as shown in Equation (5), is the measure of profitability based on the total present value of a time series of cash flows, CF_n, at any time period (n) in years from the present time with an interest rate of i. The interest rate was assumed to be the same as MARR [92]. This method converts the cash flow in the future to present values for comparison. A positive NPV indicates a viable investment. A greater and positive NPV indicates the project is competitive.

$$\text{NPV} = \sum_{n=0} \frac{CF_n}{(1+i)^n} \tag{5}$$

The IRR calculation is complementary to the NPV calculation, where it measures the discounted annual rate of return and provides a safety investment margin. The IRR is the interest rate at which NPV is equal to zero, as shown in Equation (6).

$$\text{NPV} = \sum_{n=0} \frac{CF_n}{(1+\text{IRR})^n} = 0, \tag{6}$$

The PBP is a profitability measure in terms of the length of time to recover the cost of investment. The limitation of the PBP method is that the cash flows beyond the breakeven year are no longer relevant; thus, it is not able to capture the long-term profitability of the investment. The PBP calculation used for this study was based on uneven cash flows, as shown in Equation (7). The PBP was calculated by adding the final year, n, that has a negative cumulative cash flow with the fraction of the absolute value of cumulative cash flow, CCF_n at n year to the cash flow, CF_n, at $n + 1$ year.

$$\text{PBP} = n + \frac{CCF_n}{CF_{n+1}} \tag{7}$$

ROI is a simple measure of the economic performance of the money that has been invested. The ROI is expressed as a percentage of the ratio between net profits, NP_{avg} to the TCI as in Equation (8).

$$\text{ROI (\%)} = \frac{NP_{avg}}{\text{TCI}} \times 100 \tag{8}$$

The profitability of the bio-refinery can be determined after the NPV, PBP, ROI, and IRR have been calculated. Table 6 summarizes the profitability indicators to decide whether the investment is acceptable or not. A positive NPV indicates that the earnings of the bio-refinery exceed the costs, and therefore, the bio-refinery is considered economically viable. Table 7 shows the selling price of pellet, fertilizer, xylitol, and bioethanol. The desired PBP of the investments should be under 5 years to be attractive [93]. The ROI and IRR should be more than the MARR set in this study to be profitable. The higher the ROI and IRR, the greater the returns exceed the capital cost [88].

Table 6. Profitability indicators of the bio-refinery.

Performance Criteria	Comments	Ref.
NPV	Acceptable if in positive value	[94]
PBP	Acceptable if in a short period of time	[88]
ROI	Higher than the MARR	[92]
IRR	Higher than the MARR	[88]

Table 7. Selling price of products.

Product	Cost (USD/ton)	Ref.
Pellet	90	[95]
Xylitol	5500	[96]
Bioethanol	963	[92]
Fertilizer	300	[97]

3. Results and Discussion

To examine the economic feasibility of the bio-refinery, three scenarios are evaluated in different settings. The scenario analysis is simulated based on the assumption that an existing pellet production plant's expansion produces higher-value products: xylitol and bioethanol.

Scenario 1: Economic analysis at different EFB splitting ratios to pellet production and xylitol and bioethanol production. In this scenario, the effect of EFB splitting ratios on the economic performance metrics is calculated to analyze the production process's viability.

Scenario 2: Sensitivity analysis at different EFB and product prices based on an EFB splitting ratio of 30% to pellet production and 70% to xylitol and bioethanol production. The purpose of the sensitivity analysis is to address the price fluctuations of EFB feedstock cost and market selling price of the pellet, xylitol, and bioethanol.

Scenario 3: Economic analysis of the coproduction of pellet with xylitol or pellet with bioethanol. The purpose of this analysis is to investigate whether the coproduction of pellets with xylitol or pellets with bioethanol is more economically feasible than the baseline scenario.

3.1. Profitability Analysis of Scenario 1

Figure 3 shows the profitability analysis of Scenario 1 at different EFB splitting ratios for pellet production and xylitol and bioethanol production, with a minimum acceptable return rate (MARR) of 16% and a 20-year life span. Both splitting ratios of 0% and 100% indicated that all EFB feedstock was fed into the production stream of xylitol/bioethanol and pellet, respectively. The splitting ratio of 0% had the highest net present value (NPV) of 129 million USD. Nevertheless, the NPV dropped drastically with the increase in the EFB splitting ratio to pellet production. In fact, the NPV plunged below zero at a splitting ratio of 80% and beyond and recovered slightly to the value of approximately 8 million USD at 100% EFB for pellet production only. In other words, for any EFB splitting ratio below 80%, the production of these three products is still feasible. The main reason for reducing NPV with the increase in EFB splitting ratio is the high total capital investment (TCI) and total production cost (TPC) of both xylitol and bioethanol processes that are not able to recover from revenue generation. It should be noted that xylitol products are the primary contributor to overall revenue because of their high market price of 5500 USD/ton [96]. As more EFB was diverted to the production of pellets, xylitol and bioethanol's capacity was subsequently reduced, which reduced the overall revenue. For every 10% increment in pellet production, the revenue is reduced by 19 million USD annually on average. By estimating from the slope in Figure 4, the reduction rate of revenue was much higher than TCI and TPC. The recovery of NPV to a positive value at 100% EFB for pellet production is due to the exclusion of TCI and TPC of xylitol and bioethanol, which require expensive enzymes and yeast for production.

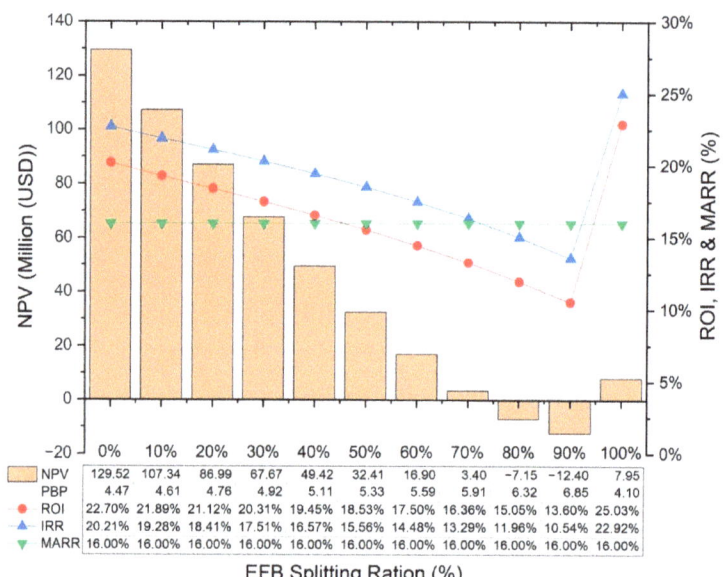

Figure 3. Economic profitability analysis at different EFB splitting ratios to pellet production.

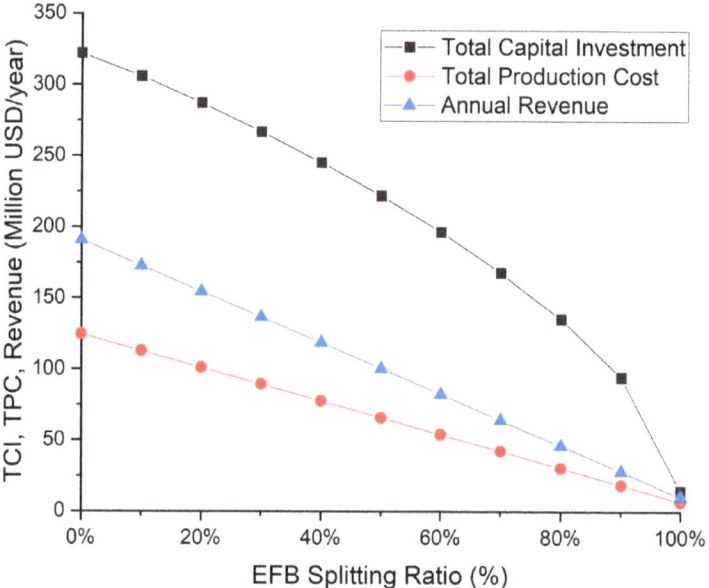

Figure 4. Changes in TCI, TPC, and annual revenue (at full capacity) at different EFB splitting ratios to pellet production.

On the other hand, the payback period (PBP) is an indicator of the length of time needed for the initial investment to break even or recover the investment cost; thus, a shorter PBP is preferred over a longer PBP. In this analysis, the payback period showed a reversed trend of NPV, where the length of PBP increased gradually from 4.5 years (0% EFB splitting ratio) to 6.9 years (90% EFB splitting ratio) and decreased to 4.1 years at a 100% EFB splitting ratio. Typically, a PBP of less than or approximately five years is favorable [93]. Therefore, an EFB splitting ratio below 40% is justifiable. Similar to the NPV analysis, the TCI and TPC for pellet production only were 96% and 94% lower than those of xylitol and bioethanol production only, respectively. If all three products are to be considered in the production, TCI and TPC contributions by xylitol and bioethanol production will be significant. The unit operations involved in pellet production only consisted of a pretreatment reactor, dryer, and pelletization mill as the main process equipment. Those were less complex and less costly than the unit operations involved in xylitol and bioethanol production only, which consisted of a hydrolysis and fermentation reactor, activated carbon adsorption column, and distillation column. Moreover, the raw materials for xylitol and bioethanol production also involved the use of expensive enzymes (6310 USD/ton) [84] and yeast (5700 USD/ton) [84], as well as large amounts of sodium hydroxide and water, which have also contributed to a higher TPC.

Both return on investment (ROI) and internal return rate (IRR) values showed a similar reduction trend of NPV from the EFB splitting ratios from 0% to 90% but rebounded strongly to an ROI value of 23% and an IRR value of 25%, higher than that at the EFB splitting ratio of 0%. The difference between ROI and IRR is that IRR takes into account the time value of money by assuming the NPV equals 0 at the end of the 20-year life span; hence, the IRR values are slightly higher than the ROI values, which are in the range of 2% to 3%. The ROI and IRR were less favorable for the coproduction of xylitol and bioethanol than for pellet production only because the TCI and TPC of xylitol and bioethanol production were much more costly than those of pellet production. The ROI and IRR were then compared to MARR. It was found that the ROI and IRR of EFB splitting ratios below 40% and 70% were above 16%, respectively, which indicated that the investment was feasible and acceptable.

Comparing these four economic performance metrics, several possible production combinations were considerable, depending on the investor's interest and the demand for the products. For example, suppose the investor has an existing pellet production plant and would like to diversify some of the EFB feedstock to xylitol and bioethanol production. In that case, it is recommended that the EFB splitting to the existing pellet production should be 40% or less. The ROI and IRR of these options may seem lower than those of the existing pellet production plant. The NPV, which is the time value of money of cumulative cash flow, is at least one order of magnitude greater than that of the existing pellet production plant. Nonetheless, the tradeoff would be a slightly longer PBP as well as a higher TCI and TPC.

3.2. Profitability Analysis of Scenario 2

While the TCI requires a large lump sum at the beginning of the investment, it is usually a one-off contribution and is not affected by the global market supply chain. Both TPC and revenue contributed by the feedstock price and product selling price play a more important role during the operation lifetime. To understand the effect of the feedstock price and product selling price on the economic analysis, we performed a sensitivity analysis using the EFB splitting ratio of 30% to pellet production as a basis, where the corresponding prices and economic performance metrics are listed in Table 8.

Figure 5 shows the changes in economic performance metrics (NPV, PBP, ROI, and IRR) with EFB feedstock and product prices. Figure 5A shows that the feedstock price did not significantly impact the economic performance metrics. The price changes in EFB feedstock only affect the TPC. With a savings of 33% of the feedstock price (4 USD/ton), the NPV was improved by 5%, the PBP was reduced by less than 1%, and the ROI and IRR were improved by approximately 1%. In contrast, if the feedstock price was increased by 33% (8 USD/ton) and by 67% (10 USD/ton), the NPV was reduced by 5% and 10%, respectively. PBP length was increased slightly by 1% to 2%, while the ROI and IRR were reduced by less than 3%. In fact, an EFB price of 10 USD/ton still generated a positive NPV value, a PBP period of 5 years was still acceptable, and both the ROI and IRR were still more significant than the MARR. The sensitivity analysis concluded that the contribution of EFB feedstock price to the total production cost is less significant.

Table 8. Baseline parameters and economic performance metrics at an EFB splitting ratio of 30% to pellet production.

Parameters	Baseline Data
NPV (Million USD)	67.67
PBP (years)	4.92
ROI (%)	17.51%
IRR (%)	20.31%
EFB Price (USD/ton)	6
Xylitol Price (USD/ton)	5500
Bioethanol Price (USD/ton)	963
Pellet Price (USD/ton)	90

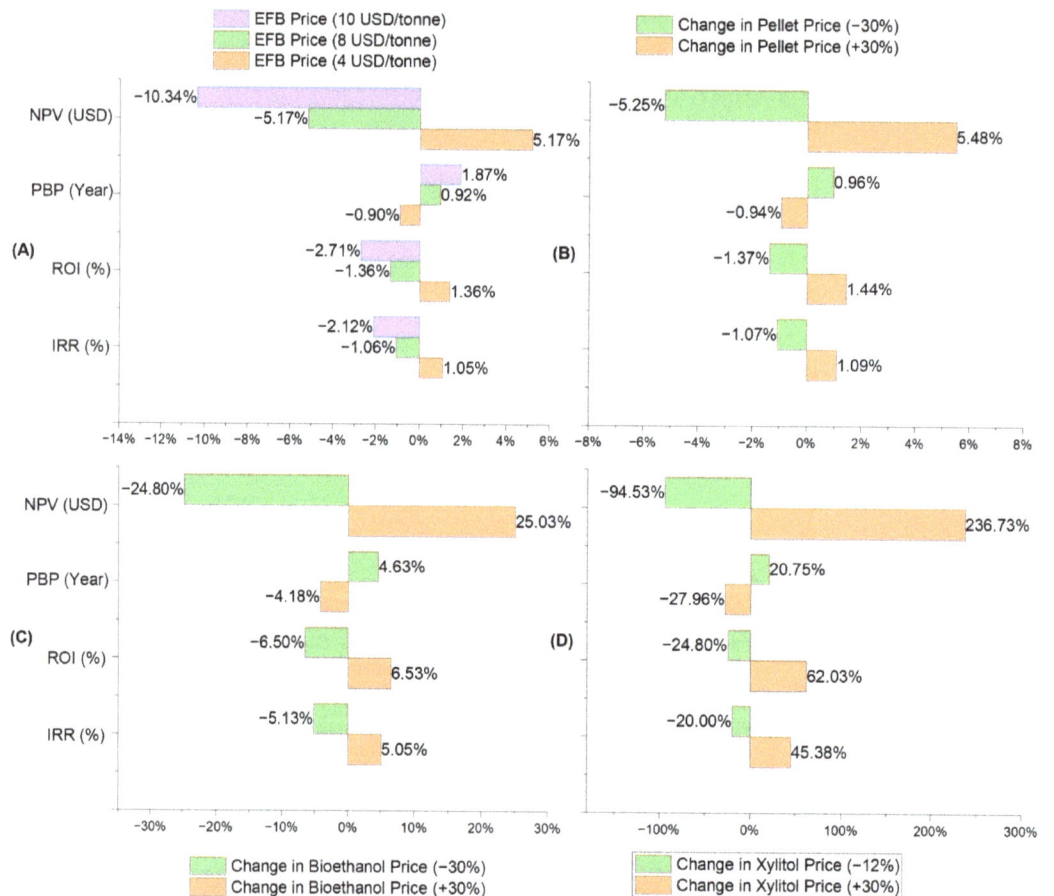

Figure 5. Sensitivity analysis of price changes in (**A**) EFB; (**B**) pellet; (**C**) bioethanol; and (**D**) xylitol at the basis of EFB splitting ratio of 30% to pellet production and MARR of 16%.

As shown in Figure 5B, the changes in the pellet price also indicate a relatively low impact on the economic performance metrics. When the pellet price (90 USD/ton) was increased by 30%, the NPV, ROI, and IRR were increased by 5%, 1.4%, and 1.1%, respectively, while the PBP was reduced by less than 1%. Conversely, when the pellet's price was reduced by 30%, the NPV, ROI, IRR, and PBP showed their corresponding opposite values. The small changes in the economic performance metrics were partly due to the pellet's selling price being much lower than that of the other two products; hence, the contribution to the changes in revenue was negligible.

The changes in bioethanol price have a greater impact on the economic performance metrics, as shown in Figure 5C. An increase of 30% in bioethanol's price improved the NPV, ROI, and IRR by 25%, 6.5%, and 5.0%, respectively, and reduced the PBP by approximately 4.2%. A positive increase in the price of bioethanol is highly desirable to offset the expensive raw materials used in production, such as enzymes and yeast, and the large quantity of sodium hydroxide and water consumption in the delignification process.

The changes in xylitol price have the most significant effect on the economic performance metrics, as shown in Figure 5D. An increase in xylitol's price by 30% increased the NPV by 237%, improved the ROI and IRR by 62% and 45%, respectively, and shortened the PBP by 28%. Nevertheless, xylitol's lower boundary price could not be lower than 4802 USD/ton, approximately 12% lower than the benchmark price of 5500 USD/ton, to maintain a positive NPV. A 12% reduction in xylitol price shrank NPV, ROI, and IRR by 95%, 24.8%, and 20.0%, respectively, and increased the PBP by 20.8%.

In summary, there are three factors that contribute to an attractive economic performance: TCI, TPC, and revenue generation. Pricing sensitivity analysis has addressed the contributions of TPC and revenue generation to the economic performance metrics. It has been shown that the feasibility of coproduction of all three products (pellet, bioethanol, and xylitol) is highly dependent on the selling price of xylitol, which offers little room for price competition. Alternatively, one should forgo one of the products to further mitigate the risk and reduce TCI. This leads to the next scenario analysis, which is the coproduction of pellets with xylitol and the coproduction of pellets with bioethanol.

3.3. Profitability Analysis of Scenario 3

In Scenario 3, the economic performances of the coproduction of pellet with xylitol and coproduction of pellet with bioethanol were compared with the coproduction of pellet with xylitol and bioethanol on the basis of an EFB splitting ratio of 30% to pellet production at a MARR of 16%. The results are presented in Figure 6. Only the coproduction of pellet and xylitol is a feasible solution compared to the coproduction of pellet and bioethanol. On the other hand, the NPV of coproduction of pellets with xylitol was increased by 31%, while the PBP was reduced slightly to less than 4.5 years. The ROI and IRR of this combination were improved to 20% and 23%, respectively, above the MARR of 16%.

The coproduction of pellets and bioethanol was a no-go option due to its expensive pretreatment process, low bioethanol yield, and selling price of bioethanol, which contributed to a negative NPV (Figure 6). From the simulation, the price of ethanol was required to be at least 1882 USD/ton to achieve a MARR of 16% or 1538 USD/ton at a lower MARR of 10%. This is consistent with Do et al. [98] on the limitation of bioethanol production from EFB. To further confirm bioethanol production's economic feasibility from EFB as a lignocellulosic source, a scenario of total conversion from EFB to bioethanol was further conducted. With the bioethanol production of 0.16 ton/ton of EFB, the production is only feasible with a bioethanol market price of 1758 USD/ton. This is in accord with the techno-economic analysis performed by Dávila et al. [75], which indicated that even with heat integration in the process that further reduced the production cost by 43%, the production price was still higher than the market price. This reflects the energy-intensive process with the heavy use of chemicals aside from the expensive enzymes for the purification and conversion involved in bioethanol production. This result also reflects that commercial production of bioethanol from lignocellulosic raw materials alone has not been widely implemented [30].

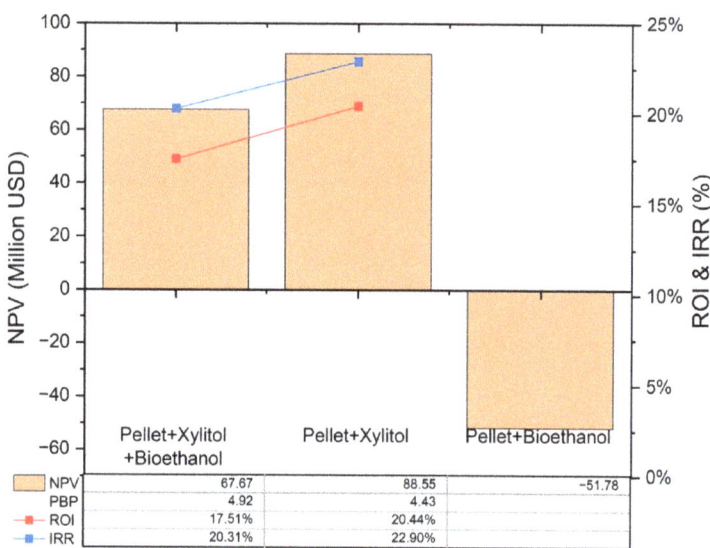

Figure 6. Economic profitability analysis of coproduction of pellet with xylitol and bioethanol, pellet with xylitol, and pellet with bioethanol at EFB splitting ratio of 30% to pellet production and MARR of 16%.

4. Conclusions

In this study, three scenarios were evaluated to determine the profitability of a bio-refinery producing pellet, xylitol, and bioethanol. The bio-refinery was found to be profitable at an EFB splitting ratio of below 40% for pellet production, resulting in a positive NPV, PBP lower than five years, and ROI and IRR higher than the MARR value of 16%. The results also showed that it is possible to produce both pellets and xylitol, which resulted in a higher NPV, shorter PBP, and higher ROI and IRR than the baseline scenario. The selling price of ethanol from either coproduction of pellets and bioethanol or bioethanol alone is still less competitive entering the market. Still, the coproduction of bioethanol with xylitol is feasible with a higher NPV than coproduction with pellets. Nevertheless, this work has successfully demonstrated that the valorization of EFB to high-value products is feasible.

Author Contributions: Conceptualization, K.L.L. and W.Y.W.; methodology, K.L.L. and W.Y.W.; S.K.L. and M.T.L.; formal analysis, N.J.R. and K.L.L.; investigation, K.L.L. and W.Y.W.; resources, S.K.L. and M.T.L.; writing—original draft preparation, N.J.R.; writing—review and editing, K.L.L. and W.Y.W.; supervision, K.L.L. and W.Y.W.; funding acquisition, K.L.L. and M.T.L. All authors have read and agreed to the published version of the manuscript.

Funding: This research was funded by TNB Research Sdn. Bhd., grant number TNBR/SF0348/2019 and by Universiti Kebangsaan Malaysia, grant number GP-2019-K017662 and PP-SELFUEL-2022.

Institutional Review Board Statement: Not applicable.

Informed Consent Statement: Not applicable.

Data Availability Statement: Not applicable.

Acknowledgments: We would like to thank for the administrative support from UKM Pakarunding Sdn. Bhd.

Conflicts of Interest: The authors declare no conflict of interest. The funders had no role in the design of the study, in the collection, analyses, or interpretation of data, in the writing of the manuscript, or in the decision to publish the results.

References

1. Ahmad Parveez, G.K.; Hishamuddin, E.; Loh, S.K.; Ong-Abdullah, M.; Mohamed Salleh, K.; Bidin, M.N.I.Z.; Sundram, S.; Azizul Hasan, Z.A.; Idris, Z. Oil Palm Economic Performance in Malaysia and R&D Progress in 2019. *J. Oil Palm Res.* **2020**, *32*, 159–190. [CrossRef]
2. Al-Qahtani, A.M.; Jebaraj, S. Oil Demand Forecasting in Malaysia in Transportation Sector Using Artificial Neural Network. *Int. J. Sci. Eng. Invent.* **2019**, *5*, 8–14. [CrossRef]
3. Chan, Y.H.; Cheah, K.W.; How, B.S.; Loy, A.C.M.; Shahbaz, M.; Singh, H.K.G.; Yusuf, N.a.R.; Shuhaili, A.F.A.; Yusup, S.; Ghani, W.A.W.A.K.; et al. An overview of biomass thermochemical conversion technologies in Malaysia. *Sci. Total Environ.* **2019**, *680*, 105–123. [CrossRef] [PubMed]
4. Hassan, N.; Idris, A.; Akhtar, J. Overview on Bio-refinery Concept in Malaysia: Potential High Value Added Products from Palm Oil Biomass. *J. Kejuruter.* **2019**, *2*, 113–124. [CrossRef]
5. Vaskan, P.; Pachón, E.R.; Gnansounou, E. Techno-economic and life-cycle assessments of biorefineries based on palm empty fruit bunches in Brazil. *J. Clean. Prod.* **2018**, *172*, 3655–3688. [CrossRef]
6. Hamzah, M.A.A.; Alias, A.B.; Him, N.R.N.; Rashid, Z.A.; Ghani, W.A.W.A.K. Characterization of food waste and empty fruit bunches (EFB) for anaerobic digestion application. *J. Phys. Conf. Ser.* **2019**, *1349*, 12132. [CrossRef]
7. MPOB. Overview of the Malaysian Oil Palm Industry 2019. Available online: https://bepi.mpob.gov.my/images/overview/Overview_of_Industry_2019.pdf (accessed on 20 September 2020).
8. Ahmad, F.B.; Zhang, Z.; Doherty, W.O.S.; O'Hara, I.M. The outlook of the production of advanced fuels and chemicals from integrated oil palm biomass biorefinery. *Renew. Sustain. Energy Rev.* **2019**, *109*, 386–411. [CrossRef]
9. Rosli, N.S.; Harun, S.; Md Jahim, J.; Othaman, R. Chemical and Phyiscal Characterization of Oil Palm Empty Fruit Bunch. *Malays. J. Anal. Sci.* **2017**, *21*, 188–196. [CrossRef]
10. Darojat, K.; Hadi, W.; Rahayu, D.E. Life Cycle Assessment (LCA) utilization of oil palm empty fruit bunches as bioenergy. *AIP Conf. Proc.* **2019**, *2194*, 020019. [CrossRef]
11. Hamzah, N.; Tokimatsu, K.; Yoshikawa, K. Solid Fuel from Oil Palm Biomass Residues and Municipal Solid Waste by Hydrothermal Treatment for Electrical Power Generation in Malaysia: A Review. *Sustainability* **2019**, *11*, 1060. [CrossRef]
12. Alaw, F.A.; Sulaiman, N.S. A Review of Boiler Operational Risks in Empty Fruit Bunch Fired Biopower Plant. *J. Chem. Eng. Ind. Biotechnol.* **2020**, *5*, 29–35. [CrossRef]
13. Poh, P.E.; Wu, T.Y.; Lam, W.H.; Poon, W.C.; Lim, C.S. Sustainability of waste management initiatives in palm oil mills. In *Green Energy and Technology*; Springer: Cham, Switzerland, 2020; pp. 57–73. [CrossRef]
14. Abdulrazik, A.; Elsholkami, M.; Elkamel, A.; Simon, L. Multi-products productions from Malaysian oil palm empty fruit bunch (EFB): Analyzing economic potentials from the optimal biomass supply chain. *J. Clean. Prod.* **2017**, *168*, 131–148. [CrossRef]
15. Garcia-Nunez, J.A.; Ramirez-Contreras, N.E.; Rodriguez, D.T.; Silva-Lora, E.; Frear, C.S.; Stockle, C.; Garcia-Perez, M. Evolution of palm oil mills into bio-refineries: Literature review on current and potential uses of residual biomass and effluents. *Resour. Conserv. Recycl.* **2016**, *110*, 99–114. [CrossRef]
16. Mohd Yusof, S.J.H.; Zakaria, M.R.; Roslan, A.M.; Ali, A.A.M.; Shirai, Y.; Ariffin, H.; Hassan, M.A. Oil palm biomass biorefinery for future bioeconomy in Malaysia. In *Lignocellulose for Future Bioeconomy*; Ariffin, H., Sapuan, S.M., Hassan, M.A., Eds.; Elsevier: Amsterdam, The Netherlands, 2019; pp. 265–285. [CrossRef]
17. James Rubinsin, N.; Daud, W.R.W.; Kamarudin, S.K.; Masdar, M.S.; Rosli, M.I.; Samsatli, S.; Tapia, J.F.; Wan Ab Karim Ghani, W.A.; Lim, K.L. Optimization of oil palm empty fruit bunches value chain in peninsular malaysia. *Food Bioprod. Process.* **2020**, *119*, 179–194. [CrossRef]
18. Tapia, J.F.D.; Samsatli, S.; Doliente, S.S.; Martinez-Hernandez, E.; Ghani, W.A.B.W.A.K.; Lim, K.L.; Shafri, H.Z.M.; Shaharum, N.S.N.B. Design of biomass value chains that are synergistic with the food–energy–water nexus: Strategies and opportunities. *Food Bioprod. Process.* **2019**, *116*, 170–185. [CrossRef]
19. Taqwa, S.A.; Purwanto, W.W. A Superstructure Based Enviro-Economic Optimization for Production Strategy of Oil Palm Derivatives. *IOP Conf. Ser. Mater. Sci. Eng.* **2019**, *543*, 012062. [CrossRef]
20. Harahap, B.M.; Mardawati, E.; Nurliasari, D. A comprehensive review: Integrated microbial xylitol, bioethanol, and cellulase production from oil palm empty fruit bunches. *J. Ind. Pertan.* **2020**, *2*, 142–157.
21. Fuad, M.A.H.M.; Faizal, H.M.; Ani, F.N. Experimental investigation on water washing and decomposition behaviour for empty fruit bunch. *J. Adv. Res. Fluid Mech. Therm. Sci.* **2019**, *59*, 207–219.
22. Novianti, S.; Zaini, I.N.; Nurdiawati, A.; Yoshikawa, K. Low Potassium Content Pellet Production by Hydrothermal-Washing Co-treatment. *Int. J. Chem. Chem. Eng. Syst.* **2016**, *1*, 28–38.
23. Tang, Y.; Chandra, R.P.; Sokhansanj, S.; Saddler, J.N. Influence of steam explosion processes on the durability and enzymatic digestibility of wood pellets. *Fuel* **2018**, *221*, 87–94. [CrossRef]
24. Chala, G.T.; Guangul, F.M.; Sharma, R. Biomass Energy in Malaysia-A SWOT Analysis. In Proceedings of the 2019 IEEE Jordan International Joint Conference on Electrical Engineering and Information Technology, JEEIT 2019—Proceedings, Amman, Jordan, 9–11 April 2019; pp. 401–406. [CrossRef]
25. Europen Pellet Council. *Statistical Pellet Report*; Europen Pellet Council: Ixelles, Belgium, 2018.
26. FutureMetrics. *Quarterly Pellet Market Report*; FutureMetrics: Bethel, ME, USA, 2018.
27. HPBA. *Market Research Reports*; Hearth, Patio & Barbecue Association: Arlington, VA, USA, 2020.

28. Brunerová, A.; Müller, M.; Šleger, V.; Ambarita, H.; Valášek, P. Bio-Pellet Fuel from Oil Palm Empty Fruit Bunches (EFB): Using European Standards for Quality Testing. *Sustainability* **2018**, *10*, 4443. [CrossRef]
29. Mohapatra, S.; Ray, R.C.; Ramachandran, S. Bioethanol from Biorenewable Feedstocks: Technology, Economics, and Challenges. In *Bioethanol Production from Food Crops*; Academic Press: Cambridge, MA, USA, 2019; pp. 3–27. [CrossRef]
30. Rosales-Calderon, O.; Arantes, V. A review on commercial-scale high-value products that can be produced alongside cellulosic ethanol. *Biotechnol. Biofuels* **2019**, *12*, 240. [CrossRef]
31. Delgado Arcaño, Y.; Valmaña García, O.D.; Mandelli, D.; Carvalho, W.A.; Magalhães Pontes, L.A. Xylitol: A review on the progress and challenges of its production by chemical route. *Catal. Today* **2020**, *344*, 2–14. [CrossRef]
32. Clauser, N.M.; Gutiérrez, S.; Area, M.C.; Felissia, F.E.; Vallejos, M.E. Alternatives of small-scale biorefineries for the integrated production of xylitol from sugarcane bagasse. *J. Renew. Mater.* **2018**, *6*, 139–151. [CrossRef]
33. Baruah, J.; Nath, B.K.; Sharma, R.; Kumar, S.; Deka, R.C.; Baruah, D.C.; Kalita, E. Recent trends in the pretreatment of lignocellulosic biomass for value-added products. *Front. Energy Res.* **2018**, *6*, 141. [CrossRef]
34. Hamzah, N.H.C.; Markom, M.; Harun, S.; Hassan, O. The Effect of Various Pretreatment Methods on Empty. *Malays. J. Anal. Sci.* **2016**, *20*, 1474–1480. [CrossRef]
35. Cocero, M.J.; Cabeza, Á.; Abad, N.; Adamovic, T.; Vaquerizo, L.; Martínez, C.M.; Pazo-Cepeda, M.V. Understanding biomass fractionation in subcritical & supercritical water. *J. Supercrit. Fluids* **2018**, *133*, 550–565. [CrossRef]
36. Michelin, M.; Romani, A.; Salgado, J.M.; Domingues, L.; Teixeira, J.A. Production of hemicellulases, xylitol, and furan from hemicellulosic hydrolysates using hydrothermal pretreatment. In *Hydrothermal Processing in Biorefineries*; Springer: Cham, Switzerland, 2017; pp. 285–315. [CrossRef]
37. Pino, M.S.; Rodríguez-Jasso, R.M.; Michelin, M.; Flores-Gallegos, A.C.; Morales-Rodriguez, R.; Teixeira, J.A.; Ruiz, H.A. Bioreactor design for enzymatic hydrolysis of biomass under the biorefinery concept. *Chem. Eng. J.* **2018**, *347*, 119–136. [CrossRef]
38. Derman, E.; Abdulla, R.; Marbawi, H.; Sabullah, M.K. Oil palm empty fruit bunches as a promising feedstock for bioethanol production in Malaysia. *Renew. Energy* **2018**, *129*, 285–298. [CrossRef]
39. Fatriasari, W.; Anita, S.H.; Risanto, L. Microwave Assisted Acid Pretreatment of Oil Palm Empty Fruit Bunches (EFB) to Enhance Its Fermentable Sugar Production. *Waste Biomass Valorization* **2017**, *8*, 379–391. [CrossRef]
40. Mardawati, E.; Herliansah, H.; Suryadi, E.; Hanidah, I.I.; Siti Setiasih, I.; Robi, A.; Sukarminah, E.; Djali, M.; Rialita, T.; Cahyana, Y. Optimization of Particle Size, Moisture Content and Reaction Time of Oil Palm Empty Fruit Bunch Through Ozonolysis Pretreatment. *J. Jpn. Inst. Energy* **2019**, *98*, 132–138. [CrossRef]
41. Medina, J.D.C.; Woiciechowski, A.; Filho, A.Z.; Nigam, P.S.; Ramos, L.P.; Soccol, C.R. Steam explosion pretreatment of oil palm empty fruit bunches (EFB) using autocatalytic hydrolysis: A biorefinery approach. *Bioresour. Technol.* **2016**, *199*, 173–180. [CrossRef] [PubMed]
42. Risanto, L. Fitria; Fajriutami, T; Hermiati, E. Enzymatic saccharification of liquid hot water and dilute sulfuric acid pretreated oil palm empty fruit bunch and sugarcane bagasse. *IOP Conf. Ser. Earth Environ. Sci.* **2018**, *141*, 012025. [CrossRef]
43. Rizal, N.F.A.A.; Ibrahim, M.F.; Zakaria, M.R.; Abd-Aziz, S.; Yee, P.L.; Hassan, M.A. Pre-treatment of Oil Palm Biomass for Fermentable Sugars Production. *Molecules* **2018**, *23*, 1381. [CrossRef]
44. Lam, P.Y.; Lim, C.J.; Sokhansanj, S.; Lam, P.S.; Stephen, J.D.; Pribowo, A.; Mabee, W.E. Leaching characteristics of inorganic constituents from oil palm residues by water. *Ind. Eng. Chem. Res.* **2014**, *53*, 11822–11827. [CrossRef]
45. Novianti, S.; Nurdiawati, A.; Zaini, I.N.; Sumida, H.; Yoshikawa, K. Hydrothermal treatment of palm oil empty fruit bunches: An investigation of the solid fuel and liquid organic fertilizer applications. *Biofuels* **2016**, *7*, 627–636. [CrossRef]
46. Kofman, P.D. New Fuels: Thermally Treated Biomass. Available online: http://www.coford.ie/media/coford/content/publications/projectreports/cofordconnects/cofordconnectsnotes/00675CCNPP40Revised091216.pdf (accessed on 20 September 2020).
47. Yaacob, N.; Rahman, N.A.; Matali, S.; Idris, S.S.; Alias, A.B. An overview of oil palm biomass torrefaction: Effects of temperature and residence time. *IOP Conf. Ser. Earth Environ. Sci.* **2016**, *36*, 012038. [CrossRef]
48. Faizal, H.M.; Jusoh, M.A.M.; Rahman, M.R.A.; Syahrullail, S.; Latiff, Z.A. Torrefaction of palm biomass briquettes at different temperature. *J. Teknol.* **2016**, *78*, 61–67. [CrossRef]
49. Ruksathamcharoen, S.; Chuenyam, T.; Stratong-on, P.; Hosoda, H.; Ding, L.; Yoshikawa, K. Effects of hydrothermal treatment and pelletizing temperature on the mechanical properties of empty fruit bunch pellets. *Appl. Energy* **2019**, *251*, 113385. [CrossRef]
50. Ahda, Y.; Prakoso, T.; Rasrendra, C.B.; Susanto, H. Hydrothermal treatment, pelletization and characterization of oil palm empty fruit bunches as solid fuel. *IOP Conf. Ser. Mater. Sci. Eng.* **2019**, *543*, 012061. [CrossRef]
51. Zhai, L.; Manglekar, R.R.; Geng, A. Enzyme production and oil palm empty fruit bunch bioconversion to ethanol using a hybrid yeast strain. *Biotechnol. Appl. Biochem.* **2019**, *67*, 714–722. [CrossRef]
52. Campioni, T.S.; Soccol, C.R.; Libardi Junior, N.; Rodrigues, C.; Woiciechowski, A.L.; Letti, L.A.J.; Vandenberghe, L.P.D.S. Sequential chemical and enzymatic pretreatment of palm empty fruit bunches for Candida pelliculosa bioethanol production. *Biotechnol. Appl. Biochem.* **2019**, *67*, 723–731. [CrossRef]
53. Kresnowati, M.; Mardawati, E.; Setiadi, T. Production of Xylitol from Oil Palm Empty Friuts Bunch: A Case Study on Biofinery Concept. *Mod. Appl. Sci.* **2015**, *9*, 206–213. [CrossRef]
54. Meilany, D.; Kresnowati, M.T.A.P.; Setiadi, T.; Boopathy, R. Optimization of xylose recovery in oil palm empty fruit bunches for xylitol production. *Appl. Sci.* **2020**, *10*, 1391. [CrossRef]

55. Sukhang, S.; Choojit, S.; Reungpeerakul, T.; Sangwichien, C. Bioethanol production from oil palm empty fruit bunch with SSF and SHF processes using Kluyveromyces marxianus yeast. *Cellulose* **2020**, *27*, 301–314. [CrossRef]
56. Kapoor, M.; Semwal, S.; Gaur, R.; Kumar, R.; Gupta, R.P.; Puri, S.K. The Pretreatment Technologies for Deconstruction of Lignocellulosic Biomass. In *Waste to Wealth*; Springer: Singapore, 2018; pp. 395–421. [CrossRef]
57. Laca, A.; Laca, A.; Díaz, M. Hydrolysis: From cellulose and hemicellulose to simple sugars. In *Second and Third Generation of Feedstocks*; Elsevier: Amsterdam, The Netherlands, 2019; pp. 213–240. [CrossRef]
58. Siregar, J.S.; Ahmad, A.; Amraini, S.Z. Effect of Time Fermentation and Saccharomyces Cerevisiae Concentration for Bioethanol Production from Empty Fruit Bunch. *J. Phys. Conf. Ser.* **2019**, *1351*, 012104. [CrossRef]
59. Sudiyani, Y.; Dahnum, D.; Burhani, D.; Putri, A.M.H. Evaluation and comparison between simultaneous saccharification and fermentation and separated hydrolysis and fermentation process. In *Second and Third Generation of Feedstocks*; Elsevier: Amsterdam, The Netherlands, 2019; pp. 273–290. [CrossRef]
60. Ghazali, N.F.; Makhtar, N.A. Enzymatic hydrolysis of oil palm empty fruit bunch and its kinetics. *Malays. J. Anal. Sci.* **2018**, *22*, 715–722. [CrossRef]
61. Mardawati, E.; Purwadi, R.; Kresnowati, M.T.A.P.; Setiadi, T. Evaluation of the enzymatic hydrolysis process of Oil Palm empty fruit bunch using crude fungal xylanase. *ARPN J. Eng. Appl. Sci.* **2017**, *12*, 5286–5292.
62. De Paula, R.G.; Antoniêto, A.C.C.; Ribeiro, L.F.C.; Carraro, C.B.; Nogueira, K.M.V.; Lopes, D.C.B.; Costa Silva, A.; Zerbini, M.T.; Pedersoli, W.R.; Do Nascimento Costa, M.; et al. New genomic approaches to enhance biomass degradation by the industrial trichoderma reesei. *Int. J. Genom.* **2018**, *2018*, 17. [CrossRef]
63. Mohd Azhar, S.H.; Abdulla, R.; Jambo, S.A.; Marbawi, H.; Gansau, J.A.; Mohd Faik, A.A.; Rodrigues, K.F. Yeasts in sustainable bioethanol production: A review. *Biochem. Biophys. Rep.* **2017**, *10*, 52–61. [CrossRef]
64. Dahnum, D.; Tasum, S.O.; Triwahyuni, E.; Nurdin, M.; Abimanyu, H. Comparison of SHF and SSF processes using enzyme and dry yeast for optimization of bioethanol production from empty fruit bunch. *Energy Procedia* **2015**, *68*, 107–116. [CrossRef]
65. Barathikannan, K.; Agastian, P. Xylitol: Production, Optimization and Industrial Application. *Int. J. Curr. Microbiol. Appl. Sci.* **2016**, *5*, 324–339. [CrossRef]
66. Dasgupta, D.; Ghosh, D.; Bandhu, S.; Adhikari, D.K. Lignocellulosic sugar management for xylitol and ethanol fermentation with multiple cell recycling by Kluyveromyces marxianus IIPE453. *Microbiol. Res.* **2017**, *200*, 64–72. [CrossRef]
67. Felipe Hernández-Pérez, A.; de Arruda, P.V.; Sene, L.; da Silva, S.S.; Kumar Chandel, A.; de Almeida Felipe, M.d.G. Xylitol bioproduction: State-of-the-art, industrial paradigm shift, and opportunities for integrated biorefineries. *Crit. Rev. Biotechnol.* **2019**, *39*, 923–924. [CrossRef]
68. Hafyan, R.; Bhullar, L.; Putra, Z.; Bilad, M.; Wirzal, M.; Nordin, N. Sustainability assessment of xylitol production from empty fruit bunch. *MATEC Web Conf.* **2019**, *268*, 06018. [CrossRef]
69. Borokhova, O.E.; Mikhailova, N.P. Microbial conversion of D-xylose. *Microbiol. Res.* **1996**, *65*, 581–588.
70. Tamburini, E.; Costa, S.; Marchetti, M.G.; Pedrini, P. Optimized production of xylitol from xylose using a hyper-acidophilic Candida tropicalis. *Biomolecules* **2015**, *5*, 1979–1989. [CrossRef]
71. Kresnowati, M.T.A.P.; Desiriani, R.; Wenten, I.G. Ultrafiltration of hemicellulose hydrolysate fermentation broth. *AIP Conf. Proc.* **2017**, *1818*, 20024. [CrossRef]
72. Agensi Inovasi Malaysia. *National Biomass Strategy 2020: New Wealth Creation for the Palm Oil Industry*; Agensi Inovasi Malaysia: Cyberjaya, Malaysia, 2013.
73. Nitsos, C.; Rova, U.; Christakopoulos, P. Organosolv fractionation of softwood biomass for biofuel and biorefinery applications. *Energies* **2018**, *11*, 50. [CrossRef]
74. Lim, M.T. Tenaga Nasional Berhad Research (TNBR), Bandar Baru Bangi, Selangor, Malaysia. Personal communication, 2020.
75. Dávila, J.A.; Rosenberg, M.; Cardona, C.A. A biorefinery approach for the production of xylitol, ethanol and polyhydroxybutyrate from brewer's spent grain. *AIMS Agric. Food* **2016**, *1*, 52–66. [CrossRef]
76. Ng, R.T.L.; Hassim, M.H.; Ng, D.K.S.; Tan, R.R.; El-Halwagi, M.M. Multi-objective design of industrial symbiosis in palm oil industry. In *Computer Aided Chemical Engineering*; Elsevier: Amsterdam, The Netherlands, 2014. [CrossRef]
77. Puig-Arnavat, M.; Shang, L.; Sárossy, Z.; Ahrenfeldt, J.; Henriksen, U.B. From a single pellet press to a bench scale pellet mill—Pelletizing six different biomass feedstocks. *Fuel Process. Technol.* **2016**, *142*, 27–33. [CrossRef]
78. Talero, G.F.; González, A.H. Use of Colombian oil palm wastes for pellets production: Reduction of the process energy consumption by modifying moisture content. In Proceedings of the 6th International Conference on Engineering for Waste and Biomass Valorisation, Albi, France, 23–26 May 2016.
79. Lee, S.C.; Park, S. Removal of furan and phenolic compounds from simulated biomass hydrolysates by batch adsorption and continuous fixed-bed column adsorption methods. *Bioresour. Technol.* **2016**, *216*, 661–668. [CrossRef] [PubMed]
80. Bukhari, N.A.; Abu Bakar, N.; Loh, S.K. Bioethanol Production by Fermentation of Oil Palm Empty Fruit Bunches Pretreated with Combined Chemicals. *J. Appl. Environ. Biol. Sci.* **2014**, *4*, 234–242.
81. Holgueras Ortega, J. Process Design of Lignocellulosic Biomass Fractionation into Cellulose, Hemicellulose and Lignin by Prehydrolysis and Organosolv Process. Master's Thesis, Wageningen University, Wageningen, The Netherlands, 2015.
82. Do, T.X.; Lim, Y.i.; Jang, S.; Chung, H.J. Hierarchical economic potential approach for techno-economic evaluation of bioethanol production from palm empty fruit bunches. *Bioresour. Technol.* **2015**, *189*, 224–235. [CrossRef]
83. Daham, A. *Basic Principles and Calculations in Chemical Engineering First Year (Report)*; FT Press: Upper Saddle River, NJ, USA, 2014.

84. Loh, S.K. Malaysian Palm Oil Board (MPOB), Bandar Baru Bangi, Selangor, Malaysia. Personal Communication, 2020.
85. Nilesh, P.P.; Vilas, S.P.; Shashikant, L.B. Molecular Sieve Dehydration: A Major Development In The Field Of Ethanol Dehydration To Produce Fuel Ethanol. *Asian J. Sci. Technol.* **2016**, *7*, 2897–2902.
86. Lemmens, S. Cost engineering techniques & their applicability for cost estimation of organic rankine cycle systems. *Energies* **2016**, *9*, 485. [CrossRef]
87. Access Intelligence, LLC. The Chemical Engineering Plant Cost Index. Available online: https://chemengonline.com/pci-home (accessed on 20 September 2020).
88. Özüdoğru, H.M.R.; Nieder-Heitmann, M.; Haigh, K.F.; Görgens, J.F. Techno-economic analysis of product biorefineries utilizing sugarcane lignocelluloses: Xylitol, citric acid and glutamic acid scenarios annexed to sugar mills with electricity co-production. *Ind. Crops Prod.* **2019**, *133*, 259–268. [CrossRef]
89. Sinnott, R.; Towler, G. Costing and Project Evaluation. In *Chemical Engineering Design*, 6th ed.; Elsevier: Amsterdam, The Netherlands, 2020; pp. 275–369. [CrossRef]
90. Sinnot, R. Costing and project evaluation. *Chem. Eng. Des.* **2005**, *6*, 243–280.
91. SPAN. *Water Rates*; Suruhanjaya Perkhidmatan Air Negara: Cyberjaya, Malaysia, 2015.
92. Bidar, B.; Shahraki, F. Energy and exergo-economic assessments of gas turbine based CHP systems: A case study of SPGC utility plant. *Iran. J. Chem. Chem. Eng.* **2018**, *37*, 209–223.
93. Ferreira da Silva, A.; Brazinha, C.; Costa, L.; Caetano, N.S. Techno-economic assessment of a Synechocystis based biorefinery through process optimization. *Energy Rep.* **2020**, *6*, 509–514. [CrossRef]
94. Hasanly, A.; Khajeh Talkhoncheh, M.; Karimi Alavijeh, M. Techno-economic assessment of bioethanol production from wheat straw: A case study of Iran. *Clean Technol. Environ. Policy* **2018**, *20*, 357–377. [CrossRef]
95. Jara, A.A.; Daracan, V.C.; Devera, E.E.; Acda, M.N. Techno-financial Analysis of Wood Pellet Prodiction in the Philippines. *J. Trop. For. Sci.* **2016**, *28*, 517–526.
96. Shankar, K.; Kulkarni, N.S.; Sajjanshetty, R.; Jayalakshmi, S.K.; Sreeramulu, K. Co-production of xylitol and ethanol by the fermentation of the lignocellulosic hydrolysates of banana and water hyacinth leaves by individual yeast strains. *Ind. Crops Prod.* **2020**, *155*, 112809. [CrossRef]
97. Lys, P.; Cachia, F. *Handbook on Agricultural Cost of Production Statistics*; Food and Agriculture Organization of the United Nations: Rome, Italy, 2016; Available online: https://www.fao.org/3/ca6411en/ca6411en.pdf (accessed on 20 September 2020).
98. Do, T.X.; Lim, Y.I.; Jang, S.; Chung, H.J.; Lee, Y.W. Process Design and Economics for Bioethanol Production Process from Palm Empty Fruit Bunch (EFB). *Comput. Aided Chem. Eng.* **2014**, *33*, 1777–1782. [CrossRef]

Article

One-Step Synthesis of High-Performance N/S Co-Doped Porous Carbon Material for Environmental Remediation

Xiaoyu Huo [1,2], Chao Jia [2], Shanshan Shi [1,2], Tao Teng [2], Shaojie Zhou [2], Mingda Hua [2], Xiangdong Zhu [2,3], Shicheng Zhang [1,2,*] and Qunjie Xu [1,*]

[1] College of Environmental and Chemical Engineering, Shanghai University of Electric Power, Shanghai 200090, China; huoxiaoyu@mail.shiep.edu.cn (X.H.); shanshanshi@mail.shiep.edu.cn (S.S.)
[2] Shanghai Technical Service Platform for Pollution Control and Resource Utilization of Organic Wastes, Shanghai Key Laboratory of Atmospheric Particle Pollution and Prevention (LAP3), Department of Environmental Science and Engineering, Fudan University, Shanghai 200438, China; 20110740026@fudan.edu.cn (C.J.); 20210740076@fudan.edu.cn (T.T.); sjzhou17@fudan.edu.cn (S.Z.); 20210740045@fudan.edu.cn (M.H.); zxdjewett@fudan.edu.cn (X.Z.)
[3] Shanghai Institute of Pollution Control and Ecological Security, Shanghai 200092, China
* Correspondence: zhangsc@fudan.edu.cn (S.Z.); xuqunjie@shiep.edu.cn (Q.X.)

Abstract: Potassium thiocyanate (KSCN), a highly efficient "three birds with one stone" activator, might work with inorganic activators to produce excellent N/S co-doped porous carbon (NSC) materials for environmental remediation. However, the effects of inorganic activators on cooperative activation are unclear. As a result, the influence of inorganic activators on the synthesis of NSC materials was investigated further. This study shows that the surface areas of the NSC materials acquired through cooperative activation by potassium salts (KOH or K_2CO_3) were considerably higher than those acquired through KSCN activation alone (1403 m^2/g). Furthermore, KSCN could cooperate with K_2CO_3 to prepare samples with excellent specific surface area (2900 m^2/g) or N/S content. The as-prepared NSC materials demonstrated higher adsorption capability for chloramphenicol (833 mg/g) and Pb^{2+} (303 mg/g) (pore fitting, complexation). The research provides critical insights into the one-step synthesis of NSC materials with a vast application potential.

Keywords: N/S co-doping; porous carbon; chloramphenicol adsorption; Pb^{2+} removal

1. Introduction

Because of the ongoing energy crisis and environmental concerns, the production and application of carbon materials derived from biomass has recently garnered a lot of interest [1–7]. As biomass resources are abundant and sustainable in the environment, they are considered as renewable carbon precursors of porous carbon materials [8]. The incorporation of heteroatoms into carbon frameworks enriches the surface functional groups and improves the chemical properties of materials' surfaces [9–11]. For example, the introduction of N elements could improve the surface polarity of carbon materials, thus improving the hydrophilicity and number of reactive sites on the materials [12–14]. Other studies have suggested that S heteroatoms could boost the chemical reactivity of carbon materials and greatly improve their affinity to heavy metals [15,16]. Benefiting from the synergistic effects of different heteroatoms, co-doping with multiple heteroatoms improves the overall performance of porous carbon materials more effectively than single-atom doping [17]. Therefore, much effort has been expended on the synthesis of heteroatom-doped porous carbon materials from biomass.

Currently, the most basic ways for creating heteroatom-doped porous carbon materials derived from biomass are self-doped technologies, which require the use of inorganic activators (such as KOH and $ZnCl_2$) to directly activate the biomass-containing heteroatom sources (e.g., N, S, or P sources) [18–20]. The performance of the resulting heteroatom-doped carbon materials critically depends on the reaction of the inorganic activators with the heteroatoms in the biomass during the activation process. However, as self-doping technology relies on the heteroatom concentration of the biomass precursor, this limits its applicability range. As a result, most techniques have concentrated on mixing inorganic activators (e.g., KOH, K_2CO_3, or $CaCl_2$) with organic dopants (e.g., urea, melamine, or thiourea), which cooperatively activate biomass to synthesize heteroatom-doped porous carbon materials [21–25]. It has been shown, for example, that K_2CO_3 and urea could be utilized as a cooperative activator to prepare N-doped porous carbon materials with a large surface area from biomass [26]. Through the complexation of inorganic activators and N-dopants, an N-containing metal oxide (e.g., KOCN) could be formed through this synthetic method, and the porosity of the samples would be improved by the subsequent carbothermal reduction reaction (KOCN + C → KCN + CO) [27]. However, the simultaneous activation of inorganic activators and organic dopants is both resource consuming and difficult, requiring excess dopants to obtain a satisfactory heteroatom-doped carbon material. In recent research, we discovered that KSCN, a "three birds with one stone" activator plays dual roles of porogen and dopant. More specifically, KSCN achieves simultaneous pore generation and N/S doping through an oxygen displacement reaction [28]. KSCN is far superior to standard organic dopants owing to its high heteroatom doping capability and minimal carbon content. Meanwhile, KSCN might also work with inorganic activators to improve the material characteristics via the carbothermal reduction process. However, the synergistic activation of biomass by inorganic activators and KSCN has been little explored in the preparation of carbon materials.

To address this need, the present study examines the impacts of several inorganic activators (KOH, K_2CO_3, and $ZnCl_2$) on the characteristics (porosity and N/S content) of NSC materials. Simultaneously, the influence of K_2CO_3 and KSCN loading contents on the porosity and N/S doping of materials was explored further. Furthermore, the potential of NSC materials to treat water was thoroughly investigated.

2. Materials and Methods

2.1. Synthesis of NSC Materials

Sawdust was obtained from rural Shanghai. Other reagents were purchased from Shanghai Aladdin Biochemical Technology Co., Ltd (Shanghai, China). The samples were synthesized as shown in Scheme 1. The NSC material was prepared by activation of sawdust (80 mesh) and mixed with inorganic activators and potassium thiocyanate (KSCN, AR = 98.5%). Then the mixture was shaken in an aqueous solution for the period of 12 h and dried at 80 °C in air for 12 h before heating to 700 °C for 2 h at a heating rate of 10 °C min^{-1} in a tube furnace. The N_2 gas flow was controlled at 100 mL min^{-1}. Various inorganic activators, including K_2CO_3 (AR = 99%), KOH (AR = 85%), and $ZnCl_2$ (AR = 98%), were used to prepare NSC material at 700 °C with an inorganic activator/KSCN/biomass weight ratio of 1:1:1. Then, the effects of loading contents of K_2CO_3 and KSCN on the porosity and doping of N/S in materials were further investigated. For this purpose, the activator-to-biomass mass ratio was adjusted to be 0, 1, 1/2, and 1, respectively. The carbonized samples were further washed with 2 M HCl to completely remove any salt residues, and were then washed repeatedly with deionized water before oven-drying at 100 °C until constant weight.

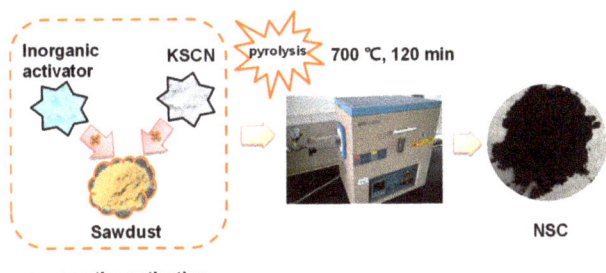

Scheme 1. Schematic of the NSC synthesis.

2.2. Characterizations of Materials

The elemental analyzer (Vario EL III) was used to perform elemental composition (C, H, N, S) analysis. The ash contents from materials were analyzed after heating materials in the air at 600 °C for 2 h. Using a Quantachrome Autosorb iQ2 apparatus, N_2 adsorption/desorption isotherms (at 77 K) was conducted to measure surface areas and pore volumes. Materials were further degassed at 120 °C for the period of 12 h before any further analysis. In this study, the Brunauer–Emmett–Teller (BET) equations were used to calculate surface areas (S_{BET}) and total pore volumes (V_T). Surface areas of micropores (S_{mic}, < 2 nm) were then determined via the t-plot analysis. The distribution of pore sizes was estimated using theoretical model such as density functional theory (DFT) [27]. The crystal structures of pristine samples were performed using powder X-ray diffractometry (XRD, X'Pert PRO), which was provided with Cu Kα radiation (40 mA, 40 kV). The 2θ range was performed from 10 to 80°. Raman spectra were further obtained by a Raman spectrometer (XploRA) with a laser of 532 nm. Five Gaussian peaks G, D, I, D', and D" at ~ 1580, ~ 1350, ~ 1220, ~ 1620, and D" ~ 1490 cm^{-1}, were fitted by parameters. The surface functionality of NSC materials containing N- and S- were measured using X-ray photoelectron spectroscopy (XPS, Thermo ESCALAB 250 XI). The N 1s spectra peaks of pyridinic, pyrrolic, quaternary, and oxidized nitrogen were fitted by deconvolution. Additionally, S_{2p} spectra included the bonding of C–SO$_3$, C–SO$_2$, C–S–C $2p_{1/2}$, and C–S–C $2p_{3/2}$. Binding energies were calibrated using the C 1s with the value of 284.6 eV. The peak data were fitted using XPS peak41 software. The morphology of the material was examined using scanning electron microscope (SEM, NovaNanoSem 450, FEI). The released CO gas at ~ 700 °C (peak 3) during cooperative activation by inorganic activator and KSCN was determined using mass spectrometer (Hiden QIC-20, MS). The m/z value for CO was set for 28.

2.3. Adsorption Experiments

The typical organic pollutant chloramphenicol (CAP) was chosen to evaluate the ability to purify water for NSC materials [29]. For the adsorption kinetic, 2.5 mg material was added to 25 mL of 120 mg/L CAP solution, and then 2 mL of the solution was collected at 2, 5, 10, 30, and 60 min for CAP concentration analysis. Batch experiments for adsorption were carried out as follows: material with 2.5 mg was dispersed in 25 mL of CAP solutions at different concentration (5–120 mg/L). The mixture was continuously shaken at the speed of 150 rpm for 12 h at 25 °C. After reaching adsorption equilibria, the supernatants were then filtered by an organic membrane with the pore size of 0.22 μm to analyze CAP concentrations. The contents of CAP were determined using an Agilent 1260 liquid chromatography, which was equipped with a C18 column at 25 °C and an ultraviolet detector at 278 nm wavelength, and the mobile phase was 6:4 (v/v) of the mixture of ultrapure water and methanol.

The toxic heavy metals such as lead(II) (Pb^{2+}) have attracted extensive attention for causing serious problems to the ecological environment and human health; thus. Pb^{2+} adsorption ability by the N/S co-doped carbon material was further evaluated [30]. For lead(II) (Pb^{2+}) adsorption isotherm, material with 20 mg was dispersed in 40 mL of Pb^{2+}

solutions, which have various concentrations ranging from 20 to 600 mg/L. Furthermore, 10 mM MES solutions were added to stabilize the pH of the solution. The mixture was constantly shaken at 150 rpm for the period of 12 h. After adsorption equilibrium, the concentration of Pb^{2+} was measured using ICP-AES (Hitachi P4010). A transmission electron microscopy (TEM-TecnaiG2F20 S-Twin FEI) was employed to explore the adsorption mechanism for NSC material. Langmuir model was used for data fitting of adsorption isotherms, and the model of pseudo-second-order was applied to interpret the adsorption mechanism, which was explicated in our previous work [31,32].

3. Results and Discussions

3.1. Effect of Inorganic Activator Type on the NSC Materials' Properties

The effect of inorganic activator types on the properties of NSC materials was investigated first. As shown in Table 1, compared with the NSC materials acquired through KSCN activation only, the surface areas of the NSC materials achieved by means of potassium salts' (KOH or K_2CO_3) cooperative activation were significantly increased. At the same time, the pore size distribution showed that the degree of porosity with the sizes less than 2 nm increased significantly, and, particularly, the supermicropore (0.7 nm < pore size < 2 nm) was increased (Figure 1a). These results indicated that potassium salts (KOH or K_2CO_3) had strong cooperative activation abilities. The KCN signals in the XRD spectra (Figure 1b) established that carbothermal reduction reactions (KOCN + C → KCN + CO) occurred during the cooperative activation by potassium salts (KOH or K_2CO_3) and KSCN, which was a widely known pore-forming reaction [33]. Moreover, the high yield for CO gas at peak 3 (greater than 700 °C) further proved the previous assumption (Figure S1a). Furthermore, it was also worth explaining that the release of CO at the low temperatures (peaks 1 and 2) were found to result from the biomass pyrolysis. However, the excessive CO generated with the carbothermal reduction could result in the enlarging of micropores. As shown in Table 1, the microporosity of KOH-derived NSC material was obviously lower than that of K_2CO_3-derived NSC material.

Table 1. NSC elemental compositions and textural characteristics of NSC produced with various types of inorganic activators. The activation conditions were set to be 700 °C, inorganic activator: KSCN: biomass mass ratio = 1:1:1.

Inorganic Activator	Yield (%)	Ash (%)	C (%)	N (%)	S (%)	S_{BET} [a] (m²/g)	S_{mic} [b] (m²/g)	V_T [c] (cm³/g)	V_{super} [d] (cm³/g)	V_{ultra} [e] (cm³/g)
/	38.7	5.65	65.7	5.03	10.01	1403	1040 (74)	1.17	0.29	0.18
K_2CO_3	26.3	4.67	71.44	2.98	9.75	2397	1970 (82)	1.24	0.63	0.13
KOH	21.2	6.05	79.35	2.28	5.67	2439	1568 (64)	1.38	0.59	0.14
$ZnCl_2$	93.9	38	38	4.46	18.09	129	- [f]	0.29	0.01	-

[a] BET surface area; [b] micropore surface area, the values in parenthesis are the S_{mic}/S_{BET} and the unit is %; [c] total pore volume; [d] supermicropore volume (0.7 nm < pore size < 2nm); [e] ultramicropore volume (<0.7 nm); and [f] not detected.

It is worth noting that the BET results showing the surface area of the carbon sample achieved by $ZnCl_2$ cooperative activation was only 129 m²/g (Table 1), indicating that the cooperative activation ability of $ZnCl_2$ was relatively low. Therefore, it was inferred that the formation of ZnS hinders the carbothermal reduction reaction. The obvious ZnS signals presented in the material obtained by $ZnCl_2$ cooperative activation further proved the above results (Figure 1b). Furthermore, compared with the NSC material obtained by the individual activation of KSCN, the N contents of the NSC samples prepared by cooperative activation of potassium salts (KOH or K_2CO_3) significantly decreased (Table 1). Therefore, it was concluded that the carbothermal reduction reaction occurs with a decrease in nitrogen content [28]. The results were further confirmed by the lower N content in the NSC materials obtained by the excessive cooperative activation of KOH than in the materials derived from K_2CO_3. Therefore, the inorganic activators were essential for the one-step synthetic pathway for N/S co-doping into carbon materials with excellent properties.

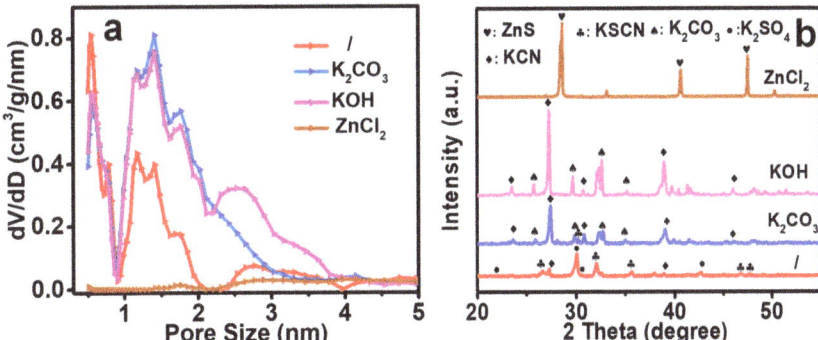

Figure 1. (a) The distribution of pore sizes for NSC, which were produced by being thermally induced using different inorganic activator, (b) XRD patterns for the unwashed NSC produced by being thermally induced using different inorganic activator. The activation conditions were set to be 700 °C, and the mass ratio of inorganic activator: KSCN: biomass was 1:1:1.

3.2. Effect of K_2CO_3 Load Content on the NSC Materials' Properties

The effects of the load content of K_2CO_3 on the pore structure and N/S doping into synthetic NSC materials were also examined. The SEM images showed that the surface of the NSC sample derived by the self-induction of KSCN was found to be less porous (Figure 2a), whereas the NSC material obtained from K_2CO_3 cooperative activation had great crosslinked pores (Figure 2b). This result indicated that K_2CO_3 cooperative activation could improve the pore structure of NSC materials. As demonstrated in Figure S2b, the singles of K_2SO_4 in the XRD spectra of the KSCN-derived NSC materials were attributed to the oxygen displacement reaction between the biomass and KSCN, which could effectively generate pores and achieve co-doping of N/S [28]. With the increasing load content of K_2CO_3, the KCN signals gradually strengthened (Figure S2b), while the yield significantly decreased as shown in Table S1. The data imply a gradually enhancing carbothermal reduction reaction (KOCN + C → KCN + CO), which was further confirmed by the increasing CO gas (peak 3 in Figure 2c). At the same time, the isotherms of N_2 adsorption (type I(b)) for the NSC materials further confirmed that high microporosity of the NSC materials could be achieved by the cooperative activation of K_2CO_3 (Figure S2a) [34]. The pore size distributions (Figure 2d) were concentrated in two regions: 0.5–0.7 nm and 1.1–2.0 nm. Furthermore, with the increasing load content of K_2CO_3, the pores accompanying with sizes less than 2 nm obviously increased (V_{super} reached 0.68 when K_2CO_3 ratio was found to be 2), thereby boosting the BET surface area of the NSC samples to 2900 m^2/g for the NSC samples (Table S1), outperforming the previously reported N/S-doped carbon materials [8].

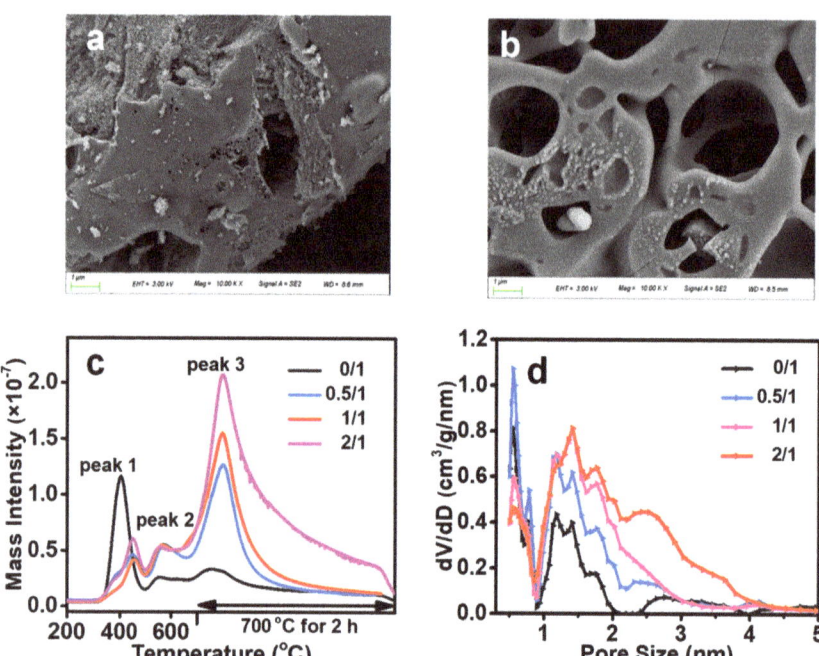

Figure 2. (a) Typical SEM images of NSC prepared by thermal activation of KSCN, (b) SEM images of NSC prepared by thermal activation of K_2CO_3 and KSCN, (c) the CO release graphs and (d) distribution of pore sizes of NSC, which were activated with various loading contents of K_2CO_3. The activation conditions were set to be 700 °C, KSCN: biomass mass ratio = 1:1, K_2CO_3: KSCN: biomass mass ratio = 2:1:1.

As shown in Table S1, the N contents of the NSC materials gradually reduced with the expanding loading content of K_2CO_3. This result also strongly indicated that the reaction of carbothermal reduction normally sacrifices some of the nitrogen content [27,35]. Four peaks were observed in the NSC samples by deconvolution of the N 1s spectrum, consisting of pyridine, pyrrole, quaternary, and oxide of nitrogen (Figure 3a). As shown in Figures 3a and S3a, the decrease in the amount of pyridinic nitrogen in the N 1s XPS spectra suggested a favorable role of this nitrogen form in the joint reactions between K_2CO_3 and KSCN. It is also noteworthy that the S content of the NSC materials remained stable with the increasing loading content of K_2CO_3 (weight ratio of K_2CO_3 and biomass < 1), inferring that S did not participate in the above reaction (Table S1). The established $C-SO_2$ and $C-SO_3$ peaks in the spectra of S 2p further confirmed the inference (Figure 3b). However, excessive K_2CO_3 activators (weight ratio of K_2CO_3 and biomass > 1), led to a considerable reduction in the S contents in the NSC materials (Table S1). Raman spectra showed that K_2CO_3 activation could also enhance the graphitization degree of the NSC materials, which was confirmed by the decreased I_D/I_G value (Figure S3b). In summary, the addition of inorganic activators in the one-step synthesis process of NSC materials definitively affected the structure of pores and N/S doping.

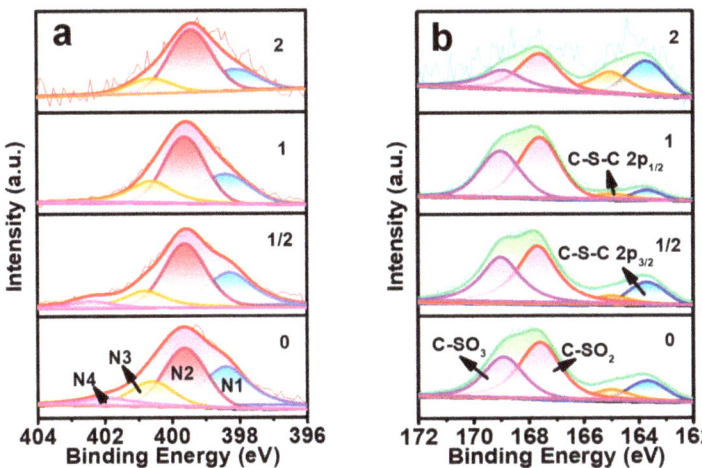

Figure 3. (a) N1s XPS spectra for different K_2CO_3 loading content-induced NSC materials (N1 pyridinic N, N2 pyrrolic N, N3 quaternary N, N4 oxidic N), (b) S 2p spectra for different K_2CO_3 loading content-initiated NSC materials. The activation conditions were set to be 700 °C, KSCN: biomass mass ratio = 1:1, various mass ratios between K_2CO_3 and biomass.

3.3. Effect of KSCN Load Content on the NSC Materials' Properties

The BET surface areas of the NSC materials achieved by collaborative initiations were significantly enhanced, and this enhancement was much more extensive than those of the porous carbon (PC) materials obtained by K_2CO_3 activation (Table S2). This result indicates that the KSCN was highly involved in the activation of biomass. When the mass ratio between KSCN and biomass grew to 1, the yields for NSC reduced but the BET surface area for the NSC improved (Table S2). The distribution curves of pore size showed that the number of pores with a size smaller than 2 nm increased (Figure 4a). These results imply that KSCN played an important role in instigating the reduction reactions (KOCN + C → KCN + CO), as further proved by the steadily raised KCN intensity in the XRD spectra of pristine NSC samples (Figure 4b). Meanwhile, the CO gas production (peak 3) increased with the increasing loading content of KSCN, further confirming the above conclusions (Figure 4c). However, extra KSCN activators (weight ratio of KSCN and biomass > 1) led to the decline in the porosity in the NSC materials, which might be attributed to the pore cracking caused by excessive cooperative activation. The V_{super} and V_{ultra} also showed the same trend (Table S2). Simultaneously, the yield of NSC was significantly improved with the increasing loading content of KSCN (weight ratio of KSCN and sawdust > 1, Table S2); thus, it could be concluded that the excess KSCN could also serve as a carbon source in initiating the carbothermal reduction for biomass. Interestingly, the deconvolution results of S in NSC proved that O could mix with sulfur to produce sulfur–oxygen bonding such as $C-SO_2$ and $C-SO_3$ (Figure S4a). The oxygen content was raised when the S content was enhanced in NSC, which complied well with the above conclusion (Table S2). In addition, the raised N/S content resulted in the slight decrease in the graphitization degree of NSC (indicated by the increased I_D/I_G value) (Figure S4b).

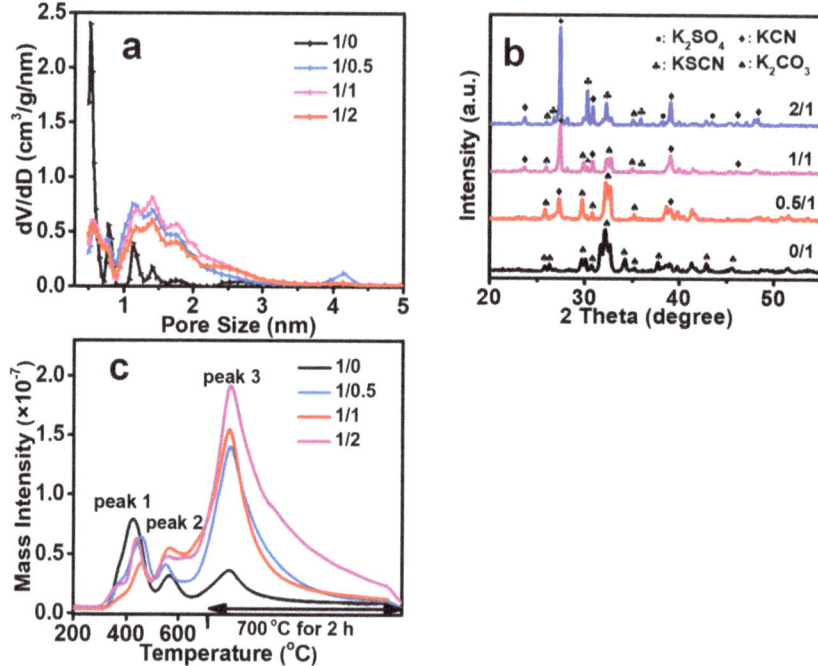

Figure 4. (**a**) Pore size in NSC materials, (**b**) XRD patterns of the unwashed NSC materials, (**c**) CO release curves of the NSC materials. The activation conditions were set to be 700 °C, K_2CO_3: biomass mass ratio of = 1:1, various mass ratios between KSCN and biomass.

3.4. The NSC Materials for Water Purification

The water purification performance of the NSC materials with different specific surface areas and N/S amounts was evaluated. The samples were labeled with A-x, in which x denotes various sample types (Table S3). As displayed in Figure 5a and Table S4, the NSC materials showed a higher adsorption capacity of CAP (maximum 833 mg/g) than that of previously reported materials [36–38]. The results of adsorption kinetics indicated that the adsorption equilibrium was reached at 30 min, which could be explained by the developed microporous structure of NSC (Figure S5a). Furthermore, the parameters for pseudo-second-order kinetic models were fitted to further reveal the adsorption process (Table S4). The adsorption capacity of CAP increased with the growing BET surface area of the NSC materials (Figure S5b,c), indicating that the pore-fitting mechanism might be essential for adsorption property. In addition, the CAP adsorption capacity for the per surface area (CAP-q_{BET}) of the NSC materials was further analyzed. The CAP-q_{BET} of the NSC materials was higher than the PC materials, suggesting that N doping played a positive role in CAP adsorption (Figure 5b).

Pb^{2+} containing wastewater poses extensive risks to the ecological environment and human health. The Langmuir model was used for fitting adsorption isotherms. The maximum adsorption capacities for Pb^{2+} varied from 100 mg/g to 303 mg/g (Figure S6 and Table S5). The significant difference in the adsorption of Pb^{2+} might be explained by changes in the sulfur contents of the NSC materials, and the results were confirmed using the positive correlation between the sulfur contents and maximum adsorption capacity (R^2 = 0.96) (Figure 6a). Therefore, the material exhibits an excellent adsorption capacity for Pb^{2+} in water. Furthermore, it was discovered that there were consistent fluctuations in the distribution and mass transfer for sulfur and Pb in TEM-EDS (Figure 6b). These observations proved that the Pb^{2+} could be fixed by complexing with the S-containing functional groups present in NSC samples. The excellent adsorption ability of materials for Pb^{2+} might be attributed to the complexation of S-containing functional groups on NSC materials.

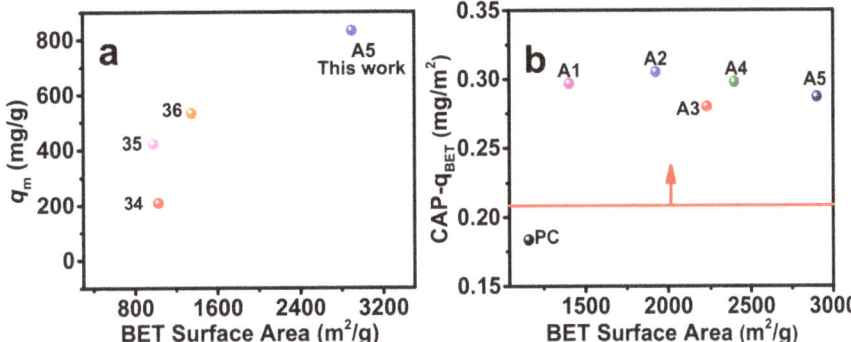

Figure 5. (a) Adsorption capacity (q_m) of CAP versus BET surface area, plotted for previously reported materials and the materials prepared in this study, (b) BET surface area of selected NSC material versus CAP-q_{BET}. $q_{BET} = q_m$/BET surface area, suggesting CAP adsorption capacity per BET surface area. PC: porous carbon material activated with K_2CO_3, activation temperature = 700 °C, K_2CO_3: biomass mass ratio = 1:1.

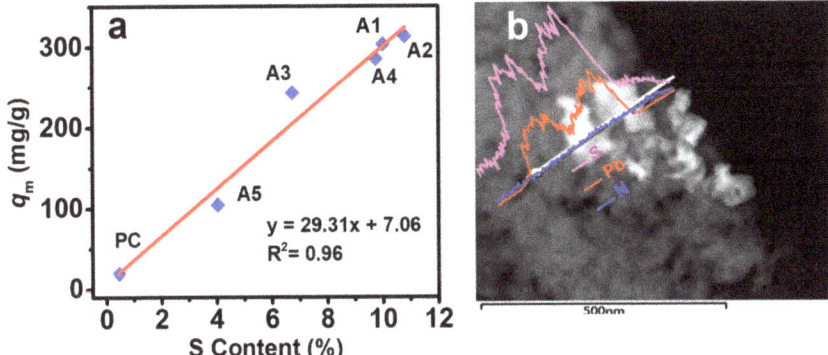

Figure 6. (a) Linear correlation between q_m and S content of NSC material, (b) representative TEM-EDS line profiles of elements in the selected NSC material (A1) after Pb^{2+} adsorption. PC: porous carbon material activated with K_2CO_3, activation temperature = 700 °C, K_2CO_3: biomass mass ratio = 1:1.

4. Conclusions

In this study, the effect of the inorganic activators on the properties of NSC materials were investigated. This study suggested that potassium salt could significantly promote the NSC materials' properties due to the strong carbothermal reduction reaction. Furthermore, the porosity and doping of nitrogen and sulfur into the obtained materials could be effectively improved by adjusting the mass ratio of KSCN to K_2CO_3. The results indicated that the as-prepared material exhibited a larger BET surface area of 2900 m^2/g or the high content of N/S. The NSC materials presented a high organic pollutant removal capacity (~833 mg/g CAP adsorption) owing to their porous properties. In addition, the doping of S greatly improved the ability of carbon materials to remove Pb^{2+} in wastewater due to the complexation effect. The results have substantial consequences for the controllable synthesis of NSC materials.

Supplementary Materials: The following supporting information can be downloaded at: https://www.mdpi.com/article/10.3390/pr10071359/s1, Figure S1. (a) Different inorganic activator on the release curves of CO during the preparation process of NSC materials. The activation conditions were set to be 700 °C, and inorganic activator, KSCN and biomass with weight ratio of 1:1:1. Figure S2. (a) N_2 adsorption–desorption isotherms of NSC materials, (b) XRD patterns of the unwashed NSC materials. The activation conditions were set to be 700 °C, KSCN and biomass with mass ratio of 1:1, various mass ratio between K_2CO_3 and biomass. Figure S3. (a) Relative ratio of pyridinic N (N1) and pyrrolic N (N2) for NSC materials, (b) Raman spectrum of NSC materials was fitted using the five Gaussian peaks (color lines). The activation conditions were set to be 700 °C, KSCN and biomass with weight ratio of 1:1, various weight ratios between K_2CO_3 and biomass. Figure S4. (a) S 2_p XPS spectra for different KSCN loading content-induced NSC materials, (b) Raman spectrum of NSC materials was fitted using the five Gaussian peaks (color lines). The activation conditions were set to be 700 °C, K_2CO_3 and biomass with weight ratio of 1:1, various weight ratios between KSCN and biomass. Figure S5. (a) Adsorption kinetics of CAP onto selected NSC, (b) correlations of the resultant NSC between maximum adsorption capacity (q_m) and BET surface area, (c) adsorption isotherms of CAP for selected NSC in aqueous solution (the adsorption isotherms were fitted with the Langmuir model). Figure S6. (a) Adsorption isotherms of Pb^{2+} for selected NSC in aqueous solution (the adsorption isotherms were fitted with the Langmuir model). Table S1. Elemental compositions and textural properties of NSC samples. The activation conditions were set to be 700 °C, KSCN and biomass with weight ratio of 1:1, various weight ratios between K_2CO_3 and biomass. Table S2. Elemental compositions and textural properties of NSC samples. The activation conditions were set to be 700 °C, K_2CO_3 and biomass with weight ratio of 1:1, various weight ratios between KSCN and biomass. Table S3. Porosity, preparation condition, and elemental compositions of selected NSC materials. The activation conditions were set to be 700 °C. Table S4. The isothermal parameters and kinetic parameters of selected NSC materials for CAP adsorption. Table S5. The isothermal parameters of selected NSC materials for Pb^{2+} adsorption.

Author Contributions: Writing—Original draft, X.H.; methodology, C.J. and S.S.; data curation, T.T., S.Z. (Shaojie Zhou), M.H. and X.Z.; funding acquisition, S.Z. (Shicheng Zhang) and Q.X. All authors have read and agreed to the published version of the manuscript.

Funding: This research was supported by the National Natural Science Foundation of China (No. 21876030), the International Cooperation Project of Science and Technology Commission of Shanghai Municipality (No. 18230710700).

Institutional Review Board Statement: Not applicable.

Informed Consent Statement: Not applicable.

Data Availability Statement: Not applicable.

Conflicts of Interest: The authors declare no conflict of interest.

References

1. Deng, J.; Li, M.M.; Wang, Y. Biomass-derived carbon: Synthesis and applications in energy storage and conversion. *Green Chem.* **2016**, *18*, 4824–4854. [CrossRef]
2. Fu, R.; Yu, C.; Li, S.F.; Yu, J.H.; Wang, Z.; Guo, W.; Xie, Y.Y.; Yang, L.; Liu, K.L.; Ren, W.C.; et al. A closed-loop and scalable process for the production of biomass-derived superhydrophilic carbon for supercapacitors. *Green Chem.* **2021**, *23*, 3400–3409. [CrossRef]
3. Park, H.; May, A.; Portilla, L.; Dietrich, H.; Münch, F.; Rejek, T.; Sarcletti, M.; Banspach, L.; Zahn, D.; Halik, M. Magnetite nanoparticles as efficient materials for removal of glyphosate from water. *Nat. Sustain.* **2020**, *3*, 129–135. [CrossRef]
4. Reza, M.T.; Rottler, E.; Tolle, R.; Werner, M.; Ramm, P.; Mumme, J. Production, characterization, and biogas application of magnetic hydrochar from cellulose. *Bioresour. Technol.* **2015**, *186*, 34–43. [CrossRef] [PubMed]
5. Li, C.; Sun, X.; Zhu, Y.; Liang, W.; Nie, Y.; Shi, W.; Ai, S. Core-shell structural nitrogen-doped carbon foam loaded with nano zero-valent iron for simultaneous remediation of Cd (II) and NAP in water and soil: Kinetics, mechanism, and environmental evaluation. *Sci. Total Environ.* **2022**, *832*, 155091. [CrossRef]
6. Creamer, A.E.; Gao, B. Carbon-Based Adsorbents for Postcombustion CO_2 Capture: A Critical Review. *Environ. Sci. Technol.* **2016**, *50*, 7276–7289. [CrossRef]
7. Jin, H.L.; Feng, X.; Li, J.; Li, M.; Xia, Y.Z.; Yuan, Y.F.; Yang, C.; Dai, B.; Lin, Z.Q.; Wang, J.C.; et al. Heteroatom-Doped Porous Carbon Materials with Unprecedented High Volumetric Capacitive Performance. *Angew. Chem. Int. Edit.* **2019**, *58*, 2397–2401. [CrossRef]

8. Demir, M.; Doguscu, M. Preparation of Porous Carbons Using NaOH, K_2CO_3, Na_2CO_3 and $Na_2S_2O_3$ Activating Agents and Their Supercapacitor Application: A Comparative Study. *ChemistrySelect* **2022**, *7*, e202104295. [CrossRef]
9. Chen, Z.; Liu, T.; Tang, J.J.; Zheng, Z.J.; Wang, H.M.; Shao, Q.; Chen, G.L.; Li, Z.X.; Chen, Y.Q.; Zhu, J.W.; et al. Characteristics and mechanisms of cadmium adsorption from aqueous solution using lotus seedpod-derived biochar at two pyrolytic temperatures. *Environ. Sci. Pollut. Res.* **2018**, *25*, 11854–11866. [CrossRef]
10. Zhao, X.C.; Wang, A.Q.; Yan, J.W.; Sun, G.Q.; Sun, L.X.; Zhang, T. Synthesis and Electrochemical Performance of Heteroatom-Incorporated Ordered Mesoporous Carbons. *Chem. Mater.* **2010**, *22*, 5463–5473. [CrossRef]
11. Zhang, Y.J.; Mori, T.; Ye, J.H.; Antonietti, M. Phosphorus-Doped Carbon Nitride Solid: Enhanced Electrical Conductivity and Photocurrent Generation. *J. Am. Chem. Soc.* **2010**, *132*, 6294–6295. [CrossRef] [PubMed]
12. Shen, Z.F.; Liu, C.L.; Yin, C.C.; Kang, S.F.; Liu, Y.A.; Ge, Z.G.; Xia, Q.N.; Wang, Y.G.; Li, X. Facile large-scale synthesis of macroscopic 3D porous graphene-like carbon nanosheets architecture for efficient CO_2 adsorption. *Carbon* **2019**, *145*, 751–756. [CrossRef]
13. Wang, L.L.; Zhu, D.Q.; Chen, J.W.; Chene, Y.S.; Chen, W. Enhanced adsorption of aromatic chemicals on boron and nitrogen co-doped single-walled carbon nanotubes. *Environ. Sci.-Nano* **2017**, *4*, 558–564. [CrossRef]
14. Wei, J.H.; Cai, W.Q. One-step hydrothermal preparation of N-doped carbon spheres from peanut hull for efficient removal of Cr(VI). *J. Environ. Chem. Eng.* **2020**, *8*, 104449. [CrossRef]
15. Saha, D.; Barakat, S.; Van Bramer, S.E.; Nelson, K.A.; Hensley, D.K.; Chen, J.H. Noncompetitive and Competitive Adsorption of Heavy Metals in Sulfur-Functionalized Ordered Mesoporous Carbon. *ACS Appl. Mater. Inter.* **2016**, *8*, 34132–34142. [CrossRef]
16. Wei, Y.; Xu, L.; Yang, K.; Wang, Y.; Wang, Z.L.; Kong, Y.; Xue, H.G. Electrosorption of Toxic Heavy Metal Ions by Mono S- or N-Doped and S, N-Codoped 3D Graphene Aerogels. *J. Electrochem. Soc.* **2017**, *164*, E17–E22. [CrossRef]
17. Chen, F.; Zhang, M.; Ma, L.L.; Ren, J.G.; Ma, P.; Li, B.; Wu, N.N.; Song, Z.M.; Huang, L. Nitrogen and sulfur codoped micro-mesoporous carbon sheets derived from natural biomass for synergistic removal of chromium(VI): Adsorption behavior and computing mechanism. *Sci. Total. Environ.* **2020**, *730*, 138930. [CrossRef]
18. Chen, H.; Yu, F.; Wang, G.; Chen, L.; Dai, B.; Peng, S.L. Nitrogen and Sulfur Self-Doped Activated Carbon Directly Derived from Elm Flower for High-Performance Supercapacitors. *ACS Omega* **2018**, *3*, 4724–4732. [CrossRef]
19. Sun, Z.J.; Liao, J.H.; Sun, B.; He, M.L.; Pan, X.; Zhu, J.P.; Shi, C.W.; Jiang, Y. Nitrogen Self-Doped Porous Carbon Materials Derived from a New Biomass Source for Highly Stable Supercapacitors. *Int. J. Electrochem. Sci.* **2017**, *12*, 12084–12097. [CrossRef]
20. Liu, W.J.; Tian, K.; Ling, L.L.; Yu, H.Q.; Jiang, H. Use of Nutrient Rich Hydrophytes to Create N,P-Dually Doped Porous Carbon with Robust Energy Storage Performance. *Environ. Sci. Technol.* **2016**, *50*, 12421–12428. [CrossRef]
21. Yue, L.M.; Xia, Q.Z.; Wang, L.W.; Wang, L.L.; DaCosta, H.; Yang, J.; Hu, X. CO_2 adsorption at nitrogen-doped carbons prepared by K_2CO_3 activation of urea-modified coconut shell. *J. Colloid Interf. Sci.* **2018**, *511*, 259–267. [CrossRef] [PubMed]
22. Zhu, Y.Y.; Chen, M.M.; Zhang, Y.; Zhao, W.X.; Wang, C.Y. A biomass-derived nitrogen-doped porous carbon for high-energy supercapacitor. *Carbon* **2018**, *140*, 404–412. [CrossRef]
23. Fuertes, A.B.; Sevilla, M. Superior Capacitive Performance of Hydrochar-Based Porous Carbons in Aqueous Electrolytes. *Chemsuschem* **2015**, *8*, 1049–1057. [CrossRef] [PubMed]
24. Sevilla, M.; Ferrero, G.A.; Fuertes, A.B. Beyond KOH activation for the synthesis of superactivated carbons from hydrochar. *Carbon* **2017**, *114*, 50–58. [CrossRef]
25. Tang, F.Y.; Wang, L.Q.; Liu, Y.N. Biomass-derived N-doped porous carbon: An efficient metal-free catalyst for methylation of amines with CO_2. *Green Chem.* **2019**, *21*, 6252–6257. [CrossRef]
26. Tsubouchi, N.; Nishio, M.; Mochizuki, Y. Role of nitrogen in pore development in activated carbon prepared by potassium carbonate activation of lignin. *Appl. Surf. Sci.* **2016**, *371*, 301–306. [CrossRef]
27. Luo, J.W.; Jia, C.; Shen, M.H.; Zhang, S.C.; Zhu, X.D. Enhancement of adsorption and energy storage capacity of biomass-based N-doped porous carbon via cyclic carbothermal reduction triggered by nitrogen dopants. *Carbon* **2019**, *155*, 403–409. [CrossRef]
28. Jia, C.; Yu, F.; Luo, J.; Chen, C.; Zhang, S.; Zhu, X. Three birds with one stone approach to superior N/S co-doped microporous carbon for gas storage and water purification. *Chem. Eng. J.* **2021**, *431*, 133231. [CrossRef]
29. Xu, L.; Wu, C.X.; Liu, P.H.; Bai, X.; Du, X.Y.; Jin, P.K.; Yang, L.; Jin, X.; Shi, X.; Wang, Y. Peroxymonosulfate activation by nitrogen-doped biochar from sawdust for the efficient degradation of organic pollutants. *Chem. Eng. J.* **2020**, *387*, 124065. [CrossRef]
30. Ma, L.J.; Wang, Q.; Islam, S.M.; Liu, Y.C.; Ma, S.L.; Kanatzidis, M.G. Highly Selective and Efficient Removal of Heavy Metals by Layered Double Hydroxide Intercalated with the MoS42- Ion. *J. Am. Chem. Soc.* **2016**, *138*, 2858–2866. [CrossRef]
31. Zhu, X.D.; Liu, Y.C.; Luo, G.; Qian, F.; Zhang, S.C.; Chen, J.M. Facile Fabrication of Magnetic Carbon Composites from Hydrochar via Simultaneous Activation and Magnetization for Triclosan Adsorption. *Environ. Sci. Technol.* **2014**, *48*, 5840–5848. [CrossRef] [PubMed]
32. Zhu, X.D.; Liu, Y.C.; Zhou, C.; Zhang, S.C.; Chen, J.M. Novel and High-Performance Magnetic Carbon Composite Prepared from Waste Hydrochar for Dye Removal. *ACS Sustain. Chem. Eng.* **2014**, *2*, 969–977. [CrossRef]
33. Sevilla, M.; Ferrero, G.A.; Diez, N.; Fuertes, A.B. One-step synthesis of ultra-high surface area nanoporous carbons and their application for electrochemical energy storage. *Carbon* **2018**, *131*, 193–200. [CrossRef]
34. Muttakin, M.; Mitra, S.; Thu, K.; Ito, K.; Saha, B.B. Theoretical framework to evaluate minimum desorption temperature for IUPAC classified adsorption isotherms. *Int. J. Heat Mass Tran.* **2018**, *122*, 795–805. [CrossRef]

35. Chen, H.; Zhou, M.; Wang, Z.; Zhao, S.Y.; Guan, S.Y. Rich nitrogen-doped ordered mesoporous phenolic resin-based carbon for supercapacitors. *Electrochim. Acta* **2014**, *148*, 187–194. [CrossRef]
36. Din, A.T.M.; Ahmad, M.A.; Hameed, B.H. Ordered mesoporous carbons originated from non-edible polyethylene glycol 400 (PEG-400) for chloramphenicol antibiotic recovery from liquid phase. *Chem. Eng. J.* **2015**, *260*, 730–739. [CrossRef]
37. Yang, J.; Ji, G.; Gao, Y.; Fu, W.; Irfan, M.; Mu, L.; Zhang, Y.; Li, A. High-yield and high-performance porous biochar produced from pyrolysis of peanut shell with low-dose ammonium polyphosphate for chloramphenicol adsorption. *J. Clean. Prod.* **2020**, *264*, 121516. [CrossRef]
38. Chen, A.; Pang, J.; Wei, X.; Chen, B.; Xie, Y. Fast one-step preparation of porous carbon with hierarchical oxygen-enriched structure from waste lignin for chloramphenicol removal. *Environ. Sci. Pollut. Res. Int.* **2021**, *28*, 27398–27410. [CrossRef] [PubMed]

Article

Molecular Dynamics Simulation for Structural Evolution of Mixed Ash from Coal and Wheat Straw

Hengsong Ji [1], Xiang Li [2,*], Mei Zhang [1], Zhenqiang Li [2], Yan Zhou [2] and Xiang Ma [3,*]

1. Institute of Energy Research, Jiangsu University, Zhenjiang 212013, China; jihengsong@ujs.edu.cn (H.J.); 18860878567@163.com (M.Z.)
2. School of Energy and Power Engineering, Jiangsu University, Zhenjiang 212013, China; 13195551761@163.com (Z.L.); YANZHOU_ZY@126.com (Y.Z.)
3. SINTEF Industry, P.O. Box 124, Blindern, 0314 Oslo, Norway
* Correspondence: xiangli@ujs.edu.cn (X.L.); Xiang.Ma@sintef.no (X.M.); Tel.: +86-157-5101-2916 (X.L.); +47-98243925 (X.M.)

Abstract: We conducted molecular dynamics (MD) simulations to investigate the structural evolution of molten slag composed of wheat straw (WS) and Shenhua (SH) coal. The content of wheat straw in the slag was varied from 0 to 100 wt%. The MD results indicated a slight reduction in the sharpness of the radial-distribution-function curve of each ion–oxygen pair and a decrease in bonding strength with increasing WS content. WS introduced many metal ions to the ash system, increasing its overall activity. The number of bridging and non-bridging oxygen atoms changed upon straw addition, which affected the stability of the system. There were relatively few highly coordinated Si ions. The number of low-coordination Si was highest for a WS content of 30%, at which the density reached a minimum value. The degree of ash polymerization was analyzed by counting the number (Q) of tetrahedra with the number (n) of the bridging oxygen atoms. With increasing WS content, Q^4 (tetrahedral Si) decreased, whereas Q^3, Q^2, Q^1, and Q^0 increased. Q^4 reached a minimum value for a WS content of 30%, at which point the degree of ion aggregation was the weakest and the degree of disorder was the strongest.

Keywords: molecular dynamics; wheat straw; Shenhua coal; ash

1. Introduction

Coal gasification is important for the high-efficiency utilization of coal, production of liquid fuel and hydrogen, reduction of iron, and other processes. Co-gasification of biomass and coal increases the energy density of the biomass and decreases the clinkering rate of coal ash owing to its high fluidity. Understanding the network structure for the co-gasification ash of biomass and coal is important when designing the gasifier and selecting the operating temperature and fuel ratio [1,2].

Shenhua (SH) coal from the China Shenhua coalfield has good surface activity, high calorific value, and low sulfur and ash content. In particular, the melting temperature of SH coal ash is lower than other coal, which is suitable for a flow bed gasifier with liquid slag discharge. Coal ash is mainly composed of metal oxides and nonmetallic oxides [3,4]. Ions can be divided into three types on the basis of their different structures in slag: network ions, modified ions, and neutral ions. The roles of SiO_2, Al_2O_3, Fe_2O_3, and CaO in melt have been studied extensively [5,6]. Because of the high SiO_2 content in coal ash, the structure of silicates is of particular importance for understanding the structure and behavior of slags. Silicate slags contain Si cations surrounded by four tetrahedrally arranged oxygen anions. These SiO_4^{4-} tetrahedra are joined in chains or rings by bridging oxygen (BO) atoms. Cations, such as Na^+, Ca^{2+}, Mg^{2+}, K^+, and Fe^{2+}, break these bonded oxygens to form non-bridging oxygen (NBO), O^-, and free oxygen, O^{2-} [7,8].

Compared with coal ash, the content of alkali metal elements such as K, Na, Mg, and Ca in biomass ash is high, and the content of network ions (Si and Al) is low [9]. The addition of biomass ash (such as wheat straw, corn straw, etc.) can reduce the melting point of coal ash [8], changing the flow characteristics of ash at high temperatures [10].

MD is an excellent tool to study the structure with classic dynamics and obtain information such as the chemical bond. This method has already been used in slag [11–16]. For example, experimental and MD simulations results indicated that the SiO_4^{4-} tetrahedron is the most stable unit in the slag, and the bonding stability decreases in the order: Si–O > Al–O > Fe–O > Mg–O > Ca–O [13]. Zhang et al. [14] also point out that alkalis (Na_2O and K_2O) have only marginal influence on the structure of molten aluminosilicates, while the Na^+ ion prefers to locate in the BO/NBO networks, and the K^+ ion tends to be present in various oxygen tri-clusters. Moreover, Si–O and Al-O networks were observed to depolymerize into a simple structure at high alkalinity, with atomic self-diffusion increasing and the viscosity decreasing with increasing alkalinity [15]. The decrease in viscosity led to an increase in the proportion of NBO in the network. In another study, an increase in the $MgO:Al_2O_3$ ratio led to the introduction of more Mg^{2+} ions to destroy the network structure, resulting in the conversion of BO to NBO [16]. Moreover, as the basicity increased, more BO was converted to free oxygen.

At present, there are plenty of studies on the structure of coal ash, but few studies on the structure of mixed ash from coal and biomass. In this study, we carried out MD simulations to investigate the effect of wheat straw (WS) content (varied from 0 to 100 wt%) on the structural properties of ash slag, including the radial distribution functions (RDF), mean square displacement (MSD), coordination numbers (CN), distribution of oxygen types, distribution of bond angles, and distribution of Q^n.

2. Methodology

2.1. Ash Samples Preparation

SH coal was supplied by the China Energy Investment Corporation; WS was collected from Bingjiang farm in Zhengjiang city. The industrial analysis of SH coal and WS was carried out using SH coal and WS powder (<100 μm) under the PRC National Standard GB/T 212-2008 [17] and GB/T28731-2012 [18], respectively. The analysis results are shown in Table 1.

SH coal ash and WS ash were prepared under the ASTM D3174-12 standard in which the final ashing temperature was set at 575 °C to minimize the loss of volatile elements such as Na [19]. The ash composition was analyzed by XRF (ARL ADVANT'X Intellipower 4200, ThermoFisher Scientific, USA), as shown in Table 2.

Table 1. Industrial analysis of SH coal and WS.

Sample	Content (wt%)				Gross Calorific Value (cal/g)
	Ash	Volatiles	H_2O	C-Fix	
SH coal	8.12	28.02	9.50	54.36	6336.8
WS	11.5	66.4	7.95	14.15	4354.4

Table 2. Composition analysis of SH coal ash and WS ash.

Sample	Content (wt%)								
	Na_2O	MgO	Al_2O_3	SiO_2	P_2O_5	SO_3	K_2O	CaO	Fe_2O_3
SH ash	1.86	2.34	19.12	58.03	0.50	5.66	1.09	8.29	3.11
WS ash	0.66	4.28	1.19	58.76	2.39	4.74	19.39	7.54	1.04

Simulation of real ash is difficult as as there are some impurities (P_2O_5, SO_3, and Fe_2O_3) in real ash, as shown in Table 2. In this simulation study, six main components in coal ash and WS ash were selected for calculation to simplify the ash system, and the molar compositions are shown in Table 3, of which the content of WS in the slag was varied from 0% to 100 wt%.

Table 3. Molar composition of different ash samples for simulation study.

Sample	Mol%					
	Na_2O	MgO	Al_2O_3	K_2O	CaO	SiO_2
SH	2.1	4.2	13.3	0.9	10.6	68.9
SH-WS(10%)	1.9	4.6	11.5	2.8	10.4	68.8
SH-WS(30%)	1.6	5.4	8.4	14.4	10.1	68.6
SH-WS(50%)	1.3	6.1	5.9	8.9	9.8	68.5
WS	0.7	7.5	0.9	14.4	9.4	68.2

2.2. Simulation Method

We used the Amorphous Cell Tools module of the Materials Studio 7.1 (MS) software to build an amorphous structural system model. The particles were randomly distributed in the system model to simulate the melt at high temperature, as shown in Figure 1. The density of the system determined the size of the cubic box.

CFF (Compass force filed) and PCFF (polymer consistent force field) in the Amorphous Cell Tools module of Materials Studio are widely used to build irregular cell structure of inorganic compounds. However, an oxygen atom cannot be added in CFF, only in the form of oxide, which is not adjusted for random distribution of atoms. In this work, the PCFF was selected for building the model as O can be added in the form of randomly distributed oxygen atoms. Other atoms were directly added to the force field, namely, Si, Na, Ca, Mg, Al, and K.

Periodic boundary conditions were applied on all sides of the model box to create an infinite system. The time step was set as 1 fs, and the data were saved each 100 steps. In the simulation system, the total number of atoms was set to 4000 and the number of different atoms was decided according to their mole fractions. The molar composition of the different ash samples is shown in Table 3. We selected the atoms with high mole fractions, i.e., Na, Mg, Al, Si, K, Ca, and O. The effective charge of ions was set as follows: Si was +1.89, Al was +1.4175, Ca was +0.945, Mg was +0.945, Na was +0.4725, K was +0.4725, and O was −0.945.

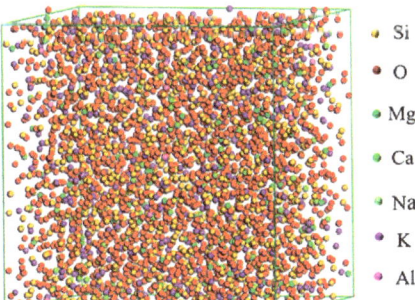

Figure 1. MD model of ash.

The number and density of each atom and the side length of the cubic model box in the system are listed in Table 4. The main components in the ash slag were feldspar (2.55–2.67 g/cm^3) and diopside (3.27–3.38 g/cm^3) [20]. Because the addition of biomass increases the diopside content in the ash, we selected a slag density of 3.0 g/cm^3. Since the sizes of Na, Mg, Al, Si, K, Ca, and O atoms are different, there are differences in the length of the cubic model box to ensure the same total atoms and slag density in different slag systems.

Table 4. Numbers of atoms, density, and side length of the cubic model box.

Sample	Atomic Number							Density (g/cm^3)	Length (Å)
	Na	Mg	Al	Si	K	Ca	O		
SH	55	53	342	884	22	135	2508	3.0	35.82
SH-WS(10%)	50	60	299	894	72	134	2491	3.0	35.93
SH-WS(30%)	42	72	223	908	161	134	2459	3.0	36.13
SH-WS(50%)	34	82	158	922	239	132	2432	3.0	36.31
WS	21	104	24	947	400	130	2374	3.0	36.67

2.3. MD Simulation Process

We used potential functions based on the Garofalini potential function [21], which includes a pair potential and three-body potential. The Born–Mayer–Huggins (*BMH*) function was used as a pair potential function, which is written as

$$V_{ij}^{BMH} = A_{ij} exp\left(\frac{-r_{ij}}{\rho_{ij}}\right) + \frac{q_i q_j}{r_{ij}} \xi\left(\frac{r_{ij}}{\beta_{ij}}\right) \quad (1)$$

where r_{ij} is the interatomic distance between atoms *i* and *j*, q_i and q_j represent the charges of atoms *i* and *j*, and ξ represents correction function. A_{ij}, ρ_{ij}, and β_{ij} represent the potential parameters of *BMH*. The parameters used are listed in Table 5. The form of the three-body potential function can be written as:

$$\varphi_{jik} = \sqrt{\lambda_{ij}\lambda_{jk}} exp\left(\frac{\gamma_{ij}}{r_{ij}-R_{ij}} + \frac{\gamma_{jk}}{r_{jk}-R_{jk}}\right)\omega_{jik} \quad (2)$$

where R_{ij} is the distance between particles *i* and *j*; R_{jk} is the distance between particles *j* and *k*; λ_{ij} and γ_{ij} are the potential parameters between particles *i* and *j*; λ_{jk} and γ_{jk} are the potential parameters between particles *j* and *k*; and ω_{jik} is a function related to the angle. The parameters of the three potentials are taken from Garofalini's work [22], as shown in Table 6. Simulations were carried out in the NVT ensemble, i.e., keeping the number of particles (N), system volume (V), and temperature (T) fixed. As shown in Figure 2, the system was thermally stabilized for 25 ps at T = 5000 K to ensure full mixing of the particles and was then cooled to 2000 K in 25 ps with a cooling rate of 1.2×10^{14} K/s. To achieve a state of equilibrium, the system was kept at 2000 K for 25 ps and then cooled to 1623 K in 25 ps with a cooling rate of 1.508×10^{13} K/s. In the final stage, the system was equilibrated at 2223 K for 25 ps to obtain a uniform ash slag in a molten state.

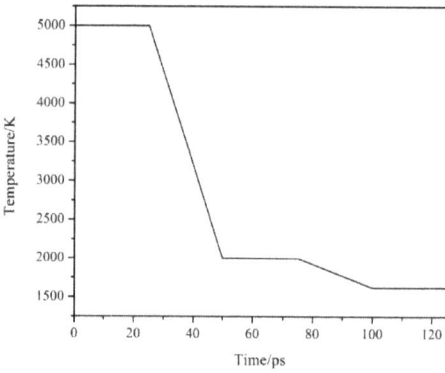

Figure 2. Temperature of the ash during the simulation.

Table 5. Two-body-potential parameters [21].

Atoms		Parameter		
i	j	A_{ij} (fJ)	β_{ij} (pm)	ρ_{ij} (pm)
Na	Na	0.2159	230	29
Na	Al	0.2178	230	29
Na	Si	0.2001	230	29
Na	O	0.3195	234	29
Mg	Mg	1.0643	231.8	29
Mg	Al	0.20846	230	29
Mg	Si	0.2216	230	29
Mg	O	0.2842	234	29
Al	Al	0.0500	235	29
Al	Si	0.2523	233	29
Al	O	0.2490	234	29
Si	Si	0.1877	230	29
Si	O	0.2962	234	29
K	K	0.99706	253.8	29
K	Al	0.4448	236.6	29
K	Si	0.4420	253.8	29
K	O	0.60802	253.8	29
Ca	Ca	0.7000	230	29
Ca	Al	0.2178	230	29
Ca	Si	0.2215	230	29
Ca	O	0.5700	234	29
O	O	0.0725	234	29

Table 6. Three-body potential parameters.

Atomtriplet	λ_{ij} (fJ)	γ_{ij} (pm)	R_{ij} (pm)	ω_{jik} (deg)
Al/Si–O–Al/Si	0.001	200	260	109.5
O–Al/Si–O	0.024	280	300	109.5

3. Results and Discussion

3.1. Radial Distribution Functions (RDFs)

From RDFs, basic structural information, including the degree of material ordering and the degree of correlation of electrons, can be obtained. The formula for the RDF can be written as [23]:

$$g_{ij}(r) = \frac{V}{N_i N_j} \sum_j \frac{\langle n_{ij}\left(r-\frac{\Delta r}{2}, r+\frac{\Delta r}{2}\right)\rangle}{4\pi r^2 \Delta r} \quad (3)$$

where V is the volume of the system; n_{ij} is the number of i atoms included in a spherical shell of thickness; Δr is located at a distance r; and N is the total number of particles.

The distributions of the RDF for each particle in the ash sample are shown in Figure 3. The location of the first peak in the RDF graph represents the average bond length of the atomic pair. For Si–O, the location of first peak in all five samples was unaffected by the addition of WS, which corresponds to a bond length of 1.61 Å. The locations of the first peaks seen in Al–O for SH, SH-WS (10%), SH-WS (30%), and SH-WS (50%) were similar, although the addition of WS shifted the peak slightly, indicating a decrease in bond length from 1.75 to 1.73 Å. The peak values of Si–O and Al–O were consistent with previous reports [21], indicating that the calculated results are reasonable.

The sharpness of the first peak in the RDF curves reflect the bonding ability and interaction between ions [24]. With increasing WS content, the width of the first peak of Si–O and Al–O increased, indicating a decrease in bond strength. The first peaks for Na–O, Mg–O, K–O, and Ca–O corresponded to bond lengths of 2.29–2.35, 2.00–2.06, 2.61–2.65, and

2.30–2.35Å, respectively. The width of the first peaks for these bonds was larger than for Si–O and Al–O, indicating weaker bonding than for Si–O and Al–O.

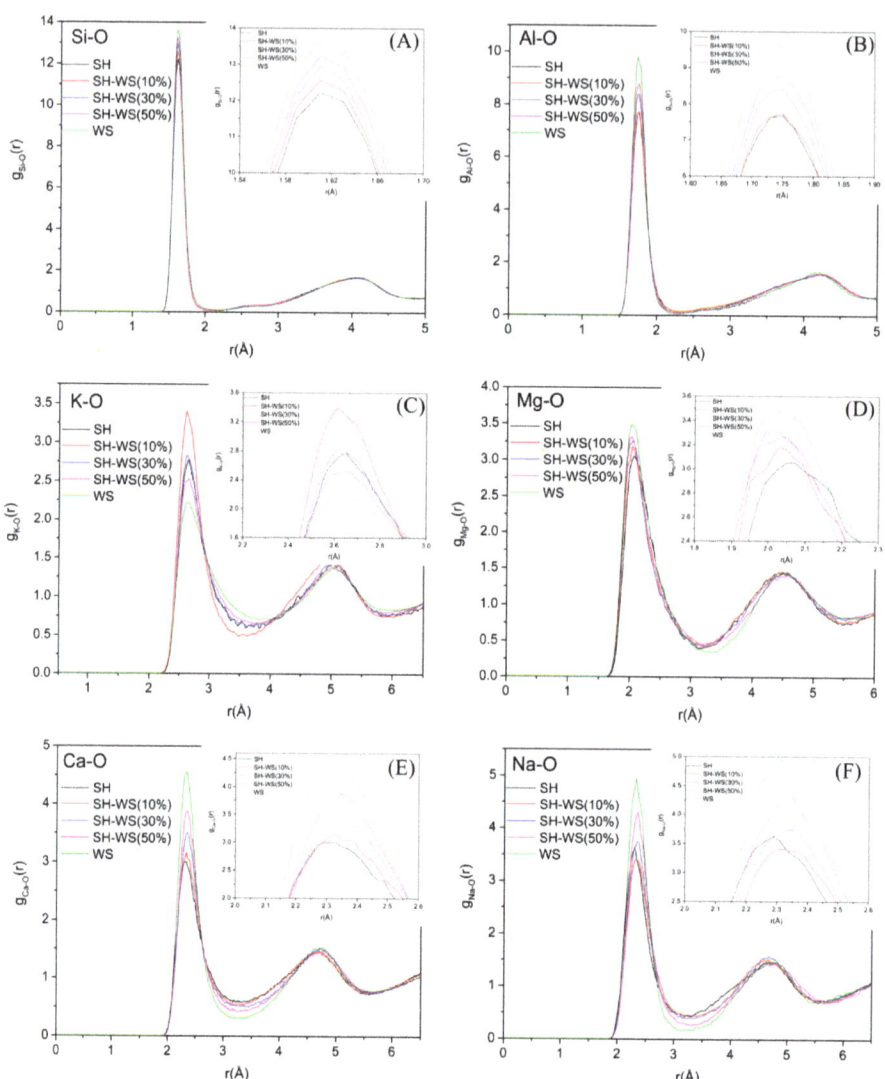

Figure 3. Radial distribution functions of Si-O (**A**), Al-O (**B**), K-O (**C**), Mg-O (**D**), Ca-O (**E**), and Na-O (**F**).

3.2. Mean Square Displacement

The atoms do not stay in a fixed position in the system; rather, they are in continuous movement. The mean square displacement (*MSD*) of each atom can be determined from its trajectory as [25]:

$$MSD = \langle \Delta r(t^2) \rangle = \frac{1}{N} \langle \sum_{i=1}^{N} |r_i(t) - r_i(0)|^2 \rangle \quad (4)$$

where $r_i(0)$ is the displacement of atom i at time zero, and $r_i(t)$ represents the displacement of atom i at time t. The MSD has a linear relationship with the diffusion coefficient (DC), which can be written as [26]:

$$DC = \lim_{t \to \infty} \frac{1}{6} \frac{d(\Delta r(t)^2)}{dt} \quad (5)$$

The MSDs of Si, Al, Ca, Mg, and K, and the total displacement of all atoms in the system are shown in Figure 4. For all samples, the activity and diffusion coefficient of Si atoms were the lowest, because Si formed the network structure of the system. However, with increasing WS content, the activity of other metal oxides (e.g., K_2O, MgO, and CaO) in the ash increased. For a WS content of 0–10 wt%, the activity of Ca was the highest among all elements in the ash. Between 30% and 100 wt%, K had the highest activity. The high activity of Ca and K is because these elements do not form the network structure and can move freely in the network. As shown in Figure 4F, the total MSD increased with increasing WS content. A possible explanation for this trend is that the addition of more oxide reduces the complexity of the system.

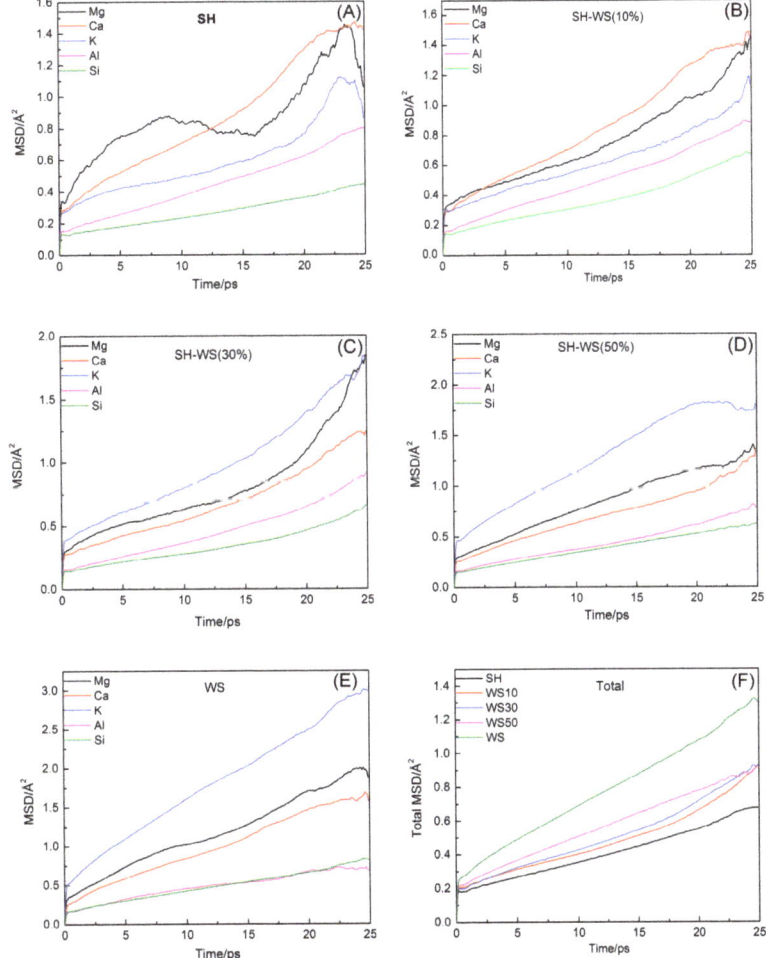

Figure 4. The mean square displacement of each atom in SH (**A**), WS (**B**), SH-WS(30%) (**C**), SH-WS(50%) (**D**), and WS (**E**). The mean square displacement of different samples (Total) (**F**).

We calculated the diffusion coefficients in the system on the basis of the law of diffusion, as shown in Figure 5. The diffusion coefficients decreased in the following order: $D_{Ca} > D_{Mg} > D_K > D_{Al} > D_{Si}$ at WS contents up to 10%. The order of diffusion coefficients was $D_K > D_{Mg} > D_{Ca}$ for WS contents between 30% and 100%, and at 100%, $D_{Si} > D_{Al}$.

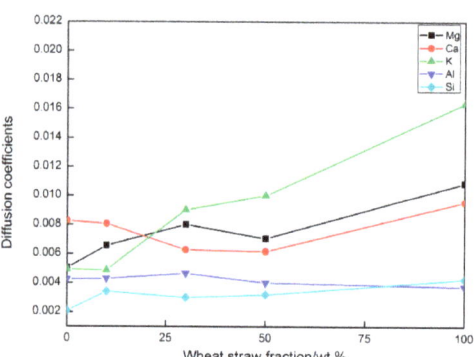

Figure 5. Diffusion coefficients of the different atoms in the ash slag system.

3.3. Coordination Numbers

The microstructure of the Na_2O-MgO-Al_2O_3-SiO_2-K_2O-CaO slag system was mainly composed of the silicoaluminate and the network modifier. The coordination numbers (CN) varied with the fraction of WS, as shown in Figure 6. It could be seen that the coordination number of Si was maintained at about four, but decreased with increasing WS content, indicating a reduction in the strength of Si–O bond. We attributed this to the depolymerization of the highly coordinated Si by metal atoms, such as Mg and K. For Al, not only four-coordinate Al–O structures but also a large number of structures with fewer than four coordination bonds formed. With increasing WS content, the fraction of more highly coordinated Al–O structures increased. This indicates that the [AlO$_4$] tetrahedral unit is not as stable as the [SiO$_4$] tetrahedron.

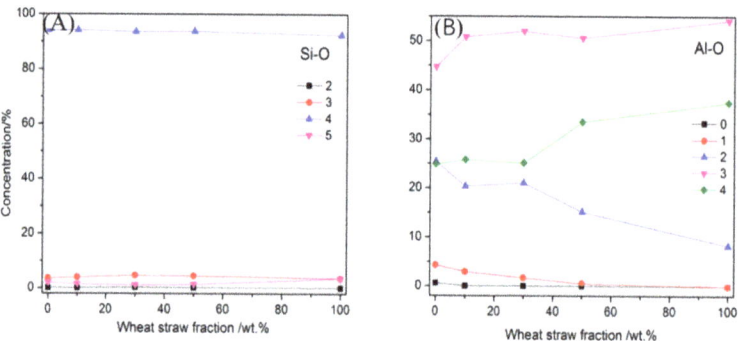

Figure 6. The variation of CN_{Si-O} (**A**) and CN_{Al-O} (**B**) against the WS content.

3.4. Distribution of Oxygen Types

The oxygen ions in this system can be classified as either BO or NBO on the basis of the cations to which they are connected. As shown in Figure 7A, the proportion of BO decreased and the proportion of NBO increased with increasing WS content. We classified BO into Si–O–Si, Si–O–Al, and Al–O–Al. There was less Al in the form of Al–O–Al than Al–O–Si, indicating the Al–O–Si bonds were more stable than Al–O–Al bonds (Figure 7B). A similar trend was not observed for Si.

For NBOs, we considered Si–O–Ca, Si–O–Na, Si–O–K, Si–O–Mg, Al–O–Mg, Al–O–Ca, Al–O–Na, and Al–O–K, the distributions of which are shown in Figure 7C. The most abundant NBOs were Al–O–K and Si–O–K, indicating that the tetrahedron is more easily destroyed by K_2O than other oxides in this network structure. Zhang et al. [23] proposed that the disparity between (Si-O-M (Ca, Mg, Na, K))/Si and (Al-O-M)/Al can indirectly reflect the proportion of charge-balancing M for tetrahedron. The ratios of (Si-O-Ca)/Si, (Al-O-Ca)/Al, (Si-O-Mg)/Si, (Al-O-Mg)/Al, (Si-O-Na)/Si, (Al-O-Na)/Al, (Si-O-K)/Si, and (Al-O-K)/Al, represent the average numbers of M, which forms non-bridging in one [SiO_4] or [AlO_4] tetrahedron. Figure 7C shows that the disparity between (Si-O-M*(Ca, Mg, Na)/Si and (Al-O-M*)/Al is much smaller than the disparity between (Si-O-K)/Si and (Al-O-K)/Al, indicating that Ca^{2+}, Mg^{2+}, Na^+ are not the charge-balancing atoms.

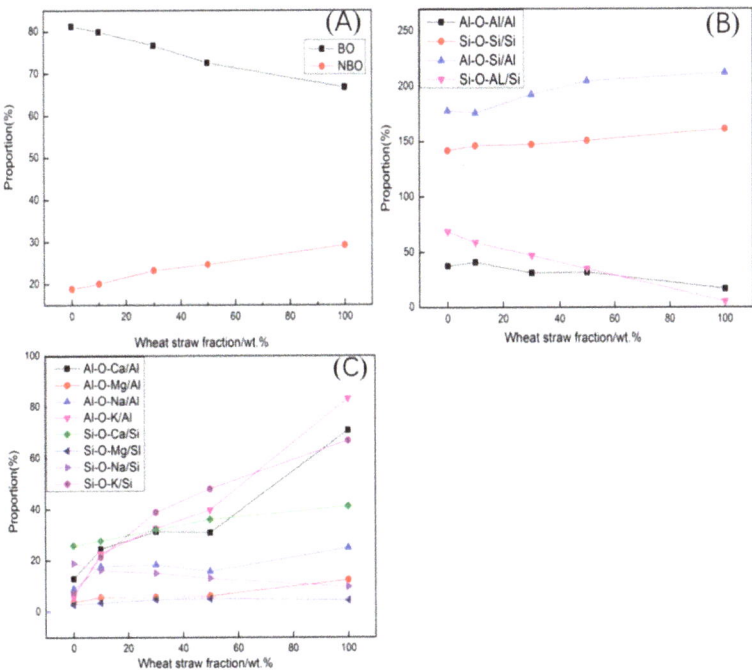

Figure 7. The influence of WS contents on the distribution of two kinds of oxygen (**A**); the influence of WS content on the distribution of bridging oxygen (**B**) and non-bridging oxygen (**C**).

3.5. Distribution of Bond Angles

The distributions of bond angles for the different samples are shown in Figure 8. The configurations of different bond angles in the Si–O and Al–O networks are shown in the insets of Figure 8A,B. The angular distributions of O–Si–O had a symmetric shape for all samples, averaging approximately 107° (Figure 8A), whereas the angular distributions of O–Al–O were asymmetric in all samples. The bond angle of O–Al–O reached a maximum when the WS content was 30%, which could be attributed to the transformation of [AlO_4] to [AlO_3]. Because of the stability of the [SiO_4] structure, the O–Si–O angle was close to the ideal tetrahedral angle (109.5°). However, at least four forms of O–Al–O (each with different bond angles) were identified owing to the presence of one-, two-, and three-coordinated Al atoms. The peak of the O–Al–O angle distribution was much wider than that of O–Si–O because of the contribution of the non-tetrahedral forms of O–Al–O.

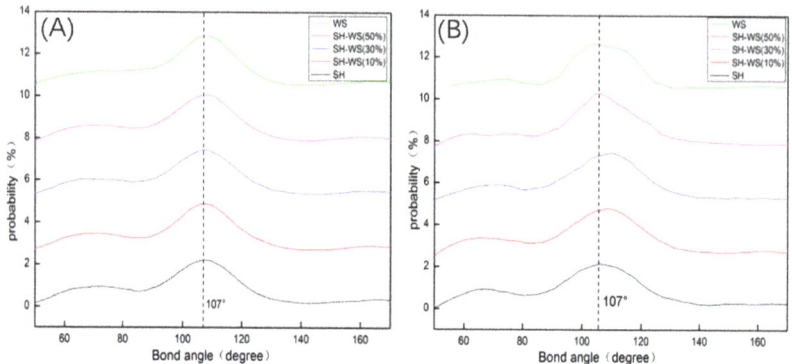

Figure 8. The influence of the WS contents on the O-Si-O angle (**A**) and O-Al-O angle (**B**).

3.6. Distribution of Q^n

The degree of polymerization is an important parameter for characterizing the integrity of the melt network structure. This can be analyzed by counting the number (Q) of tetrahedra with n bridging oxygen atoms. The influence of the WS fraction on Q^n for Si is shown in Figure 9. With increasing WS content up to 30%, Q^4 and Q^5 decreased, whereas Q^0, Q^1, Q^2, and Q^3 increased. We attribute these trends to the depolymerization of the network structure, which results in the ash having good fluidity.

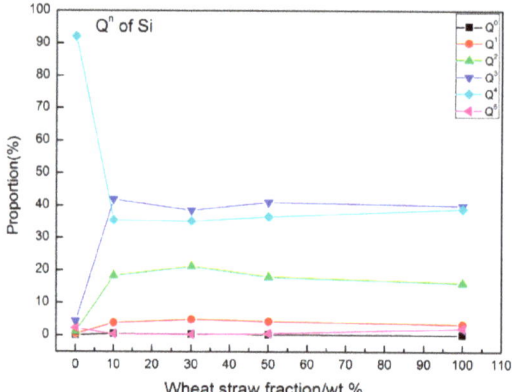

Figure 9. Q^n distribution of Si tetrahedra for different WS contents.

4. Conclusions

In conclusion, we investigated the structure of ash slags via MD simulations. The content of WS in the slag was varied from 0% to 100 wt%. The simulations provided structural information, such as the RDF, MSD, coordination number, bond angle, distribution of oxygen types, and Q^n. We draw the following conclusions:

(1) By analyzing the RDFs of the ion–oxygen pair, we found that the WS had no obvious influence on the bond length of Si–O. The widths of the first peaks in other ion–oxygen pairs were larger than that of Si–O, indicating a weaker bond strength than Si–O. The coordination number of Al was influenced by the addition of WS, demonstrating that WS influences the stability of the network structure. Moreover, the [AlO$_4$] tetrahedron was found to be less stable than the [SiO$_4$] tetrahedron.

(2) The concentration of BO decreased and that of NBO increased with increasing WS content. This indicates that the network structure can be destroyed by metal ions, resulting in the conversion of BO to NBO.

(3) The O–Si–O bond angle was constant at approximately 107.5°. The O–Al–O bond angles were wide and varied with increasing WS content.

(4) With increasing WS content, the diffusion coefficients of Si^{4+} varied only slightly with no obvious trend. The sum of all diffusion coefficients increased with WS content, indicating an increased diffusion capacity of the system.

(5) For a WS content of 30%, the Q^4 and Q^5 structural units transformed into Q^3, Q^2, Q^1, and Q^0 structural units to the greatest extent, which corresponded to the lowest degree of polymerization and highest ash fluidity.

Author Contributions: All six authors participated in the research and in the writing of this paper. All authors have read and agreed to the published version of the manuscript.

Funding: This research was supported by the Zhenjiang Science and Technology Bureau (High-tech Research Key laboratory of Zhenjiang, project No. SS2018002; Key Research and Development Program, project No. SH2020007), Department of Human Resources and Social Security of Jiangsu Province (Jiangsu Postdoctoral Research Funding Program, project No. 2021K180B), Taizhou Science and Technology Bureau (Technology Support Programme, project No. SSF20210131), and Jiangsu University (High-level Scientific Research Foundation for the introduction of talent, project No. 18JDG015).

Institutional Review Board Statement: Not applicable.

Informed Consent Statement: Not applicable.

Data Availability Statement: Not applicable.

Conflicts of Interest: The authors declare no conflict of interest.

References

1. Dudley, B. *BP Statistical Review of World Energy*; Whitehouse Associates: London, UK, 2019.
2. Jiang, N.; Li, L.; Wang, S.; Li, Q.; Dong, Z.; Duan, S.; Zhang, R.; Li, S. Variation tendency of pollution characterization, sources, and health risks of PM 2.5 -bound polycyclic aromatic hydrocarbons in an emerging megacity in China: Based on three-year data. *Atmos. Res.* **2019**, *217*, 81–92. [CrossRef]
3. Steinberg, M.; Cheng, H.-C. Modern and prospective technologies for hydrogen production from fossil fuels. *Int. J. Hydrog. Energy* **1989**, *14*, 797–820. [CrossRef]
4. Mahgagaokar, U.; Krewinghaus, A.-B. Coal conversion processes (gasification). *Encycl. Chem. Technol.* **1992**, *6*, 541–568.
5. Dai, J.; Cui, H.; Grace, J.-R. Biomass feeding for thermochemical reactors. *Prog. Energy Combust.* **2012**, *38*, 716–736. [CrossRef]
6. Yun, S.; Dai, Z.-H.; Zhou, Z.-J.; Chen, X.-L.; Yu, G.-S.; Wang, F.-C. Rapid co-pyrolysis of rice straw and a bituminous coal in a high-frequency furnace and gasification of the residual char. *Bioresour. Technol.* **2012**, *109*, 188–197. [CrossRef] [PubMed]
7. Ilyushechkin, A.-Y.; Shwe, H.-S.; Chen, X.; Roberts, D.-G. Effect of sodium in brown coal ash transformations and slagging behaviour under gasification conditions. *Fuel Process Technol.* **2018**, *179*, 86–98. [CrossRef]
8. Magdziarz, A.; Gajek, M.; Nowak-Wozny, D.; Wilk, M. Mineral phase transformation of biomass ashes—Experimental and thermochemical calculations. *Renew. Energy* **2018**, *128*, 446–459. [CrossRef]
9. Wei, J.; Gong, Y.; Ding, L.; Yu, J.; Yu, G. Influence of Biomass Ash Additive on Reactivity Characteristics and Structure Evolution of Coal Char–CO_2 Gasification. *Energy Fuels* **2018**, *32*, 10428–10436. [CrossRef]
10. Fang, X.; Jia, L. Experimental study on ash fusion characteristics of biomass. *Bioresour. Technol.* **2012**, *104*, 769–774. [CrossRef]
11. Li, K.-J.; Bouhadja, M.; Khanna, R.; Zhang, J.-L.; Liu, Z.-J.; Zhang, Y.-P.; Yang, T.-J.; Sahajwalla, V.; Yang, Y.-D.; Barati, M. Influence of SiO_2 reduction on the local structural order and fluidity of molten coke ash in the high temperature zone of a blast furnace. A molecular dynamics simulation investigation. *Fuel* **2016**, *186*, 561–570. [CrossRef]
12. Zhang, S.; Zhang, X.; Liu, W.; Lv, X.; Bai, C.; Wang, L. Relationship between structure and viscosity of CaO–SiO_2–Al_2O_3–MgO–TiO_2 slag. *J. Non-Crys. Solids* **2014**, *402*, 214–222. [CrossRef]
13. Xuan, W.; Wang, H.; Xia, D. Deep structure analysis on coal slags with increasing silicon content and correlation with melt viscosity. *Fuel* **2019**, *242*, 362–367. [CrossRef]
14. Li, K.; Khanna, R.; Bouhadja, M.; Zhang, J.; Liu, Z.; Su, B.; Yang, T.; Sahajwalla, V.; Singh, C.-V.; Barati, M. A molecular dynamic simulation on the factors influencing the fluidity of molten coke ash during alkalization with K_2O and Na_2O. *Chem. Eng. J.* **2017**, *313*, 1184–1193. [CrossRef]
15. Wu, T.; Wang, Q.; Yu, C.-F.; He, S.-P. Structural and viscosity properties of CaO-SiO_2-Al_2O_3-FeO slags based on molecular dynamic simulation. *J. Non-Crys. Solids* **2016**, *450*, 23–31. [CrossRef]
16. Jiang, C.; Li, K.; Zhang, J.; Qin, Q.; Liu, Z.; Sun, M.; Wang, Z.; Wang, L. Effect of MgO/Al_2O_3 ratio on the structure and properties of blast furnace slags: A molecular dynamics simulation. *J. Non-Crys. Solids* **2018**, *502*, 76–82. [CrossRef]

17. PRC National Standard GB/T 212-2008. *Methods for Industrial Analysis of Coal*; China Quality and Standards Publishing: Beijing, China, 2008.
18. PRC National Standard GB/T 28731-201. *Methods for Industrial Analysis of Solid Biomass Fuel*; China Quality and Standards Publishing: Beijing, China, 2012.
19. ASTM Standard D3174-2012. *Standard Test Method for Ash in the Analysis Sample of Coal and Coke from Coal*; ASTM International: West Conshohocken, PA, USA, 2012.
20. Ma, C.; Skoglund, N.; Carlborg, M.; Brostrm, M. Viscosity of molten $CaO-K_2O-SiO_2$ woody biomass ash slags in relation to structural characteristics from molecular dynamics simulation. *Chem. Eng. Sci.* **2020**, *215*, 115464. [CrossRef]
21. Dai, X.; Bai, J.; Huang, Q.; Liu, Z.; Bai, X.; Cao, R.; Wen, X.; Li, W.; Du, S. Viscosity temperature properties from molecular dynamics simulation: The role of calcium oxide, sodium oxide and ferrous oxide. *Fuel* **2019**, *237*, 163–169. [CrossRef]
22. Litton, D.-A.; Garofalini, S.-H. Atomistic structure of sodium and calcium silicate intergranular films in alumina. *J. Mater. Res. Technol.* **1999**, *14*, 1418–1429. [CrossRef]
23. Zheng, K.; Zhang, Z.; Yang, F.; Sridhar, S. Molecular Dynamics Study of the Structural Properties of Calcium Aluminosilicate Slags with Varying Al_2O_3/SiO_2 Ratios. *ISIJ Int.* **2012**, *52*, 342–349. [CrossRef]
24. Liu, L.; Xu, Z.; Li, R.; Zhu, R.; Xu, J.; Zhao, J.; Wang, C.; Nordlund, K.; Fu, X.; Fang, F. Molecular dynamics simulation of helium ion implantation into silicon and its migration. *Nucl. Instrum. Meth. B* **2019**, *456*, 53–59. [CrossRef]
25. Wang, B.; Cormack, A.-N. Molecular dynamics simulations of Mg-doped beta"-alumina with potential models fitted for accurate structural response to thermal vibrations. *Solid State Ionics* **2014**, *263*, 9–14. [CrossRef]
26. Mongalo, L.; Lopis, A.-S.; Venter, G.-A. Molecular dynamics simulations of the structural properties and electrical conductivities of $CaO-MgO-Al_2O_3-SiO_2$ melts. *J. Non-Crys. Solids* **2016**, *452*, 194–202. [CrossRef]

Review

Assessment of Manure Compost Used as Soil Amendment—A Review

Elena Goldan [1], Valentin Nedeff [1,2], Narcis Barsan [1], Mihaela Culea [3], Mirela Panainte-Lehadus [1,*], Emilian Mosnegutu [1], Claudia Tomozei [1], Dana Chitimus [1] and Oana Irimia [1,*]

1 Faculty of Engineering, Vasile Alexandri University of Bacau, Calea Marasesti, No. 157, 600115 Bacau, Romania
2 Gheorghe Ionescu Sisesti, Academy of Agricultural and Forestry Sciences, 6 Marasti Blvd., 011464 Bucharest, Romania
3 Faculty of Letters, Vasile Alexandri University of Bacau, Calea Marasesti, No. 157, 600115 Bacau, Romania
* Correspondence: mirelap@ub.ro (M.P.-L.); oana.tartoaca@ub.ro (O.I.)

Abstract: Organic waste management is an important concern for both industries and communities. Proper management is crucial for various reasons, such as reducing greenhouse gas emissions, promoting sustainability, and improving public health. Composted manure is a valuable source of nutrients and organic matter that can be used as a soil amendment in agriculture. Some important benefits of using composted manure in agriculture include: improves soil fertility, enhances soil structure, reduces soil erosion, suppresses plant diseases, and reduces reliance on synthetic fertilizers. Composted manure represents one of the most effective methods of organic waste valorization. Its macronutrients and micronutrients content can increase plant yield, without any reported negative or toxic effects on the soil and plants at various application rates. However, improper use of farmyard manure can have negative effects on the environment, such as air pollution from greenhouse gas emissions, soil acidification, and contamination of surface water and groundwater by nitrates and phosphates. The properties of the soil, including aeration, density, porosity, pH, water retention capacity, etc., can be improved by the structure and composition of manure. The slow-release source of nutrients provided by the nutrient content of compost can determine proper plants growth. However, it is crucial to use compost in moderation and regularly test soil to prevent excessive nutrient application, which can have adverse effects on plants and the environment.

Keywords: waste management; manure compost; soil amendment

1. Introduction

Waste management is a global concern, which uses valuable resources and determines the implementation of restrictive policies regarding waste valorization/recovery/reuse, etc. [1]. Organic waste results in large quantities and can cause significant pollution levels and environmental problems in the absence of rapid control [2]. More than 50% of this waste could be recycled or reused to turn waste into a usable resource [1,3].

Livestock production in the European Union generates an annual output of around 1400 million tons of manure [4]. Livestock farming in Europe remains responsible for a wide range of environmental problems, and public pressure for improvement is unlikely to disappear in the future [5]. The livestock sector is an important user of natural resources and has a significant influence on air quality, global climate, soil quality, biodiversity, and water quality by modifying the biogeochemical cycles of nitrogen, phosphorus, and carbon. The size of global livestock production is the result of human food development, which is based on products of animal origin [6].

The livestock sector must respond to the global demand for food, but under certain conditions, it can generate environmental problems and climate change [2]. Worldwide, about 65 billion chickens, 1.5 billion pigs, 1 billion goats and sheep, and about 330 million cattle and buffaloes are bred for meat production. The number of cattle used for milk production reaches almost 234 million, while the egg production sector has 7.6 billion specimens [6].

Organic wastes intended for use on agricultural land can be processed by different methods so as to retain as many nutrients as possible (e.g., nitrogen N, or phosphorus P) and to increase their capacity for agricultural use while minimizing their impact on the environment. Depending on the technology used, the processing of organic waste can add agronomic, economic, and ecological value to the final product [7].

Proper management of organic waste as a soil amendment can improve essential services provided by the soil, such as water filtration, food production, and climate regulation. The use of waste as a soil amendment can produce positive effects on the organic matter content and trace elements, reducing the need for inorganic fertilizers. In addition to this, it can help with forming soil macroaggregates and improving soil structure, water infiltration, and water retention capacity in the soil [1,8].

Organic fertilizers (amendments) are important sources of nutrients for sustainable agricultural production and, in combination with soil microorganisms and fauna, can bring a significant contribution to improving soil structure and favorable plant-growing conditions [9–11].

From the literature analysis, some knowledge gaps related to manure composting can be identified. Optimal composting conditions for manure depend on factors such as: type of manure, climate, and other variables. Ongoing research is being conducted to determine the ideal conditions for manure composting, including temperature, moisture content, aeration, and carbon–nitrogen ratio. While composting can reduce pathogen levels in manure, there is still uncertainty about the effectiveness of different composting methods in reducing pathogen contamination, particularly for emerging antibiotic-resistant bacteria. Although the literature generally mentions that composting can reduce greenhouse gas emissions from manure, there is limited knowledge about the overall impact of composting on emissions and the potential for composting to mitigate climate change [2–12].

This article aims to present the main aspects specific to the use of manure in agriculture, while also highlighting the main benefits of composting. Some important parameters were also discussed by considering the main objective of the manure compost used as a soil amendment to ensure an environmentally friendly product, which is beneficial for the growth and development of plants, with a low production cost compared to other methods.

2. Some Perspectives of Manure Used as Soil Amendment

The livestock production results in large amounts of waste, which can become a barrier to development if not disposed of properly [12]. In Europe, the livestock sector is currently responsible for around 80% of total ammonia emissions, 10–17% of greenhouse gas (GHG) emissions, 40–50% of diffuse nitrogen, and 70% of inorganic phosphorus. Several government policies have been implemented by the European Union, contributing to the development of manure treatment technologies, while some Member States have increased the use of manure in agriculture and have, therefore, reduced the impact on the environment [13].

2.1. Manure Management

The management of animal waste (manure) includes primarily composting for agricultural applications, combustion used for the production of heat and electricity, and the conversion of livestock waste into bioenergy through biological or thermochemical processes [12].

Manure disposal becomes a problem due to the increase in its volume and the risk of potential contamination of the soil, air, surface water, and groundwater caused by the draining of this organic waste from storage sites and by the odors released [14]. There are many problems with the storage and use of untreated manure, such as odor, emissions, leaching of hazardous substances and the appearance of health risks, the loss of nutrients, and the difficulty of handling this waste [15].

The use of improper disposal practices could cause serious environmental problems, which could include the addition of potentially harmful metals, inorganic salts, and pathogens to the soil, and would lead to increased soil nutrient loss through leaching and to increased emissions of hydrogen sulfide, ammonia, and other toxic gases into the air [16].

Alternative environmentally friendly disposal methods with potential financial benefits are manure processing technologies, which provide energy and products from manure [17]. Although manure is a resource for preserving soil fertility, its management has become one of the main environmental problems [4]. Even though the main objective of manure processing is to reduce the impact on the environment, not all technologies achieve the reduction of pollution, and most technologies are considered too costly [17]. In order to assess the economic and environmental sustainability of manure management methods and to support decision making, different types of methods are used, based on mathematical programming or simulation methods [17].

Since the early 1990s, the European Union legislation has regulated animal production and, indirectly, the use of manure [18]. This legislation requires that management criteria, such as the best application rates and timing, be adapted to specific local conditions (soil, climate, culture, type of manure) [18].

Overall, effective manure management is important for both environmental and agricultural sustainability. By implementing best practices for manure management, we can reduce the negative impacts of manure while maximizing its potential benefits as a nutrient-rich fertilizer. Additionally, if we refer to the manure management, we can mention the practices used to handle and process manure related to the current concepts on manure management such as: nutrient management, which is a critical component of manure management. Nutrients in manure, such as nitrogen, phosphorus, and potassium, can be valuable fertilizers for crops, but if not managed properly, they can also contribute to environmental problems such as nutrient pollution. Anaerobic digestion is a process that breaks down organic matter in manure in the absence of oxygen, producing biogas (a mixture of methane and carbon dioxide) and a nutrient-rich digestate that can be used as a fertilizer. This process can help reduce greenhouse gas emissions and odors from manure. Composting is another way to process manure that can help reduce odors and pathogens. The process involves allowing manure to decompose naturally with the help of microorganisms, turning it periodically to ensure adequate aeration. Manure storage and handling can help to reduce environmental and health risks associated with manure. This can involve using covered storage structures, proper ventilation, and appropriate equipment for handling and spreading manure. Water management can also be mentioned and involves managing water excess, a consequence that can lead to nutrient runoff and contamination of surface and groundwater. Proper management of water can help minimize these risks [12–18].

2.2. The Principal Techniques Used as Manure Treatment for Use in Agriculture

2.2.1. Anaerobic Digestion

Anaerobic digestion is the process of organic material degradation by microorganisms in the absence of oxygen, producing a biogas composed mainly of methane and carbon dioxide (CH_4 and CO_2) [4]. Manure used for anaerobic digestion becomes a compound rich in nutrients, called digestate, which makes it a potential substitute for chemical fertilizers [4]. Anaerobic digestion is a widely used process for stabilizing waste, controlling pollution, improving manure quality, and biogas production [17]. Biogas plants come with many significant benefits, including the reduction of methane emissions, the production of electricity, and renewable heat, resulting in reduced odor and CO_2 [19].

2.2.2. Mechanical/Physical Separation of Manure

Manure separation produces two fractions: a liquid fraction, which contains a small amount of dry matter, and a solid fraction [17]. The purpose of separation is to reduce the volume of manure and to obtain a solid fraction that can be used for the fertilization of crops [17]. The performance of this process can be determined by the degree of clarification of the supernatant produced, which is then used for irrigation. If activities around the farm are considered, such as washing manure channels, a high degree of clarification is necessary, but without producing large volumes of diluted sludge. Usually, mechanical/physical separation processes succeed one another. A single separation process rarely removes suspended matter and produces a solid phase rich in dry matter [20].

2.2.3. Aerobic Treatment

The purpose of aerobic treatment is to remove nitrogen by nitrification and denitrification, which is achieved by alternating the anoxic and aerobic phase, or by low levels of aeration [4]. Aerobic treatment was initially used to reduce nitrogen excess and was then supplemented with a mechanical separation to manage phosphorus excess [4]. This process results in nitrogen emissions and, sometimes, under adverse conditions, nitrogen oxide can form, and the resulting sediment can be mechanically separated and used for the fertilization of crops or for composting [4]. For example, biological nitrogen removal can only be obtained by combining anaerobic and aerobic treatments. Therefore, effective integrated anaerobic/aerobic treatments can only be achieved through a better management of electron flows. Several authors have reported experimental and full-scale applications of the combined anaerobic–aerobic process configuration [21].

2.2.4. Pyrolysis

During pyrolysis, manure is decomposed in an oxygen-deficient environment to produce gas, liquid, and coal [22]. The resulting coal is a solid residue, composed mainly of carbonic and inorganic materials (ash), and can be applied as a soil amendment and for the production of activated carbon [22]. In many situations, pyrolysis is applied in several stages for a better evolution of the product regarding its properties and temperature [23].

With regard to the use of the resulting compound after using pyrolysis as a soil amendment, it can be stated that if the pyrolysis is carried out at temperatures between 400–550 °C, there is no major impact on the pH [24].

2.2.5. Composting

An alternative approach to manure management is composting, which involves stabilizing organic matter, suppressing weeds and pathogens, deodorizing, improving the handling of the finished product, and the possibility of safe storage and transport [15].

Composting is an ecological and economical alternative for organic waste treatment, which turns manure into fertilizer/organic amendment [25]. Composting is not considered a new technology, but among the waste management strategies this method is considered a suitable option for manure management due to the economic and environmental benefits [26]. The composting process, if carried out correctly, converts wet and odorous organic waste into a dry, odorless, decomposed, and reusable product [27].

There are four stages in the composting process: the initial mesophilic phase, where mesophilic bacteria and fungi degrade simple organic compounds, such as sugars, amino acids, proteins, etc., by rapidly increasing the temperature; the thermophilic phase, when the composting material reaches its maximum temperature (>40 °C), this being the fastest stage of the decomposition process; the cooling phase, which is a decrease in temperature due to the reduction of microbial activity; the maturation phase, representing a long stabilization period of time meant to help obtaining high-quality stabilized, matured, and humified compost [28].

Composting is a method that can be used to reduce the amount of organic waste through recycling, because during the composting process up to 30% of the volume of waste can be reduced, resulting in a product that can have beneficial effects on the soil [27,29]. Composting significantly reduces the volume of manure through biochemical mineralization and partial humification of organic compounds, also reducing the mass, water content, and many of the undesirable elements present in manure, such as pathogens, parasites, and weed seeds [30–32]. The composting process can be influenced by many factors, including: oxygen content/level, humidity, biochemical composition of manure, pH, and temperature, which ultimately affect the quality of the final product and the efficiency of composting [28].

Both pathogens and weed seeds in the raw material are suppressed by: high temperatures, microbial antagonism and/or competition for nutrients, toxicity produced by by-products through organic matter degradation (e.g., ammonia, sulfides, organic acids, and phenolic compounds), and enzymatic degradation [32].

As a result of this process, the risk of spreading pathogens, parasites, or weed seeds is eliminated or reduced, and a stabilized final product is obtained, which can be used to improve and maintain the quality and fertility of the soil [26]. All these present environmental and economic advantages, such as more efficient transport and storage compared to the original raw material [32]. Composting reduces the environmental risks that can occur if untreated manure is stored (degradation of water, soil, and air quality) and improves soil quality in nutrient-deficient areas [30].

The advantages of using compost compared to the use of untreated manure are: pathogens and weeds elimination/reduction, microbial stabilization, volume and humidity reduction, odors elimination and control, ease of storage, transport and use, the production of organic fertilizers, or good-quality substrates [26]. Another advantage is the homogeneous and fragmented structure of the compost, which leads to easier scattering on the soil surface compared to uncomposted manure [32].

3. Manure and Compost as Soil Amendment

3.1. Some Positive Aspects Related to the Use of Manure and Compost as Soil Amendment

Manure has been used on farmland for many centuries, not only for the disposal of this waste, but also as a fertilizer/amendment [33]. Before the advent of inorganic fertilizers, manure was the main source of nutrients added to the soil for plant growth [33]. Manure from cattle was used centuries ago to improve soil fertility due to an increase in the content of organic matter in the soil and the improvement of its physical, chemical, and biological properties [34].

The application of manure to the soil has been shown to improve soil structure and adsorption properties by reducing volumetric density while increasing porosity, infiltration rate, hydraulic conductivity, and stability of the aggregate [9]. Used as a soil amendment, manure adds nutrients to soils (e.g., organic nitrogen or ammonia) and also improves soil structure, thereby increasing nutrient retention, the amount of organic matter, and the water retention capacity [9,35].

The beneficial effects of the use of manure on agricultural land are generally based on the ability to favorably alter soil properties, such as the availability of nutrients for plants, soil pH, cation-exchange capacity, water retention capacity [36,37]. Manure generally contains bicarbonates, organic anions, and basic cations such as Ca^{2+} or Mg^{2+}, which can buffer and neutralize soil acidity [9]. The increase of the soil cation-exchange capacity (CEC) after the application of manure is attributed to the increase in organic matter and carbon in the soil [9]. Since manure has a high content of organic matter, its application to the soil often contributes to the restoration of organic matter in degraded areas, and most of the nutrients are in organic form, which causes a gradual release of nutrients over time [35]. This waste added to the soil can contribute to the development of soil biodiversity in terms of species number, abundance, and diversity [9]. Soil biodiversity can contribute to the suppression of diseases through a variety of mechanisms, including the reduction of the abundance of certain pathogens and pests, releasing allochemics, increasing pH, and increasing the presence of soil microbial antagonists, such as *Actinobacteria* [9].

In general, the manure compost can improve the physical, chemical, and biological properties of soil as the other types of compost. So, we refer to the three types of properties that can be mentioned [33–38]:

- Physical properties of manure compost can vary depending on the specific mixture of organic materials used in the composting process. However, there are several physical properties for manure compost such as: texture, color, odor, and moisture.
- The chemical properties of manure compost are an important factor in determining its effectiveness as a fertilizer. Some key chemical properties of manure compost are nutrient content, organic matter, pH, salinity, and heavy metals.
- Biological properties of manure compost refer in general to microorganisms, a variety of microorganisms, including bacteria, fungi, and protozoa, which can help to break down organic matter and release nutrients. These microorganisms can also help to suppress plant pathogens and improve soil structure. Additionally, at the biological properties of manure, beneficial insects can be mentioned because manure compost can attract beneficial insects, such as earthworms, which can help to improve soil structure and promote healthy plant growth.

Properly used manure is a valuable source of plant nutrients and improves the soils' quality and productivity [38]. The increase in plant yield following the application of manure is due to the nutrients supplied to the soil, to the organic matter, to increased water retention in the soil, and due to the general improvement in the physico-chemical properties of the soil [16,39,40]. The use of an organic source of nutrients, such as manure from cattle, has become an alternative used to prevent a reduction in crop productivity, improving soil quality at the same time [37]. Soils modified with manure tend to have a higher pH and improved productivity resulting from increasing the availability of N, P, K, Ca, and Mg for plants [9]. This organic waste contains numerous essential elements necessary for plant growth, and its long-term use can have beneficial effects on plants, even at low application rates [9,41].

Compost contributes to the formation of soil structure, can improve soil physical properties by improving soil structure, water retention capacity, and aeration, while improving soil chemical properties by providing nutrients and adjusting soil pH. Compost also improves soil biological properties by promoting beneficial microorganisms and enhancing soil fertility [30,42].

Compost used in the soil can improve soil properties such as organic matter, water and nutrient retention capacity, infiltration, aeration, resistance to compaction and erosion, and resistance to soil-borne diseases [43]. Compost obtained from manure provides significant benefits when incorporated into the soil, since the organic matter in manure acts as a nutrient reservoir, improves the nutrient cycle, increases the cation-exchange capacity (CEC), pH, and also improves the physical properties of the soil, such as aggregation, friability, density, porosity, root infiltration, water retention capacity, and water infiltration [9,44,45].

Compost is favorable for the development of soil macrofauna, which plays an important role in improving soil quality. Furthermore, compost slowly releases nutrients that can be taken up by plants and thus contributes to improving crop productivity [43]. Compost contains both macroelements (mainly nitrogen, phosphorus, and potassium) and microelements, which are essential for plant growth; therefore, its use contributes to improving soil fertility [42]. Compost stimulates the activity of microorganisms, thus increasing the availability of nutrients for plants and produces hormone-like substances that can contribute to the growth of crops [42].

Finished compost contains highly active microbial communities that can stimulate soil biota and microbial community structure (this can alter the function of microorganisms involved in the biogeochemical cycle), suppress diseases, help plant development, increase nutrient availability, and increase fertilizer use efficiency and plant production [9]. Compared to the chemical ones, animal waste-based fertilizers (manure) can limit pest growth by increasing the content of micronutrients in the vegetal tissue and/or by increasing the production of defensive secondary metabolites [45].

It is very important to establish the cattle manure application rates, so as to ensure the necessary nutrients to the plants and to have beneficial effects on the properties of the soil [40].

3.2. Some Risks Related to the Use of Manure as Soil Amendment

Manure collection, storage, processing, and application on agricultural fields causes losses of nitrogen, phosphorus, and carbon [46]. These losses have an impact on the environment and human health, including climate change, soil acidification, water eutrophication, and the formation of particles in the air [46]. The use of manure as fertilizer also has several disadvantages for the environment, which are mainly related to the contamination of water sources with nitrates and phosphates [47]. For example, a part of the N and P supplied in excess of the plant requirements can leak out of the root area and thus contaminate the soil and affect both surface and ground waters [15,40,48].

The direct use of untreated manure in the soil at high application rates diminishes its role as soil amendment, and it is often seen as a problem of waste disposal rather than as a valuable source of nutrients [30]. It is very important to properly manage and use the manure application rates in order to avoid massive leaching of contaminants in groundwater [47]. Very large amounts of manure can cause water pollution and eutrophication by leaching and/or nitrate/phosphate leakage/and air pollution by greenhouse gas emissions and ammonia [4].

Following cattle manure overapplication, a possible decrease in crop yield and a negative impact on the environment may occur [40]. Manure in an untreated form, used as a soil amendment, can cause certain environmental and food safety problems [49]. This untreated organic waste is an important source of ammonia, which may lead to negative effects on crop growth and is, at the same time, an important source of pathogens and nitrates that can be transferred to surface and groundwater [49].

The levels of pathogens in manure depend on the type of animal, on its state of health and on how the manure was stored before use [50]. Pathogens can persist in manure for a long time depending on the storage conditions, type of manure mixture, storage temperature, and type of pathogen [50].

Some of the risks related to the use of manure as a soil amendment are presented in Table 1.

The inadequate use and inefficient recycling of manure from animals, especially in regions with high animal density, have exerted several negative effects on the environment [13]. This can contribute to increased climate change through emissions of methane and nitrogen dioxide, while ammonia emitted by manure can affect air, soil, and water quality [48]. Ammonia has undesirable direct and indirect effects on natural ecosystems, on greenhouse gas emissions, and on human health. For example, NH_3 contributes to the formation of small particles less than 2.5 µm in the atmosphere, and when inhaled, these particles, which are less than 2.5 µm, reach the deepest areas of the lungs, causing serious health problems. The nitrogen is mainly emitted as ammonia, nitrogen dioxide, nitrogen monoxide, harmless inhaler gas (N_2), and nitrate (NO_3^-), which drain and accumulate in soils, surface water, and groundwater [46]. Phosphorus, which is not absorbed by the crops, is retained in the soil and is susceptible to the leaching process [46]. Carbon, which contributes to climate change, is mainly emitted as methane and carbon dioxide. Emissions of N and P cause low efficiency in nutrient use. Reducing N emissions and increasing the efficiency of nitrogen use leads to a low environmental impact [46].

Table 1. Some of the risks related to the use of manure as a soil amendment.

Risks	Risks Description	Ref.
Nutrient balance	Manure contains a variety of nutrients such as nitrogen, phosphorus, and potassium, but the nutrient content can vary depending on the animal type, diet, and storage practices. Over-application of manure can lead to an imbalance in soil nutrients, resulting in excessive growth of plants, soil acidity, and soil nutrient pollution. For example, pig manure is typically high in nitrogen. So, if too much nitrogen is applied to the soil in the form of pig manure, it can lead to excessive vegetative growth, reduced fruit and seed production, and increased susceptibility to pests and diseases.	[46]
Pathogens	Manure can contain harmful pathogens such as bacteria, viruses, and parasites that can cause human and animal illnesses. If manure is not properly handled or stored, these pathogens can spread to crops, soil, water, and air, posing a health risk to humans and animals. For example, the main pathogen risks associated with bovine manure are *E. coli*, *Salmonella*, and *Cryptosporidium*.	[32,50]
Contamination	Manure can contain heavy metals, antibiotics, and hormones, which can contaminate the soil and water. These contaminants can accumulate in the soil over time and pose a risk to human and animal health if they enter the food chain. One significant example of contamination refers to antibiotic residues from chicken manure that can reduce its effectiveness as a fertilizer, as the residues can inhibit the growth of beneficial microorganisms in the soil that are important for nutrient cycling and soil health.	[40,48]
Odors	Manure can emit a strong odor that can cause discomfort and annoyance to nearby residents and nature in general. This can lead to complaints and even legal action against farmers who use manure as a soil amendment.	[51]
Environmental Impact	The improper use of manure can have negative environmental impacts, such as eutrophication, soil erosion, and greenhouse gas emissions. Excessive use of manure can lead to the runoff of nutrients and pathogens into water bodies, leading to algae blooms and fish kills.	[13,48]

Manure treatment is recommended so as to reduce greenhouse gas and ammonia emissions [4]. The volatilization of certain organic compounds from manure that cause unpleasant odors can represent yet another problem [51].

Figure 1 presents some aspects related to the manure and manure compost uses.

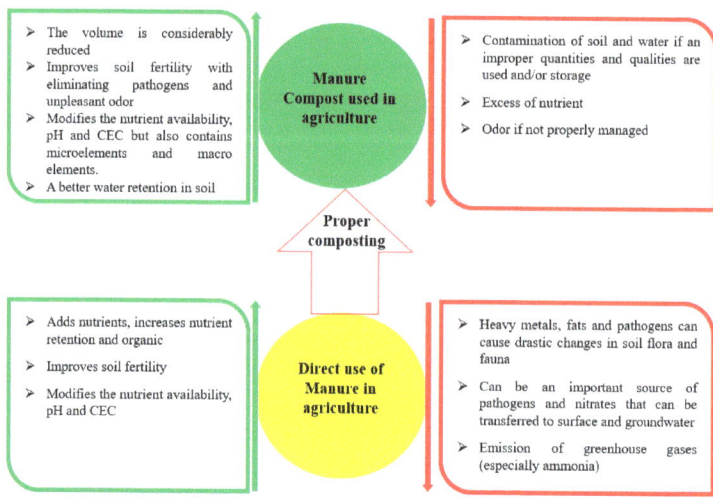

Figure 1. Some aspects related to the use of manure and manure compost used as soil amendment.

3.3. The Influence of Manure Compost on the Soil and Plants

Compost has high organic content (90–95%), but compared to chemical fertilizers, it generally includes low concentrations of nitrogen, phosphorus, potassium, and macro- and micronutrients [27].

Among the organic amendments, manure has been widely used on agricultural fields, and the composted form of this organic waste is preferred to eliminate the risk of nitrogen loss through leaching and surface leakage, as well as for suppressing pathogens, mitigating greenhouse gas emissions, and increasing organic matter in the soil. The use of compost to increase crop productivity is associated with minimizing the risk of spreading pathogens and weeds and improving soil quality and fertility [31].

Table 2 presents experiments from specialized literature, carried out under different experimental conditions, in which the fertilizing potential of manure was tested.

Table 2. Some examples of cattle manure compost.

Experimental Conditions	Raw Materials/Application Rate	Effect on the Soil	Effect on Plants	Ref.
The experiments were conducted in plastic containers, in outdoor conditions, for a time span of 300 days. Two experiments were conducted, each with 3 types of soil collected from different agricultural land plots. Experiment 1: 50% humidity. Experiment 2: 95% humidity.	vineyard soil vineyard soil + CMC compost (cattle manure compost) potato crops soil potato crops soil + CMC compost orchard soil orchard soil + CMC compost Application rate: 9 t/ha	Compost increased pH, EC, organic matter content, cation-exchange capacity, and nutrients. In the case of soil–compost mixtures maintained at 95% WHC moisture content, the EC, CEC, N, K, and Na values were lower than in the mixtures from the experiment with 50% moisture content.	No data.	[52]
The experiment was conducted in field conditions. Rice seedlings (*Oryza sativa Japonica*) were transplanted to the flooded field.	control soil cattle manure compost CMC, 5 t/ha swine manure compost, 6.35 t/ha.	Both types of compost increased pH, nutrient availability (C, N, and P), microbial biomass and enzymatic activities. The increase of these parameters was more significant in the case of cattle manure compost.	Plant growth parameters recorded significant values in the case of both types of compost. The yield of rice plants recorded maximum values in the case of cattle manure compost.	[53]
An experiment was conducted in field conditions for 210 days. Study plant: onions.	control soil soil + cattle manure compost and (chaff) rice husk Application rate: 20, 40, 60, and 80 t/ha.	Organic matter and soil pH increased in accordance with the increase of application rates. Compared to soil control, compost increased the amount of nutrients.	Growth parameters of onion plants recorded high values in the case of all application rates.	[54]
The experiment was conducted in greenhouse conditions, in a time span of 3 months. Study plant: French marigolds. Compost was produced from cattle manure and wood splinters.	commercial peat (control variant) 100% synthetic aggregate 20% CMC compost (cattle manure compost) + 80% synthetic aggregate 40% CMC compost +60% synthetic aggregate 60% CMC compost +40% synthetic aggregate 100% CMC compost.	pH and EC increased depending on the increase in the concentration of compost. Compost significantly increased N, K, P, C, Mg, and Ca. High concentrations of compost increased the values of Cu, Zn, Cr, Mn, and Pb, but these were lower than the limits imposed by the law.	A maximum yield of French marigolds was recorded in the case of the substrate with 40% compost. K, Mg, Ca, and P had the highest values in the substrate with 100% compost. Cu, Mn, Zn, Cd, Cr, and Pb were much lower than the phytotoxic levels in specialized literature.	[55]
The experiment lasted 8 weeks in greenhouse conditions and two types of soil were used. Study plant: spinach. Cattle manure was composted in poplar leaves mixture, ratios: 1:0; 1:1; 1:2, and 1:3 (manure: leaves).	sandy soil + compost (1:0, manure: leaves) sandy soil + compost (1:1) sandy soil + compost (1:2) sandy soil + compost (1:3) clay soil + compost (1:0) clay soil + compost (1:1) clay soil + compost (1:2) clay soil + compost (1:3) Application rate: 20 t/ha.	Compost increased the content of K and P. Cu, Zn, Fe, and Cd decreased in proportion to the increase in the amount of poplar leaves in the compost. An increase in the quantity of poplar leaves fom the compost caused increases in pH and EC of the soil.	Spinach biomass increased in the case of compost with a large quantity of leaves. Plant biomass was higher in the sandy soil. Increasing the ratio of leaves in the compost reduced N, Zn, Fe, Cu, and Cd and increased the content of P and K in spinach.	[29]

Table 2. Cont.

Experimental Conditions	Raw Materials/Application Rate	Effect on the Soil	Effect on Plants	Ref.
The study was conducted between 2006/2007 and 2007/2008, in field conditions. Study plant: maize.	control soil soil+ poultry manure compost soil+ cattle manure compost soil + urea soil + chemical fertilizer (Calcium Ammonium Nitrate—CAN) soil+ ammonia sulphate (AS) Application rate: 60 and 90 kg N ha^{-1}	Organic matter, organic carbon, total nitrogen, phosphorus, pH, EC and Ca, Mg, K, Na, recorded high values in the case of the two compost types compared to the soil modified with inorganic fertilizer.	The number of gray leaves decreased in the case of composts. The yield of plants reached the highest values in the case of cattle manure compost samples.	[56]
The experiment was conducted between 2010 and 2014 on a cultivated land rotating wheat (*Triticum aestivum* L.) with maize (*Zea mays* L.) crops.	control soil soil + 100% inorganic fertilizer (IF) soil + 25% CMC + 75% (IF) soil + 50% CMC + 50% (IF) soil + 75% CMC +25% (IF) soil + 100% CMC.	The inorganic fertilizer produced a decrease in the water content and the total N content, but instead these parameters increased in accordance with the increase in the amount of compost from cattle manure.	The average annual yield of wheat and maize plants increased in all treatments. The highest yield was obtained in the treatment with 25% CMC + 75% IF compost.	[57]
The experiment was conducted between April 2002 and May 2003, in field conditions. Study plant: maize.	control soil soil+ CMC compost +IF soil + IF.	Compost treatment resulted in more significant increases of pH, EC, organic matter, and nutrient content. Cr, Ni, Pb, and Cd were similar in both treatments and were not significantly higher than the values in the control soil.	The yield of maize grain production did not vary significantly. Ca, Mg, Mn, Fe, Cu, Zn, and B did not vary significantly. Cr, Ni, Pb, Cd, and Hg in maize grain were lower than the detection limit.	[15]
Experiment conducted in a vineyard in a time span of 5 years. Study plant: grapevine.	control soil, soil+ CMC compost applied between rows soil + grapevine compost applied between rows/to the inter-rows areas soil + grapevine applied under rows Application rate: 4 t/ha.	The use of compost resulted in an increase of soil pH, organic matter, total nitrogen, and microbial biomass compared to the control variant.	The vegetative growth of the vine was best stimulated by the compost from cattle manure. The number of grapes and their weight were similar in the case of both types of compost.	[58,59]
Experiment in greenhouse conditions for 35 days. Study plant: spinach.	control soil, soil + 5% cotton compost soil + 10% cotton compost soil + 5% CMC compost soIl + 10% CMC compost	Compared to cotton compost, cattle manure compost significantly increased the amount of nutrients and organic matter.	Both types of compost had a positive effect on spinach plants. The productivity of spinach plants was significantly improved by the use of cattle manure compost.	[43]
Experiment conducted in greenhouse conditions, for a period of 90 days. Study plant: autumn barley	control soil soil + 0% CMC compost, 100% sewage sludge biochar soil + 10% compost, 90% biochar soil + 20% compost, 80% biochar soil + 30% compost, 70% biochar soil + 40% compost, 60% biochar soil + 50% compost, 50% biochar soil + 60% compost, 40% biochar; soil + 70% compost, 30% biochar soil + 80% compost, 20% biochar soil + 90% compost, 10% biochar soil + 100% compost, 0% biochar Application rate: 5 and 30 t/ha.	Organic matter, organic carbon, and soil organic content increased due to the application of compost mixed with biochar for both application rates, but a more significant increase was recorded in the case of the application rate of 30 t/ha. ATR-FTIR spectra showed that the chemical composition of the soil did not change as a result of applying compost–biochar mixtures to the soil	No data.	[11]

Table 2. Cont.

Experimental Conditions	Raw Materials/Application Rate	Effect on the Soil	Effect on Plants	Ref.
Experiment conducted between August–November 2016, for a period of 90 days. Study plant: autumn barley	control soil soil + 0% cattle manure compost, 100% sewage sludge biochar soil + 10% compost, 90% biochar soil+ 20% compost, 80% biochar soil + 30% compost, 70% biochar sol + 40% compost, 60% biochar soil + 50% compost, 50% biochar soil + 60% compost, 40% biochar; soil + 70% compost, 30%biochar soil + 80% compost, 20% biochar soil + 90% compost, 10% biochar; soil + 100% compost, 0% biochar Application rates: 5 and 30 t/ha.	Compost–biochar mixtures used at an application rate of 30 t/ha significantly increased the pH, soil respiration, and electrical conductivity in the soil.	A more significant increase in plant height, number of shoots, and dry biomass was determined at application rates of 30 t/ha of compost–biochar mixtures, especially in the case of mixtures with a high concentration of compost.	[60]
Experiment conducted in greenhouse conditions, for a period of 90 days, having as study plant autumn barley.	control soil soil +0% cattle manure compost, 100% sewage sludge biochar. soil + 10% compost, 90% biochar; soil+ 20% compost, 80% biochar; soil + 30% compost, 70% biochar; soil + 40% compost, 60% biochar; soil + 50% compost, 50% biochar; soil + 60% compost, 40% biochar; soil + 70% compost, 30% biochar; soil + 80% compost, 20% biochar; soil + 90% compost, 10% biochar; soil + 100% compost, 0% biochar Application rates: 5 and 30 t/ha.	Pb and Cd concentrations recorded an increase in the case of mixtures with 100% sewage sludge biochar. Cu concentration increased at application rates of both 5 t/ha and 30 t/ha in accordance with the increase in cattle manure compost concentration in the mixtures.	No data.	[3]

As it is presented in Table 2, composting can indeed increase pH, electrical conductivity (EC), and organic matter content of the composted material. During the composting process, microorganisms break down organic matter, releasing carbon dioxide as a byproduct. This can lead to a decrease in acidity (lowering of pH) in the composted material. Additionally, the decomposition of organic matter can release nutrients and minerals, which can contribute to an increase in EC [52]. Additionally, the extent of the increase in organic matter and soil pH will depend on the amount and quality of the compost applied, as well as the starting properties of the soil. Generally, the higher the application rate of compost, the greater the increase in organic matter and soil pH, up to a certain point where additional compost may not result in further increases. It is important to note that the effects of adding compost to soil may take some time to fully manifest, as the organic matter must decompose and integrate into the soil ecosystem [54,58,59].

Biotests are usually used to estimate ecotoxicity and include a leachate analysis as well as direct tests using organisms of different taxonomic and trophic levels [61]. For ecotoxicological analysis, different methods (contacts and elutriates/leachates) are recommended, using terrestrial and aquatic organisms from different trophic levels [61].

The use of compost on the soil surface depends on its maturity and stability, which can be assessed by measuring physico-chemical characteristics and phytotoxicity [25]. Phytotoxicity is one of the most important criteria for assessing the quality of compost used for agricultural purposes [62]. Phytotoxicity is mainly caused by increased solubility of heavy metals, or by the production of phytotoxic substances such as ammonia, ethylene oxide, and organic acids [62]. The germination index (GI) is widely used to assess the phytotoxicity of compost, given that a high germination index indicates a decrease in phytotoxicity, and the obtained results should be carefully interpreted, as they are affected by the type of seeds used and the source of the compost [62].

Earthworms and other animal species such as enchytraeids, collembola, soil mites, isopods, nematodes, and protozoa also indicate the possibility to assess soil ecotoxicity [62].

Table 3 presents toxicity experiments in specialized literature, regarding toxicity testing of cattle manure. From the literature data, it can be observed that the toxicity testing of cattle manure is an important step in assessing its safety as a soil amendment. Manure can contain various contaminants, including pathogens, antibiotics, heavy metals, and hormones, which can potentially harm human health and the environment if not properly managed. For example, one common method of toxicity testing for manure is a bioassay. Additionally, plant germination and growth tests can be used to assess the chronic toxicity of manure, as plants are sensitive to long-term exposure to contaminants. Chemical analyses can also be conducted to assess the concentration of various contaminants in the manure.

Table 3. Experiments using sewage sludge biochar/sewage sludge.

Experimental Conditions	Raw Materials/Application Rate	Effects on Plants/Test Organisms	Ref.
The experiment was conducted on a field cultivated with wheat and corn by rotation. At two temporal intervals (June and October 2014) from each plot, a cube of soil was sampled, and earthworms of *Eisenia foetida* and *Pheretima guillelmi* species were sorted manually.	control soil, soil + 100% inorganic fertilizer, soil + 25% CMC compost + 75% inorganic fertilizer, soil + 50% CMC compost + 50% inorganic fertilizer; soil + 75% CMC compost + 25% inorganic fertilizer, soil + 100% CMC compost.	Treatment with 100% inorganic fertilizer had a negative effect on earthworms. The total density and biomass of earthworms of *Eisenia foetida* species increased in proportion to the increase of the compost concentration. Treatment with 75% compost +25% inorganic fertilizer had a positive effect on *P. guillelmi* earthworms.	[57]
Folsomia candida species was used in the first test conducted in a time span of 28 days, under laboratory conditions, at a temperature of 20–22 °C, in the dark. In the second test, *Eisenia Andrei* was used as the test organism. The containers were kept at a temperature of 20 °C for 14 days.	artificial soil soil + 0% CMC compost, 100% sewage sludge biochar soil + 10% compost, 90% biochar soil + 20% compost, 80% biochar soil + 30% compost, 70% biochar soil + 40% compost, 60% biochar soil + 50% compost, 50% biochar soil + 60% compost, 40% biochar soil + 70% compost, 30%biochar soil + 80% compost, 20% biochar soil + 90% compost, 10% biochar soil + 100% compost, 0% biochar Application rates: 5 and 30 t/ha	The number of juveniles of *Folsomia candida* determined at a 30 t/ha application rate of compost–biochar mixtures, did not exceed the number detected at application rates of 5 t/ha. In the case of the test in which the *Eisenia Andrei* was used, the compost from cattle manure, used at a concentration of 100% in the mixture, produced a significant increase in the biomass of the earthworms.	[10]
The *Tetrahymena pyriformis* species was chosen as an indicator of toxicity. The samples were incubated for 36 days.	0 Compost: control soil soil + 12.5 g leachate soil + 25 g leachate soil + 37.5 g leachate soil + 50 g leachate 25 g Compost: control soil, soil+ 12.5 g leachate + compost soil+ 25 g leachate + compost soil + 37.5 g leachate + compost soil+ 50 g leachate + compost. 50 g Compost: control soil soil + 12.5 g leachate + compost soil+ 25 g leachate + compost soil + 37.5 g leachate + compost soil+ 50 g leachate + compost.	Increases in compost rates had the effect of increasing pH and enzymatic activities. Increases in the amounts of leachate in the soil produced an increase in the toxicity of the samples. A remarkable decrease in toxicity was observed following the addition of cattle manure compost.	[63]

3.4. Future Research Directions

After a literature evaluation and interpretation, some future research directions can include:
- Conducting comparative studies regarding the effects of chemical fertilizers and the effects of treated organic waste on the soil and plants.
- Analyzing the use of other treated organic waste, generated by animal husbandry (poultry, horses, pigs, sheep, etc.), to reduce the amount of organic waste, including their use in a way that is beneficial to the environment, and to reduce the use of chemical fertilizers.
- Analyzing long-term studies to identify the efficiency of treated organic waste and its persistent effects in the soil.
- Comparative studies that examine the production costs of chemical and organic fertilizers and determine the effects that may occur during the production and long-term use of organic and chemical fertilizers.

4. Conclusions

The use of manure in agriculture is considered an optimal method for valorizing this waste, because the nutrients in this waste are recycled and reused in a beneficial way for degraded soils. The improvement of the physical, chemical, and biological properties of the soil due to the use of manure as a fertilizer is widely known. Similarly, this organic waste increases crop productivity due to the high content of organic matter and nutrients necessary for plant growth and development.

Manure, if used improperly, can cause disadvantages for the environment, for example, air pollution by greenhouse gas emissions, soil acidification, contamination of surface and groundwater by nitrates and phosphates.

Composting of manure allows obtaining a humified product, which contains organic matter and has the capacity to favorably modify the properties of the soil, thus contributing to the recovery of degraded agricultural soils following intensive agriculture.

The structure and composition of manure improve the properties of the soil, such as: aeration, density, porosity, pH, electrical conductivity, water retention capacity, etc. Additionally, the nutrient content of compost can ensure the growth and development of plants by providing a slow-release source of nutrients that can be taken up by plants as they need them. However, it is important to use compost in moderation and to test soil regularly to avoid over-application of nutrients, which can be harmful to plants and the environment.

The content of nutrients in the compost ensures the growth and development of plants over long periods of time, without the need to apply another type of fertilizer for a period of 2–3 years, thus reducing the number of chemical fertilizers. Additionally, by creating a healthy, balanced soil ecosystem, compost can reduce the need for chemical pesticides and herbicides.

Compost from cattle manure, used in various studies, under different experimental conditions, had a positive impact on the soil due to improving soil quality and productivity.

In general, the use of manure compost as a soil amendment can provide some benefits such as: improving soil fertility, waste reducing, improving soil health, slow-release source of nutrients for plants grows, and reducing environmental impacts.

Author Contributions: Conceptualization, E.G. and V.N.; methodology, E.G.; validation, C.T. and M.P.-L.; formal analysis, M.C.; investigation, E.G.; writing—original draft preparation, E.G.; writing—review and editing, N.B.; visualization, E.M., D.C. and O.I.; supervision, V.N. All authors have read and agreed to the published version of the manuscript.

Funding: This research received no external funding.

Institutional Review Board Statement: Not applicable.

Informed Consent Statement: Not applicable.

Conflicts of Interest: The authors declare no conflict of interest.

References

1. Renaud, M.; Chelinho, S.; Alvarenga, P.; Mourinha, C.; Palma, P.; Sousa, J.P.; Natal-da-Luz, T. Organic wastes as soil amendments—Effects assessment towards soil invertebrates. *J. Hazard. Mater.* **2017**, *330*, 149–156. [CrossRef] [PubMed]
2. Zhang, X.; Zhao, Y.; Zhu, L.; Cui, H.; Jia, L.; Xie, X.; Li, J.; Wei, Z. Assessing the use of composts from multiple sources based on the characteristics of carbon mineralization in soil. *Waste Manag.* **2017**, *70*, 30–36. [CrossRef]
3. Westerman, P.W.; Bicudo, J.R. Management considerations for organic waste use in agriculture. *Bioresour. Technol.* **2005**, *96*, 215–221. [CrossRef]
4. Loyon, L. Overview of manure treatment in France. *Waste Manag.* **2017**, *61*, 516–520. [CrossRef]
5. Loyon, L.; Burton, C.; Misselbrook, T.; Webb, J.; Philippe, F.; Aguilar, M.; Doreau, M.; Hassouna, M.; Veldkamp, T.; Dourmad, J.; et al. Best available technology for European livestock farms: Availability, effectiveness and uptake. *J. Environ. Manag.* **2016**, *166*, 1–11. [CrossRef]
6. Tullo, E.; Finzi, A.; Guarino, M. Review: Environmental impact of livestock farming and Precision Livestock Farming as a mitigation strategy. *Sci. Total Environ.* **2019**, *650*, 2751–2760. [CrossRef] [PubMed]
7. Case, S.D.C.; Oelofse, M.; Hou, Y.; Oenema, O.; Jensen, L.S. Farmer perceptions and use of organic waste products as fertilisers—A survey study of potential benefits and barriers. *Agric. Syst.* **2017**, *151*, 84–95. [CrossRef]
8. Goldan, E.; Nedeff, V.; Barsan, N.; Culea, M.; Tomozei, C.; Panainte-Lehadus, M.; Mosnegutu, E. Evaluation of the Use of Sewage Sludge Biochar as a Soil Amendment—A Review. *Sustainability* **2022**, *14*, 5309. [CrossRef]
9. Abbott, L.; Macdonald, L.; Wong, M.; Webb, M.; Jenkins, S.; Farrell, M. Potential roles of biological amendments for profitable grain production—A review. *Agric. Ecosyst. Environ.* **2018**, *256*, 34–50. [CrossRef]
10. Goldan, E.; Nedeff, V.; Barsan, N.; Mosnegutu, E.; Sandu, A.V.; Panainte, M. The effect of biochar mixed with compost on heavy metal concentrations in a greenhouse experiment and on Folsomia candida and Eisenia Andrei in laboratory conditions. *Rev. Chim.* **2019**, *70*, 809–813. [CrossRef]
11. Goldan, E.; Nedeff, V.; Sandu, I.; Barsan, N.; Mosnegutu, E.; Panainte, M. The use of biochar and compost mixtures as potential organic fertilizers. *Rev. Chim.* **2019**, *70*, 2192–2197. [CrossRef]
12. Zhu, L.D.; Hiltunen, E. Application of livestock waste compost to cultivate microalgae for bioproducts production: A feasible framework. *Renew. Sustain. Energy Rev.* **2016**, *54*, 1285–1290. [CrossRef]
13. Hou, Y.; Velthof, G.L.; Case, S.D.C.; Oelofse, M.; Grignani, C.; Balsari, P.; Zavattaro, L.; Gioelli, F.; Bernal, M.P.; Fangueiro, D.; et al. Stakeholder perceptions of manure treatment technologies in Denmark, Italy, the Netherlands and Spain. *J. Clean. Prod.* **2018**, *172*, 1620–1630. [CrossRef]
14. Cao, H.; Xin, Y.; Wang, D.; Yuan, Q. Pyrolysis characteristics of cattle manures using a discrete distributed activation energy model. *Bioresour. Technol.* **2014**, *172*, 219–225. [CrossRef] [PubMed]
15. Gil, M.V.; Carballo, M.T.; Calvo, L.F. Fertilization of maize with compost from cattle manure supplemented with additional mineral nutrients. *Waste Manag.* **2008**, *28*, 1432–1440. [CrossRef] [PubMed]
16. Gomez-Brandon, M.; Lazcano, C.; Dominguez, J. The evaluation of stability and maturity during the composting of cattle manure. *Chemosphere* **2008**, *70*, 436–444. [CrossRef]
17. Gebrezgabher, S.A.; Meuwissen, M.P.M.; Oude Lansink, A.G.J.M. A multiple criteria decision making approach to manure management systems in the Netherlands. *Eur. J. Oper. Res.* **2014**, *232*, 643–653. [CrossRef]
18. Zavattaro, L.; Bechini, L.; Grignani, C.; van Evert, F.K.; Mallast, J.; Spiegel, H.; Sandén, T.; Pecio, A.; Cervera, J.V.G.; Guzmán, G.; et al. Agronomic effects of bovine manure: A review of long-term European field experiments. *Eur. J. Agron.* **2017**, *90*, 127–138. [CrossRef]
19. Tsai, W.T.; Lin, C.I. Overview analysis of bioenergy from livestock manure management in Taiwan. *Renew. Sustain. Energy Rev.* **2009**, *13*, 2682–2688. [CrossRef]
20. Burton, C.H. The potential contribution of separation technologies to the management of livestock manure. *Livest. Sci.* **2007**, *112*, 208–216. [CrossRef]
21. Bortone, G. Integrated anaerobic/aerobic biological treatment for intensive swine production. *Bioresour. Technol.* **2009**, *100*, 5424–5430. [CrossRef] [PubMed]
22. Yuan, X.; He, T.; Cao, H.; Yuan, Q. Cattle manure pyrolysis process: Kinetic and thermodynamic analysis with isoconversional methods. *Renew. Energy* **2017**, *107*, 489–496. [CrossRef]
23. Ábrego, J.; Arauzo, J.; Sánchez, J.L.; Gonzalo, A.; Cordero, T.; Rodriguez-Mirasol, J. Structural Changes of Sewage Sludge Char during Fixed-Bed Pyrolysis. *Ind. Eng. Chem.* **2009**, *48*, 3211–3221. [CrossRef]
24. Atienza-Martínez, M.; Ábrego, J.; Gea, G.; Marías, F. Pyrolysis of dairy cattle manure: Evolution of char characteristics. *J. Anal. Appl. Pyrolysis* **2020**, *145*, 104724. [CrossRef]
25. Huang, J.; Yu, T.; Gao, H.; Yan, X.; Chang, J.; Wang, C.; Hu, J.; Zhang, L. Chemical structures and characteristics of animal manures and composts during composting and assessment of maturity indices. *PLoS ONE* **2017**, *12*, e0178110. [CrossRef]
26. Bernal, M.P.; Alburquerque, J.A.; Moral, R. Composting of animal manures and chemical criteria for compost maturity assessment. A review. *Bioresour. Technol.* **2009**, *100*, 5444–5453. [CrossRef]
27. Khater, E.S.G. Some Physical and Chemical Properties of Compost. *Int. J. Environ. Waste Manag.* **2015**, *5*, 172. [CrossRef]

28. Kim, S.Y.; Jeong, S.T.; Ho, A.; Hong, C.H.; Lee, A.H.; Kim, P.J. Cattle manure composting: Shifts in the methanogenic community structure, chemical composition, and consequences on methane production potential in a rice paddy. *Appl. Soil Ecol.* **2018**, *124*, 344–350. [CrossRef]
29. Anwar, Z.; Irshad, M.; Mahmood, Q.; Hafeez, F.; Bilal, M. Nutrient uptake and growth of spinach as affected by cow manure co-composted with poplar leaf litter. *Int. J. Recycl. Org.* **2017**, *6*, 79–88. [CrossRef]
30. Larney, F.J.; Hao, X. A review of composting as a management alternative for beef cattle feedlot manure in southern Alberta, Canada. *Bioresour. Technol.* **2007**, *98*, 3221–3227. [CrossRef]
31. Arriaga, H.; Viguria, M.; López, D.M.; Merino, P. Ammonia and greenhouse gases losses from mechanically turned cattle manure windrows: A regional composting network. *J. Environ. Manag.* **2017**, *203*, 557–563. [CrossRef] [PubMed]
32. Viaene, J.; Van Lancker, J.; Vandecasteele, B.; Willekens, K.; Bijttebier, J.; Ruysschaert, G.; De Neve, S.; Reubens, B. Opportunities and barriers to on-farm composting and compost application: A case study from northwestern Europe. *Waste Manag.* **2016**, *48*, 181–192. [CrossRef] [PubMed]
33. Kumar, R.R.; Park, B.J.; Cho, J.Y. Application and environmental risks of livestock manure. *J Korean Soc Appl Biol Chem.* **2013**, *56*, 497–503. [CrossRef]
34. Indraratne, S.P.; Hao, X.; Chang, C.; Godlinski, F. Rate of soil recovery following termination of long-term cattle manure applications. *Geoderma* **2009**, *150*, 415–423. [CrossRef]
35. Mosebi, P.; Truter, W.F.; Madakadze, I. Manure from cattle as fertilizer for soil fertility and growth characteristics of Tall Fescue (Festuca arundinacea) and Smuts Finger grass (Digitaria erianta). *Livest. Res. Rural. Dev.* **2015**, *27*, 190.
36. Thangarajan, R.; Bolan, N.S.; Tian, G.; Naidu, R.; Kunhikrishnan, A. Role of organic amendment application on greenhouse gas emission from soil. *Sci. Total Environ.* **2013**, *465*, 72–96. [CrossRef]
37. Hariadi, Y.C.; Nurhayati, A.Y.; Hariyani, P. Biophysical Monitoring on the Effect on Different Composition of Goat and Cow Manure on the Growth Response of Maize to Support Sustainability. *Agric. Agric. Sci. Procedia* **2016**, *9*, 118–127. [CrossRef]
38. Lazcano, C.; Gomez-Brandon, M.; Dominguez, J. Comparison of the effectiveness of composting and vermicomposting for the biological stabilization of cattle manure. *Chemosphere* **2008**, *72*, 1013–1019. [CrossRef]
39. Torrellas, M.; Burgos, L.; Tey, L.; Noguerol, J.; Riau, V.; Palatsi, J.; Antón, A.; Flotats, X.; Bonmatí, A. Different approaches to assess the environmental performance of a cow manure biogas plant. *Atmos. Environ.* **2018**, *177*, 203–213. [CrossRef]
40. Schlegel, A.J.; Assefa, Y.; Bond, H.D.; Haag, L.A.; Stone, L.R. Changes in soil nutrients after 10 years of cattle manure and swine effluent application. *Soil Tillage Res.* **2017**, *172*, 48–58. [CrossRef]
41. Liu, C.; Guo, T.; Chen, Y.; Men, Q.; Zhu, C.; Huang, H. Physicochemical characteristics of stored cattle manure affect methane emissions by inducing divergence of methanogens that have different interactions with bacteria. *Agric. Ecosyst. Environ.* **2018**, *253*, 38–47. [CrossRef]
42. Pergola, M.; Piccolo, A.; Palese, A.M.; Ingrao, C.; Di Meo, V.; Celano, G. A combined assessment of the energy, economic and environmental issues associated with on-farm manure composting processes: Two case studies in South of Italy. *J. Clean. Prod.* **2018**, *172*, 3969–3981. [CrossRef]
43. Xu, C.; Mou, B. Short-term Effects of Composted Cattle Manure or Cotton Burr on Growth, Physiology, and Phytochemical of Spinach. *HortScience* **2017**, *51*, 1517–1523. [CrossRef]
44. Forján, R.; Rodríguez-Vila, A.; Cerqueira, B.; Covelo, F.E. Comparison of the effects of compost versus compost and biochar on the recovery of a mine soil by improving the nutrient content. *J. Geochem. Explor.* **2017**, *183*, 46–57. [CrossRef]
45. Rowen, E.; Tooker, J.F.; Blubaugh, C.K. Managing fertility with animal waste to promote arthropod pest suppression. *Biol. Control* **2019**, *134*, 130–140. [CrossRef]
46. De Vries, J.; Hoogmoed, W.; Groenestein, C.; Schröder, J.; Sukkel, W.; De Boer, I.; Koerkamp, P.G. Integrated manure management to reduce environmental impact: I. Structured design of strategies. *Agric. Syst.* **2015**, *139*, 29–37. [CrossRef]
47. Franco, A.; Schuhmacher, M.; Roca, E.; Domingo, J.L. Application of cattle manure as fertilizer in pastureland: Estimating the incremental risk due to metal accumulation employing a multicompartment model. *Environ. Int.* **2006**, *32*, 724–732. [CrossRef]
48. He, Z.; Pagliari, P.H.; Waldrip, H.M. Applied and Environmental Chemistry of Animal Manure: A Review. *Pedosphere* **2016**, *26*, 779–816. [CrossRef]
49. Cao, H.; Xin, Y.; Yuan, Q. Prediction of biochar yield from cattle manure pyrolysis via least squares support vector machine intelligent approach. *Bioresour. Technol.* **2016**, *202*, 158–164. [CrossRef]
50. Venglovsky, J.; Martinez, J.; Placha, I. Hygienic and ecological risks connected with utilization of animal manures and biosolids in agriculture. *Livest. Sci.* **2006**, *102*, 197–203. [CrossRef]
51. Flessa, H.; Dörsch, P.; Beese, F.; König, H.; Bouwman, A.F. Influence of cattle wastes on nitrous oxide and methane fluxes in pasture land. *J. Environ. Qual.* **1996**, *25*, 1366–1370. [CrossRef]
52. Gil, M.V.; Calvo, L.F.; Blanco, D.; Sánchez, M.E. Assessing the agronomic and environmental effects of the application of cattle manure compost on soil by multivariate methods. *Bioresour. Technol.* **2008**, *99*, 5763–5772. [CrossRef]
53. Das, S.; Jeong, A.T.; Das, S.; Kim, P.J. Composted Cattle Manure Increases Microbial Activity and Soil Fertility More Than Composted Swine Manure in a Submerged Rice Paddy. *Front. Microbiol.* **2017**, *8*, 1702. [CrossRef] [PubMed]
54. Lee, J. Evaluation of Composted Cattle Manure Rate on Bulb Onion Grown with Reduced Rates of Chemical Fertilizer. *HortTechnology* **2012**, *22*, 798–803. [CrossRef]

55. Jayasinghe, G.Y.; Arachchi, I.D.L.; Tokashiki, Y. Evaluation of containerized substrates developed from cattle manure compost and synthetic aggregates for ornamental plant production as a peat alternative. *Resour. Conserv. Recycl.* **2010**, *54*, 1412–1418. [CrossRef]
56. Lyimo, H.J.F.; Pratt, R.C.; Mnyuku, R.S.O.W. Composted cattle and poultry manures provide excellent fertility and improved management of gray leaf spot in maize. *Field Crops Res.* **2012**, *126*, 97–103. [CrossRef]
57. Guo, L.; Wu, G.; Li, Y.; Li, C.; Liu, W.; Meng, J.; Liu, H.; Yu, X.; Jiang, G. Effects of cattle manure compost combined with chemical fertilizer on topsoil organic matter, bulk density and earthworm activity in a wheat–maize rotation system in Eastern China. *Soil Tillage Res.* **2016**, *156*, 140–147. [CrossRef]
58. Gaiotti, F.; Marcuzzo, P.; Belfiore, N.; Lovat, L.; Fornasier, F.; Tomasi, D. Influence of compost addition on soil properties, root growth and vine performances of Vitis vinifera cv Cabernet sauvignon. *Sci. Hortic.* **2017**, *225*, 88–95. [CrossRef]
59. Rayne, N.; Aula, L. Livestock manure and the impacts on soil health: A review. *Soil Syst.* **2020**, *4*, 64. [CrossRef]
60. Goldan, E.; Nedeff, V.; Sandu, I.G.; Mosnegutu, E.; Panainte, M. Study of greenhouse use of biohazard wastewater and manure compost. *Rev. Chim.* **2019**, *70*, 169–173. [CrossRef]
61. Kuryntseva, P.; Galitskaya, P.; Selivanovskaya, S. Changes in the ecological properties of organic wastes during their biological treatment. *Waste Manag.* **2016**, *58*, 90–97. [CrossRef] [PubMed]
62. Tang, J.C.; Maie, N.; Tada, Y.; Katayama, A. Characterization of the maturing process of cattle manure compost. *Process Biochem.* **2006**, *41*, 380–389. [CrossRef]
63. Yang, L.; Chen, Z.; Liu, T.; Jiang, J.; Li, B.; Cao, Y.; Yu, Y. Ecological effects of cow manure compost on soils contaminated by landfill leachate. *Ecol. Indic.* **2013**, *32*, 14–18. [CrossRef]

Disclaimer/Publisher's Note: The statements, opinions and data contained in all publications are solely those of the individual author(s) and contributor(s) and not of MDPI and/or the editor(s). MDPI and/or the editor(s) disclaim responsibility for any injury to people or property resulting from any ideas, methods, instructions or products referred to in the content.

Review

Composting Processes for Agricultural Waste Management: A Comprehensive Review

Muhammad Waqas [1,†], Sarfraz Hashim [2,†], Usa Wannasingha Humphries [3,*], Shakeel Ahmad [4], Rabeea Noor [5], Muhammad Shoaib [5], Adila Naseem [6], Phyo Thandar Hlaing [1] and Hnin Aye Lin [1]

1. The Joint Graduate School of Energy and Environment (JGSEE), King Mongkut's University of Technology Thonburi, Bangkok 10140, Thailand
2. Department of Agricultural Engineering, MNS University of Agriculture, Multan 66000, Pakistan
3. Department of Mathematics, Faculty of Science, King Mongkut's University of Technology Thonburi, Bangkok 10140, Thailand
4. College of Environmental Science and Engineering, Nankai University, Tianjin 300350, China
5. Department of Agricultural Engineering, Bahauddin Zakariya University, Multan 60000, Pakistan
6. Department of Food Science and Technology, Bahauddin Zakariya University, Multan 60000, Pakistan
* Correspondence: usa.wan@kmutt.ac.th
† These authors contributed equally to this work.

Abstract: Composting is the most adaptable and fruitful method for managing biodegradable solid wastes; it is a crucial agricultural practice that contributes to recycling farm and agricultural wastes. Composting is profitable for various plant, animal, and synthetic wastes, from residential bins to large corporations. Composting and agricultural waste management (AWM) practices flourish in developing countries, especially Pakistan. Composting has advantages over other AWM practices, such as landfilling agricultural waste, which increases the potential for pollution of groundwater by leachate, while composting reduces water contamination. Furthermore, waste is burned, open-dumped on land surfaces, and disposed of into bodies of water, leading to environmental and global warming concerns. Among AWM practices, composting is an environment-friendly and cost-effective practice for agricultural waste disposal. This review investigates improved AWM via various conventional and emerging composting processes and stages: composting, underlying mechanisms, and factors that influence composting of discrete crop residue, municipal solid waste (MSW), and biomedical waste (BMW). Additionally, this review describes and compares conventional and emerging composting. In the conclusion, current trends and future composting possibilities are summarized and reviewed. Recent developments in composting for AWM are highlighted in this critical review; various recommendations are developed to aid its technological growth, recognize its advantages, and increase research interest in composting processes.

Keywords: composting; biodegradability; decomposing; organic waste; agricultural waste management

Citation: Waqas, M.; Hashim, S.; Humphries, U.W.; Ahmad, S.; Noor, R.; Shoaib, M.; Naseem, A.; Hlaing, P.T.; Lin, H.A. Composting Processes for Agricultural Waste Management: A Comprehensive Review. *Processes* **2023**, *11*, 731. https://doi.org/10.3390/pr11030731

Academic Editor: Andrea Petrella

Received: 21 October 2022
Revised: 2 December 2022
Accepted: 8 December 2022
Published: 1 March 2023

Copyright: © 2023 by the authors. Licensee MDPI, Basel, Switzerland. This article is an open access article distributed under the terms and conditions of the Creative Commons Attribution (CC BY) license (https://creativecommons.org/licenses/by/4.0/).

1. Introduction

Waste production is proportional to the number of human inhabitants worldwide. Thus, the increasing global population and continually growing human demands have resulted in massive waste production. With a population of 212 million in 2019 [1], Pakistan generates more than 20 million tons of waste annually [2]. On average in Pakistan, waste generation per capita is 0.612 kg/day; from this amount, 60 to 65% of waste is organic and biodegradable [3]. Organic matter (OM) in Pakistani soil is <1% [4]. Agricultural waste (AW) are leftovers of agricultural activity on agricultural land. Owing to lack of access to disposal sites in Pakistan, agricultural waste is frequently mismanaged; thus, most AW is burned or destroyed [3]. To protect the environment and ensure sustainable agriculture, resilient rural regions, and productive farming, it is vital to pursue the appropriate use and development of AW management (AWM). Among the different methods for managing organic waste, such as landfilling and incineration, biological decomposition of AW is considered the most effective. Composting is a low-cost method of biological decomposition. Micro-organisms control the composting process. This process influences the physical–chemical parameters of heat, aeration, water content, C:N ratio, and pH [5,6]. Composting is an alternative AWM approach and the resulting compost can be recycled into valuable products. This method is considered the most effective—it is environmentally friendly and agronomically sound since the resulting compost can be utilized as a natural, organic fertilizer and soil nutrient source [7]. Composting has been defined by Ayilara et al. (2020) as a form of recycling in which organic waste is digested by microbial activity under regulated conditions to create valuable, ecofriendly, and environmentally friendly goods [8]. The microbial population, which includes bacteria, fungi, and worms, can also stabilize degradable OM in the compost. In addition, the features of the microbial population rely on the substrate and physical conditions, which include the substrate's wetness, temperature, and aeration. Composting is only appropriate for agricultural waste, so the performance of this procedure also depends on the properties of the waste [9]. Composting has numerous benefits, including lowering the waste volume, weight, and water content, and producing dormancy in harmful organisms [10]. The compost can therefore contribute to the enhancement of soil nutrient levels, which is required for plant growth and significantly minimizes the need for synthetic fertilizers [11]. As a result of its ability to boost the soil's organic carbon content, compost application can revitalize soils in dire need of revitalization. In addition, as a soil amendment, compost improves soil structure, water retention capacity, and tilth [12]. Composting is initiated and managed under regulated environmental conditions instead of a natural and uncontrolled process. Composting is distinguished from decomposition by its controlled process [13]. Composting requires a longer preparation time, emits a foul stench, requires a long time to mineralize, and may contain diseases that can tolerate high temperatures to some extent, i.e., thermotolerant pathogens, and contains insufficient nutrients. All of these factors have deterred farmers from employing composting as a method of sustainable agriculture. Following this, there has been abundant evidence for the invention of composting processes to manage AW.

Numerous studies have investigated various composting processes, including vermicomposting (VC), aerobic composting (AC), and anaerobic composting (AnC), to convert farm waste into farm manure [14,15]. AC is the breakdown of OM by oxygen-dependent bacteria. Composting bacteria occur naturally and thrive on the moisture that surrounds OM. Airborne oxygen diffuses into the moisture and is absorbed by the bacteria [16]. Mehta and Sirari (2018) stated that AC is the most efficient decomposition type, producing compost that matures quickly. The biological breakdown and stability of OM under conditions favorable to the multiplication and activity of thermophilic microbes results in a solid, pathogen-free product ideal for forestry and agriculture [15]. AnC is a "no oxygen" technique in which biodegradable materials are stacked in an enclosed environment. Typically, digesters are used. Anaerobic micro-organisms dominate the AnC process. These microbes produce intermediate chemicals, including hydrogen sulfide, methane, and acids, while leaving pathogens and weed seeds untouched [14]. VC is the process of using earthworms to compost biodegradable organic materials. By substantially eating all types of OM, earthworms can degrade the OM. Earthworms can consume their body weight daily e.g., earthworms weighing 0.1 kg can consume 0.1 kg of waste daily [16]. According to Barthod et al. (2018), it is a globally adopted, low-cost biological treatment procedure for the generation of biofertilizers for agricultural uses. Worms and micro-organisms are the primary agents in composting for recycling nutrients, controlling soil processes, and preserving soil fertility [10]. Recently, many investigators employed these composting processes for AWM, including Karak et al. (2013) who investigated composting rice straws, wheat straws, potato plants, and mustard stovers with fishpond substrate [7]. This process was carried out for 56 days utilizing a heap as a compost box. For all compost preparations, the compost temperature on the first day varied from 24 to 26.8 °C and climbed to 81 °C before persistence. Initial pH values ranged from 6.76 to 7.68, while total N concentration was between 14.56 and 21.57 g/kg; the content of heavy metals was below the Indian Agriculture Ministry and Cooperation's limit. After composting, the C:N ratio ranged from 11–18 [17]. Qasim et al. (2018) improved the carbon-to-nitrogen (C/N) ratio to achieve a high composting and aeration rate and to create favorable circumstances for the process [13]. Azim et al. (2018) conducted a literature assessment of the most critical startup, monitoring, and maturity criteria for various composting techniques and input materials [12]. Farmers in developing countries must be aware of the process' aspects, effectiveness, and efficiency of composting. It is challenging to decide which composting technique is effective regarding all of elements used to maintain soil health.

Composting of AW is strongly encouraged in Pakistan as a massive amount of rubbish fills our overflowing landfills. Numerous researchers have investigated organic waste's physical and chemical features during the different composting processes. Consequently, this review article examines composting as an alternate AWM strategy. Therefore, the key objectives of the current review are to (a) highlight the best prominent features of the composting method via its phases and prominence in various wastes; (b) assist farmers, researcher, and scientists in the selection of treatments for different crops substrates and help them select a composting technique by providing a comparison between different techniques; (c) provide the comparison of composting techniques on the basis of nutrients; and (d) compare two-stage composting (AnC followed by AC) with AC, AnC, or VC alone.

2. Composting

Composting is the biological conversion of the solid waste of plant and animal organic materials into a fertile matrix through numerous micro-organisms, including actinomycetes, bacteria, and fungi, in the presence of oxygen. The addition of diverse microorganism in a solid waste can convert it into compost or many by-products, .g., heat, water, and CO_2 [17,18]. Humus is the solid and stable matrix after the microbiological process that can be usefully applied to land as an organic fertilizer to increase the fertility and structure of the soil. In ancient history, i.e., pre-Columbian Indians of Amazonia or ancient Egyptians and numerous prehistoric cultures used composting as a primitive technique for the betterment of soil. In the previous four decades, the composting technique has flourished, and its beneficial impact is illustrated with scientific research. The vulnerability and interconnection of various competing factors regarding the knowledge and process engineering of a composting matrix have been established [19–21].

Composting innovative processes were developed and employed by large- or medium-scale farmers, but they are expensive for small-scale farmers because the techniques require high-tech equipment for composting. Despite discrete processes/techniques, the crucial key points of the composting processes were indistinguishable each time, like natural, chemical, and physical characteristics. Appropriateness of distinctive input supplies and alterations and their fitting structure, substrate degradability, dampness management, energy, porosity, air space, energy adjustment, deterioration, and stabilization are needed to study and distinguish compost and composting processes [19,22].

2.1. Composting Stages

Composting processes undergo four stages: mesophilic, thermophilic, cooling, and finally ending with compost maturation; these stages can happen concomitantly rather than consequently [23] (Figure 2). Each stage duration depends on the mixture's inceptive framework, water content, air circulation, and microbiological composition [24,25]. During the mesophilic phase, a combination of bacteria, fungi, and actinomycetes induce the rapid metabolism of C-abundant substrates. Moreover, this is accomplished by selecting tolerable temperatures, generally within 15–40 °C, because aerobic metabolism will produce heat. Transforming the matter and air circulation decreases the temperature, for the time being, reducing the rapid decay of other organic matter. Thus, the temperature rises once again, as shown in Figure 1. In the thermophilic phase (2nd stage), temperature increases to around 40 °C, favoring mostly thermophilic bacteria, e.g., bacillus. When C compounds are produced after substrate reduction, a modest temperature fall occurs followed by the cooling phase.

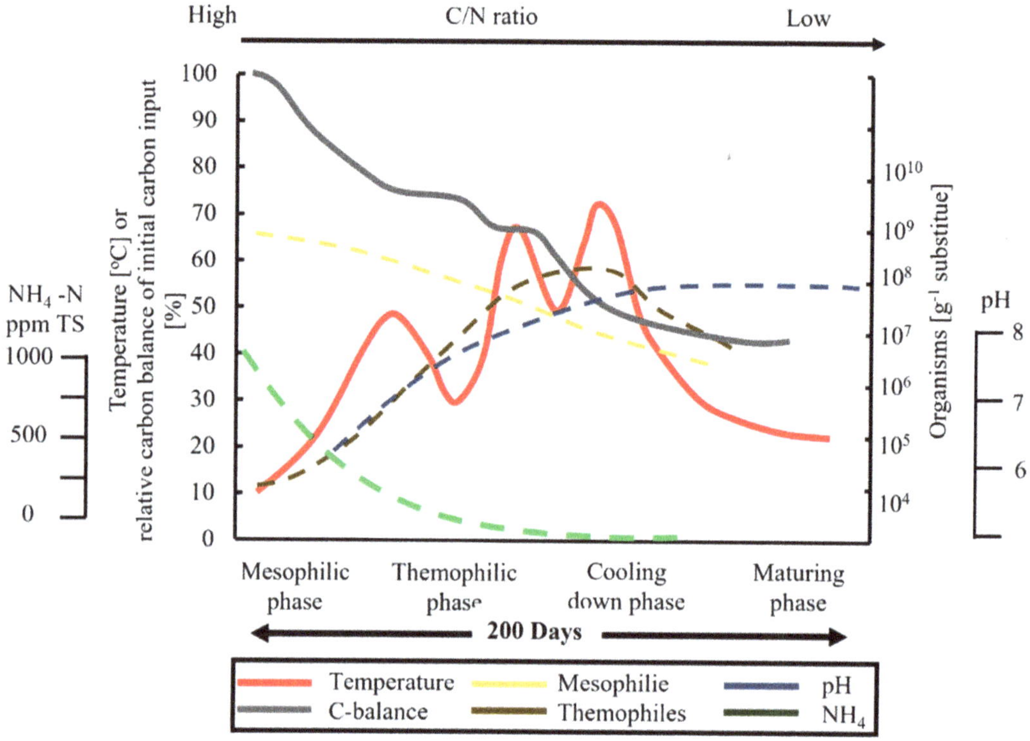

Figure 1. Time, temperature, the progression of compost biota, and further processes during discrete stages in composting.

Fungi break down more complex structures and more resistant components like lignin and cellulose molecules. Additionally, actinomycetes play a crucial role in forming humic compounds via condensation processes and breakdown [25]. Using aerobic bacteria, the final composting maturity is characterized by lower oxygen uptake rates and temperatures < 25 °C. During this final phase, the breakdown of various organic components continues, and macrofauna and soil organisms enter. By metabolizing phytotoxic chemicals, the organisms of this phase have a favorable effect on compost maturation, e.g., plant disease suppression [26].

Consequently, compost quality improves primarily during maturation (final stage) [27]. The final product of composting is characterized by pH and a lower C/N ratio of 15 to 20 compared to the initial substrate composition. It may contain a significant amount of plant-available NO_3^-, but NO_4^+ levels are low. Moreover, the intensity of the compost odor is significantly diminished [28]. However, it appears that the OM has stabilized, retaining recalcitrant C compounds [25]. Table 1 explains the favorable and sustainable application of different crop residues' influence on numerous biological, chemical, and physical aspects during the different processes. The outcomes showed which method is best with respect to input residues and the desired output products.

2.2. Discrete Waste Composting

In contrast to landfilling, which elevates the pollution risk for groundwater, discrete waste composting techniques are environment friendly and avoid groundwater contamination since chemical pollutants and bacteria are reduced during composting. Composting permits persistent organic pollutants and endocrine disruptors to remain in the soil while beneficial bacteria break down the toxins. The elimination of these harmful chemicals has not been simple. Although numerous methods have been attempted to eradicate them, there is no agreed success rate. A thorough application can increase agricultural and environmental sustainability. It also improves soil OM content and enhances agricultural productivity [29] due to the availability of plant-growth-promoting organisms and sufficient nutrients in the composted debris [30] and significantly contributes to the certification of food safety. Compost is helpful for bioremediation [6], weed control [31], plant disease control [32], pollution anticipation [33], and erosion management, in addition to its use as fertilizer. Composting also increases soil biodiversity and reduces environmental risks associated with synthetic fertilizers [34].

Composting is a fundamental aspect of a comprehensive AWM strategy. The key strategy for practical integrated AWM is nutrition level improvement. Compost is rich in essential plant nutrients, e.g., nitrogen (N), phosphorus (P), potassium (K), sulfur (S), carbon (C), and magnesium (Mg), as well as various essential trace elements [26,35]. Consequently, compost can be described as an assortment of nutrient-rich organic fertilizers [36]. Compost processing parameters and organic feedstocks determine its key chemical features, e.g., C/N ratio and pH, as well as the content of other nutrients (Table 2). Total N, P, and K levels could contribute to soil fertility when used as soil amendment agents. By adequately combining these organic components, nutrient-rich compost substrates can be produced and used in agriculture in place of commercial mineral fertilizers. This aspect is discussed in the following subsections.

- Crop residue waste

Global agricultural waste production is substantial, and crop leftover management is imperative [37]. In addition, waste disposal pollution necessitates research into eco-friendly methods for managing agricultural wastes as the increase in agricultural waste exacerbates aesthetic, health, and environmental issues. Consequently, research into secure disposal methods is necessary. Composting has evolved into an eco-friendly, cost-effective, and secure treatment technology; it is a productive method for intensifying and preserving agricultural products [38]. Biodegradable wastes, e.g., wood shavings, pine needles, dry leaves, sawdust, and coir pith, are commingled to maintain appropriate and durable humus [39]. However, lignin-rich plant products are difficult to decompose. Lime is used to accelerate the breakdown process in the garbage. These components are mixed at a ratio of 5 kg (lime) per 1000 kg (plant materials) to produce high-quality compost. Lime mixed with water may result in the formation of a semi-solid substance or a dry powder. Lime boosts humification of plant wastes by decreasing lignin structure and improving humus content [40]. Likewise, usable compost substrates can be generated from various crop leftovers using a suitable process and quality control procedures (Table 1).

- Municipal solid waste (MSW)

Increasing population, industrialization, and urbanization has elevated the levels of MSW, which has become a problematic responsibility in Pakistan and worldwide [41].

The most well-known biodegradable waste procedures are microbiological stabilization and composting [42]. Due to the high organic content of MSW, composting is theoretically one of the most suitable AWM technologies for MSW management [43–45]. In addition, it generates a soil layer known as a conditioner with agronomic benefits, and is an economically viable and valuable method for offsetting the organic part of the trash. It also reduces the disposed waste, remarkably decreasing the residual waste's pollution capacity and volume for landfilling. As a result, numerous developing Asian countries are turning to compost to manage their MSW. Picking, contaminant separation, sizing and mixing, biological decomposition, and other functions are all part of the modern MSW management composting system. Figure 2 shows the schematic flow diagram of the distinct method of MSW management from source to utmost disposal. To weigh Pakistan's Lahore compost waste intake, a weighbridge having a capacity of 75 tons is located at the Mahmood Booti open dumping site operated by the City District Government of Lahore [46]. Composting is primarily a small-scale industry in Bangladesh and the Maldives. MSW composting in Indore and a large-scale aerobic device in Mumbai were installed in India in 1994 to control 500 metric tons of MSW [44]. These are the two examples of operational large-scale composting ingenuities in India [47]. By 2008, composting had been used to treat 9% of India's MSW [44]. The average cost ranged from $25 to $30 per ton, while the market value per metric ton ranged from $33.5 to $42. India intends to add other plants in near future [48].

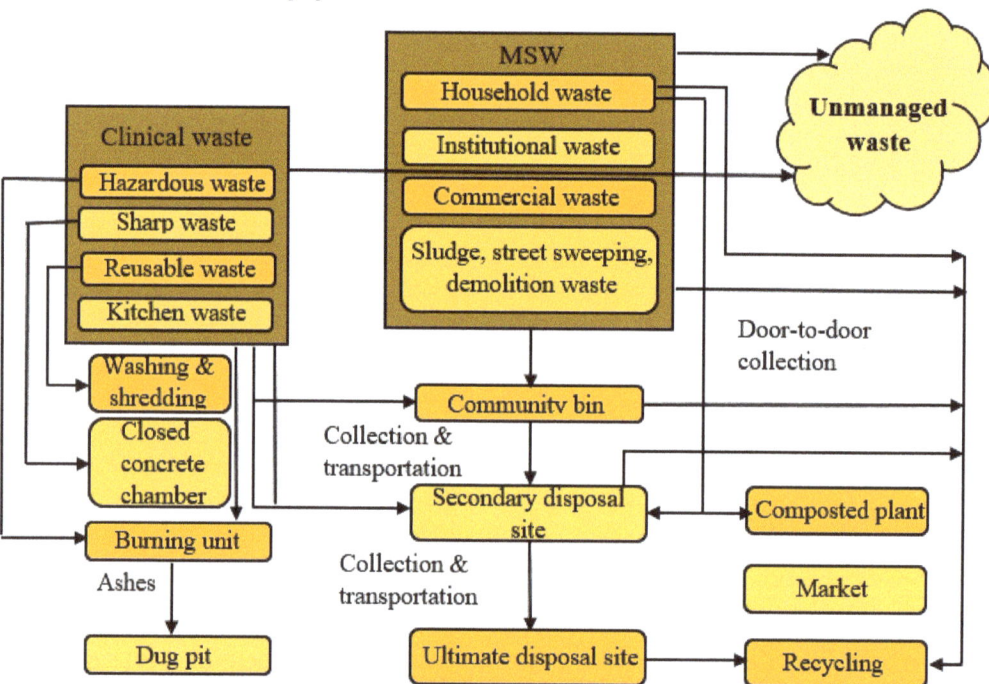

Figure 2. Schematic flow diagram of MSW management by composting.

- Biomedical waste (BMW)

The waste produced through the diagnosis, immunization, and treatment of human beings, research practices, and animal is organic BMW. In Pakistan, hospitals make approximately 2.07 kg of BMW per bed per day [49]. If BMW is not managed properly, it may cause serious environmental and health issues [50]. Therefore, safe disposal techniques need to be investigated, and composting is a sustainable option. Neem and tobacco extracts are commercially cost-effective for local small farmers and provide the best degradation of organic BMW. Thus, these extracts can be employed for conversion of organic BMW into potential fertilizer [51]. Previous research revealed that the BMW must be similarly treated with 5% sodium hypochlorite (NaOCl) at the disposal location [52]. It can be exposed to an initial decomposition process by mixing it with cow dung slurry, and then VC can be utilized to treat it further. Several epigeal species of worms may be used for this purpose. By using this approach to handle BMWs, these worms are more effective in decomposition. VC and proper handling of BMW can be energy-efficient and sustainable methods of eliminating and recycling this hazardous waste [52]. Meanwhile, the composting processes of various wastes come in discrete modes. The most utilized techniques are conventional composting, i.e., AC, AnC, and VC, and emerging composting, e.g., two-stage composting, as described below.

Table 1. Treatment methodologies of different types of crop residues.

Waste	Physicochemical Characteristics	Method	Quality Control	Final Products and Uses	Outcomes	References
Barley waste	Composting in an open-air pile that was rotated 7 times in 105 d. Average temperatures of 65–68 °C with relative humidity of 45–65%.	Maximum temperatures of 65–68 °C with humidity of 45–65%.	Composting	Fertilizer	Micronutrient absorption favored at lowest doses. Doses >10 mg/L inhibited it and depressed growth at highest levels.	[53]
Barley straw waste	Conductance (compost to water, w/w: 1:3), pH (in water and 0.01 M CaCl) Quality of dry matter (% fw, 105 °C) Ash content (% dw, 480 °C/16 h) in triplicate.	Heterotrophic mesophilic bacteria.	Composting	Composting of cow and swine waste with barley straw.	1—C/N ratio declined from 22.6–28.5 to 12.7 during composting. 2—Approximately 11–27% and 13–23% of total C and N were lost after 7 d of intensive composting and 62–66% and 23–37% for whole composting, respectively.	[54]
Barley waste	Final compost pH was 8.7 and C/N ratio was 13. No. of seeds germinated in co-compost depending on grains used.	Total OM was estimated by weight loss on ignition at 540 °C/16 h, and moisture on drying at 105 °C/24 h.	1—Composting 2—VverC	OM composition was high in barley wastes and solid poultry manure.	OM content of barley waste was high (86.3% dw) and had N deficiency.	[55]
Wheat straw waste	Compost contributed 10% of its total N for plant growth during growing season.	During growing season, compost supplied 10% of available N to plants.	1—Mature composting 2—Immature composting	Additional fertilizer	1—At 126.5 h, total H yield of 68.1 mL H/g TVS was 136-fold higher than raw wheat straw wastes. 2—Substrate pretreatment was essential in turning wheat straw wastes into biohydrogen by composts producing hydrogen.	[56]
Rice straw	Lowest C/N ratios found (17–24). Pathogenic micro-organisms were extracted from rice straw by heating at 62 °C/48 h.	Micro-organisms respiration behavior was determined on separate initial C/N (17, 24, and 40) raw materials.	Composting	Development of paper, building materials, soil incorporation, manure, energy supply, and animal feed.	Rice straw residues was rich in OM (80%), oxidizable organic C (34%), and C/N ratio (very volatile and average of 50), suggesting a potential C supply for micro-organisms that can tolerate composting conditions.	[57]
Wheat straw waste	Overall C and N of materials was estimated. Wheat straw has C/N ratio of 100 and cover-grass hay has C/N ratio of 15.	Weight loss of compost samples oven-dried at 80 °C/24 h to assess water content.	1—C1- Automatic NC analyzer connected to isotope mass spectrometer measured total N and C. 2—NH_4 and NO analysis—Traditional calorimetric approaches of flow-injection analysis.	Fertilizer	1—pH ranged 7.6–8.9, with highest values after 3–4 weeks. 2—Weight loss after weeks of composting reduced by 44–45% of original weight. 3—After 7 1/2 weeks, weight loss was 61–63% of actual weight. 4—4% N rose from 2.8 to 4.6%.	[58]

Table 1. Cont.

Waste	Physicochemical Characteristics	Method	Quality Control	Final Products and Uses	Outcomes	References
Wheat straw waste	pH = 6.9. Negligible $CaCO_3$ content. Organic C content of 11.0 g C/kg dry soil	Three types of UWC were applied 1—Bio-waste compost (BIO) from green waste and source-separated organic fraction. 2—Co-compost from mixture of 70% green waste and 30% sewage sludge. 3—Municipal solid waste compost.	1—CERES model 2—Parameter modelling	Soil conditioner or fertilizer	1—Simulated N fluxes indicated that organic amendments resulted in additional leaching of up to 8 kg N/ha/year. 2—After many years, composts mineralized 3–8% of their original organic N content. Composts with slower N release delivered more N to crops. 3—CERES used to help choose best time to apply compost.	[59]
Rice flakes	pH = 7	Aspergillus spp.	Composting	Edible products	1—As opposed to inorganic N, organic N contributed to higher enzyme production. 2—Optimum enzymatic activity was observed at 55 °C/pH 5. 3—Presence of Ca increased enzyme activity, while EDTA presence had opposite effect.	[60]
Rice straw	Temp., air circulation, moisture, and nutrients should all be appropriately managed. Initial optimal composting ratio of C/N was 25–30.	Psychrophilic and mesophilic micro-organisms.	AnC	Combination of swine manure and rice straw as fertilizer.	1—Organic compound biodegradation caused temperature increase to 40–50 °C. 2—pH in all composts were constant and steady.	[61]
Rice straw	Gravimetric approach to assess moisture content. In-house approach was used to evaluate P and K amounts.	Composting in shaded environment on premium Agro products premises. Two therapies: compost piles with EM (C1) and without EM (C2).	Composting	Final compost in matured stage range could be used without limitation.	Compost treated with EM produces more N, P, and K ($p\ 0.05$) than compost without EM treatment.	[62]
Rice straw	Individually homogenized substrates and inoculum were deposited at 4 °C for further use.	Effect of characteristics on bio gasification was calculated using Box-Behnken experimental design combined with response surface methodology.	AnC	Research contributes to understanding of intertwined symptoms and microbial activity of Alzheimer's disease.	Bio-gasification of SS-AD of composting RS had significant interactive impact on temperature, ISC, and C/N ratio. Highest biogas output achieved at 35.6 °C with 20% ISC and 29.6:1 C/N ratio	[63]

Table 2. Physiographical properties of organic feedstock materials or different wastes.

Properties	Total Organic C (g/kg)	Total N (g/kg)	C/N ratio	pH	Total P (g/kg)	Total K (g/kg)	Reference
Household waste	368	21.7	17	4.9			[64]
Manure	330	22	15	9.4	3.9	23.2	[65]
Wood chips	394	14.3	28	7.4	3.5		[66]
Sawdust	490	1.1	446	5.2	0.1	0.4	[65]
Canola	457	1.9	24	6.3	1.1	-	[67]
Rice	412	8.7	47	6.8	1.1	-	[67]
Soybean	440	23.8	18	6.3	0.9	-	[67]
Pea	436	35.0	12	6.3	4.6	-	[67]
Rice straw	39.20 [1]	0.64 [1]	61.3	7.6	0.21 [1]	1.12 [1]	[62]
Rape straw		6.52	59.8	7.11	0.99	31.64	[68]
Wheat chaff		5.24	73.8	6.93	0.62	19	[68]
Maize chaff		9.41	46.5	7.03	0.93	22.93	[68]
Rice chaff		8.51	49.1	7.82	0.88	25.31	[68]
Wheat straw biochar	-	1.38 [1]	38	7.03	0.45 [1]	1.06 [1]	[69]

[1] Values in percentage. Total N = Total concentration of N. Total P = Total concentration of P. Total k = Total concentration of K.

3. Conventional Composting

In the recent few decades, conventional composting processes, i.e., VC, AC, and AnC, have become commonly used globally. Many investigations illustrated that AWM utilization in the field in the form of composting enhanced the soil texture, and structure and has many other beneficial impacts on the field. Researchers focused on improving the composting structure, providing bioavailable components, enhancing the product consistency, and the economic and environmental effects due to the advancement of approaches and green development. This section outlines the overall features of traditional composting and the resulting compost consistency.

3.1. Vermicomposting (VC)

VC is defined as utilizing organic waste from several earthworm species [70–72], and it occurs in a bin/tub. A bin is prepared with a perforated bottom made of adjacent layers of 0.5 mm and 1 cm sieve sizes of nylon and aluminum to facilitate compost tea infiltration. For instance, cow manure was placed at the bottom with worms, including Eudrillus Eugineae/Eisenia Fetida, on top, and shredded kitchen waste was placed over the worms. One bucket of water was added daily for the survival and multiplication of worms. Water, when it leaches down, can be used as compost tea. Compost tea is a liquid fertilizer enriched in nutrients that can be applied for plant growth enhancement. The VC procedures are illustrated in the schematic diagram in Figure 3.

Several different species of worms have been used with different combinations of organic materials (waste) with the purpose of their degradation and conversion into a value-added product. Table 3 shows discrete composting parameters utilizing VC. Earthworms can decay various types of OM, including sewage sludge [73–75], cattle farm waste [76,77], poultry waste [78,79], bagasse [80], industrial waste [81–83], and residential waste [84]. Sludge is the most widely studied for VC, followed by household waste (Table 3). Worms, including Eisenia Fetida, Eudrillus eugineae, Perionyx Sansibaricus, Pontoscolex Corethrurus, Megascolex Chinensis, and Lampito Mauritii are quite effective for VC.

Comparison of compost production from organic waste between different earthworm species, including Eisenia Fetida vs. Lampito Mauritii and Eisenia Fetida vs. Eudrillus eugineae, is shown in Table 4. This study was conducted in a semi-arid climate in Jodhpur, India. The amount of N, P, and K increased while the amount of C/N and C/P decreased as VC preceded. The ideal temperature, moisture content, and pH of Eisenia Fetida were 25 °C, 75%, and 6.5, respectively, for optimum growth, while those of Lampito Mauritii were 30 °C, 60%, and 7.5, respectively. For optimum growth of the earthworm species, ideal temperature, moisture content, and pH were 25 °C, 75%, 6.5 for Eisena Fetida and 30 °C, 60%, and 7.5 for Lampito Mauritti, respectively. The results showed that Eisenia Fetida produced nutrient-rich compost more effectively and efficiently than Lampito Mauritii [85]. Performance evaluation of Eisena Fetida was accessed for six different poultry waste combinations, cow dung, and food industry waste in the semi-arid climate of Hirsa, India, and the results showed an increase in N, P, and K and a decrease in the C/N ratio. Eisena Fetida performed best when cow dung was mixed with poultry waste and food industry waste in a ratio of 2:1:1 compared to cow dung alone [86]. In another study in Kolkata, India, a combination of Eisenia Fetida with micro-organisms, including N-fixing, K-fixing, and P-solubilizing bacteria, was utilized for compost formation using sawdust, paddy straw, and water hyacinth as compost feedstocks. The results showed that not only the time for compost production was reduced, but the percentage of nutrients was also increased in the final product (Table 4). Paddy straw and water hyacinth provided better results than sawdust in compost formation [87]. Comparative studies between Eisenia Fetida vs. Eudrillus eugineae in compost production revealed that in 100 gm compost, 250 worms of the local species, Eisenia Fetida, or Eudrillus eugineae yielded 7, 11, and 17 cocoons and 460, 227, and 540 juveniles per 100 gm, respectively. Around a 40-fold increase in Eudrillus eugineae was achieved, while there was only a 10-fold increase in the local earthworms. Eudrillus eugineae produced compost within 40 days, while local species took 50 days to prepare the final compost [88,89]. Combinations of earthworms with micro-organisms (N-fixing, K-fixing, and P-solubilizing bacteria) minimized the time duration for composting and the finished compost was more highly enriched in nutrients (N, P, K, Ca, Mg, Zn) than conventional compost. As reported, adding P-solubilizing, N-fixing, and K-fixing bacteria increased the amount of N, P, and K in the final compost [90].

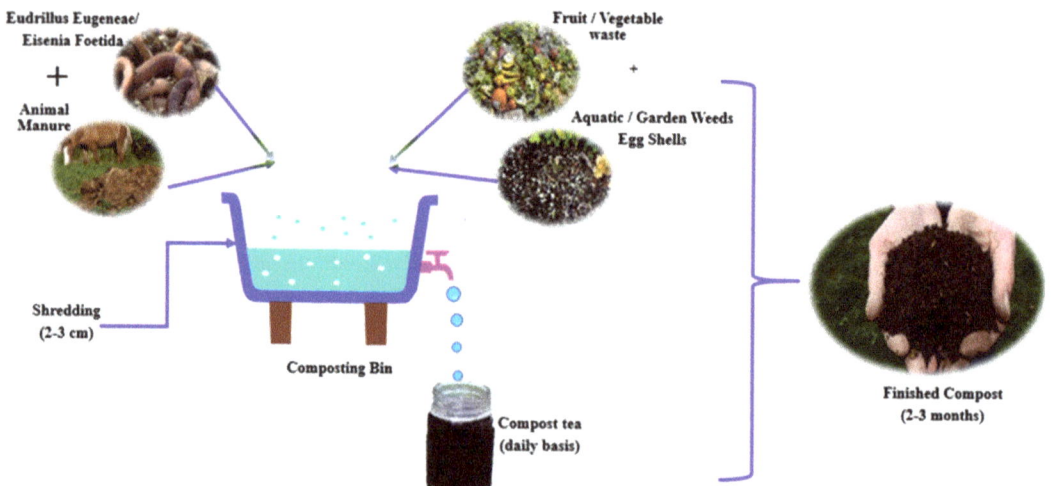

Figure 3. Schematic of vermicomposting production process.

In Kerala, India, banana, cassava, and cowpea composting materials were inoculated with N-fixing, K-fixing, and P-solubilizing bacteria. The results showed that Eudrillus eugineae performed better than other earthworms in synthesizing the nutrient-enriched compost. Using prepared Vermicompost as fertilizer provided the best results in accelerating root growth by increasing nutrient uptake and enhancing total yield [89]. Peppermill sludge, solid pulp, and cow dung were fed to Eudrillus eugeneae to reduce pollution and convert waste into value-added products. The earthworms survived and resulted in enhanced N and P content and a reduction in the C and N ratio, demonstrating the efficiency and effectiveness of Eudrillus eugeneae [91]. Four different feedstocks, namely seaweed, sugarcane trash, coir pith, and vegetable waste, were used for composting with Eudrillus eugineae. Composting for 50 days revealed that different parameters, including pH, OMC,

TOC, N, P, K, cellulose, and lignin, were decreased compared with the C/N ratio. An upsurge in the nutrients in vermicompost showed its development. It was concluded that if cow dung was added to a mixture of materials, it would enrich the nutrients in the final compost. The reproduction and growth rate of Eudrillus eugineae increased as the amount of C/N ratio increased [88]. Eudrillus eugineae was the best in forming nutrient-enriched products in less time and at a greater reproductivity rate than Eisenia fetida and local worms. Different combinations of worms, bacteria, and organic waste generated higher quality vermicompost when utilized together.

Table 3. Discrete composting parameters utilized in the vermicomposting process.

Type of Waste	Factor	Range	References
Sludge from Tannery and cattle dung	C/N ratio	19.00	[92]
Cattle dung and tannery sludge	pH	9.02	[92]
Newspaper and sawdust	pH	7.23	[93]
Distillery industry sludge	pH	6.70	[94]
Distillery industry sludge	C/N ratio	19.50	[94]
Household waste	pH	7.43	[94]
Household waste	C/N ratio	9.89	[94]
Sludge from WWT plants	EC (mS/cm)	1.81	[74]
Wastewater treatment plant's sludge	pH	6.9	[74]
Mixed (farmyard manure, agriculture, and MSW)	C/N	18.6	[95]

Table 4. Chemical properties of different worms in vermicomposting.

Worms	Composting Materials	N	P	K	C/N	C/P	Reference
Eisenia Fetida Lampito Mauritii	Sawdust Straw Biogas slurry Cow waste	3.32-fold increase	1.61-fold increase	1.13-fold increase	2.79-fold decrease	1.35-fold decrease	[85–87]
Eisenia Fetida	Kitchen waste Cow waste Poultry waste Food waste	1.6–3.6-fold increase	33.7%–54% increase	39.5%–50% increase	10.7–12.7 decrease	N/A	[85–87]
Eisenia Fetida Trichoderma viride (M) Bacillus polymixa (M) Azotobacter Bacillus firmus (M) chrococcum (M)	Water hyacinth Paddy straw Sawdust Food waste	52–72% increase	34–80% increase	45–80% increase	lowest from initial	N/A	[87,89]
Eisenia Fetida Eudrillus eugineae (B) Perionyx sansibaricus Pontoscolex corethrurus Megascolex chinensis	Banana Cassava Cowpea	62% increase	20% increase	38% increase	11 points	N/A	[87,89]
Eudrillus eugineae	Solid pulp Paper sludge Cow dung	63.31% increase	2–11-fold increase	N/A	9.6 points	N/A	[89,91]
Eudrillus eugineae	Seaweed Sugarcane trash Coir pith Vegetable waste	63.75% increase	31.58% increase	42.55% increase	23.91 for seaweed	46.04 for seaweed	[88]

3.2. Aerobic Composting (AC)

AC is the degradation of OM with micro-organisms by utilizing oxygen and it takes place in the open atmosphere as a pile or pit [96–99]. For instance, green and brown materials are shredded by a chopper and to a size of 2–3 cm or smaller to help in rapid decomposition. The shredded material is then arranged in a pile/windrow with a specific moisture content. Frequent turnings are employed with sufficient moisture for proper mixing and provide aeration to ensure micro-organisms' survival. The micro-organisms multiply in organic material with sufficient water and air and decompose organic material. After seven to eight turnings, the material becomes fine and changes its color to dark brown (depending upon the material used for composting) with reduced odor. Now the compost is ready to use as organic fertilizer. The schematic flow diagram of this process is shown in Figure 4.

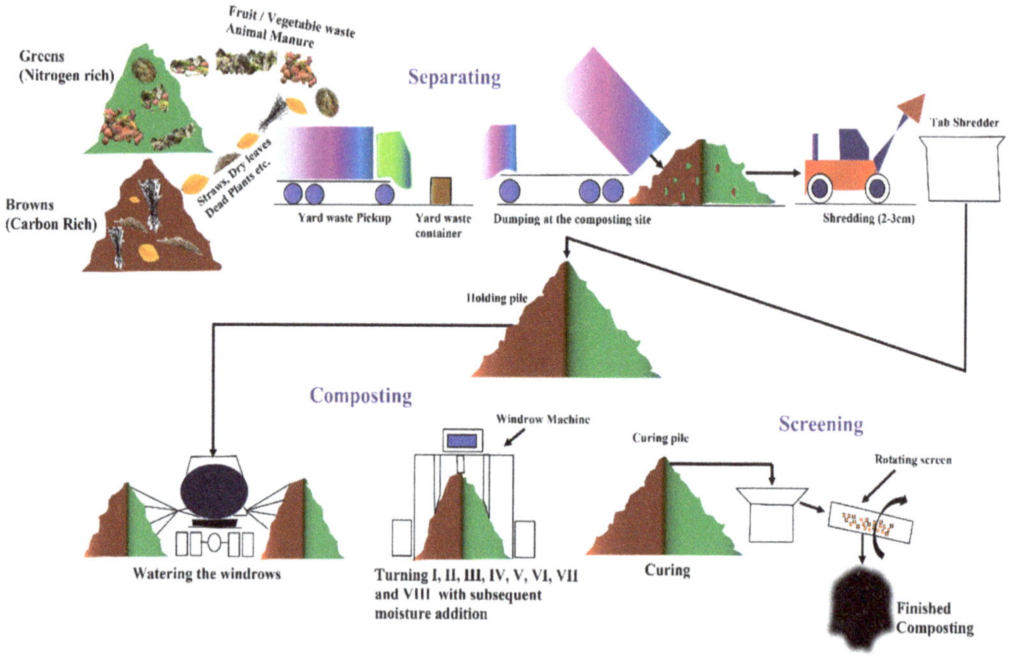

Figure 4. Schematic flow diagram of AC.

A couple of windrows are developed, aerated and turned with an air pump, and mechanically turned by a tractor installed with a bucket loader. Compost prepared with both methods has the same characteristics and organization (60 days). Both composting processes have the same temporal changes in temperature, biological, physical, and chemical parameters, as exhibited in Table 5. However, thermophilic micro-organisms eliminated the harmful bacterium fecal coliform due to increasing temperature [100]. A mixture of poultry waste, feed waste, wood chips, and feathers was used in compost formation under aerated piles, and the effect of the produced compost on soil and crop production was assessed. The results showed that several changes occurred during compost formation, e.g., temperature rise due to mesophilic and thermophilic micro-organisms and a change in OM, N, P, and K levels even in piles that were aerated but not turned. Composted poultry litter had significantly more OM than un-composted poultry litter. Thus, with the availability of OM, crop fields are less susceptible to loss [101]. Another study was performed to determine the suitability of aerated and turned piles using olive husks as the compost material. The outcome showed that both piles reached their maturity stage simultaneously, while the thermophilic phase of turned piles was achieved earlier and had slightly higher OM than the aerated pile (94% versus 84%). The variations in chemical and biological parameters were negligible in both piles, as shown in Table 5. For large-scale applications, the mechanical turning method is best to convert waste into a valuable resource as higher temperatures are achieved through mechanical turning [102].

A comparative study was carried out between aerated and turned windrows to evaluate their effectiveness, using olive mill waste mixed with grape stalks and sheep litter as composting materials. Both methods evaluated efficiency based on pH, temperature, OM, and total N. The results showed that several drawbacks were associated with the aerated composting process due to the physical properties of olive husks. The prepared compost from both methods had similar characteristics, while the thermophilic duration of the turned compost lasted longer and had higher humification than the aerated compost [103]. Different studies were conducted utilizing chicken manure, wheat straw, and bamboo biochar (Table 5). The physicochemical processes, biological parameters, and gas emissions were periodically measured to assess compost quality. Adding biochar enhanced porosity and stabilized the composting rate, accelerating the process and improving the finished compost's quality. Biochar improves looseness, provides better material degradation, and reduces GHG emissions (CO_2, N_2O, NH_3, CH_4) [104]. The composting experiment was performed in two bins with different composting materials, and the effects of low C/N ratio on the final product, including several parameters are shown in Table 6. Less straw and more swine manure were added into bin one, which had a low thermophilic duration and took longer to mature than bin two which had more straw and less swine manure. It was recommended that 172 kg of straw could be treated with one ton of swine manure. A low C/N ratio was recommended for composting rice straw with swine manure [61]. In-vessel composting offers fewer complications than windrow- or pile-composting due to reduced bioaerosols, and better AWM and control over leachate.

The composting material consisted of three different types of waste: green waste, paper waste, and bio-solids in bins 1, 2, and 3, respectively. Better results were attained from bin 3 with maximum temperature achieved and a more humified final product (Table 6). The active compost was carried out in bins, whereas the compost was taken out and matured by successive turnings to complete the maturity phase. However, precautions must be taken during the maturity phase to reduce pollution and effective AWM [105]. Several sleeves were used to prepare the compost with an equal ratio of green waste and sewage sludge. The concentration of oxygen was maintained through a perforated pipe, which was inserted at the bottom of the sleeve. Controlled moisture content and thermophilic temperature were maintained at >45 °C throughout the composting. Harmful bacteria, including Fecal coliform and E. coli, dominated initially but were subsequentially reduced, and the traces of these micro-organisms in the final product were negligible. The active phase of composting was performed in the sleeves while the maturity phase was carried out in the open. The final product was non-toxic and used as a beneficial soil additive [106].

Composting in sleeves results in reduced odor and attracted fewer insects with better leachate management compared to open windrows/piles. The composting materials included green waste and olive mill wastewater, in which the green waste remained soaked for one night. The oxygen level was maintained by the addition of a perforated PVC pipe through which air was injected into the sleeve. The temperature throughout the entire sleeve was maintained at >45 °C (Table 6). The final compost had no toxicity and basil and ornamental plant growth were tested using prepared compost. In this way, wastewater was beneficially utilized and converted into a valuable resource [107]. In another study, the performance of a closed bioreactor (In-vessel) was evaluated in terms of various physicochemical parameters, including C/N ratio, NH_3-N, pH, moisture, N content, etc. Mixtures of different food wastes were collected from several locations and placed in a bioreactor with the recommended initial standards. The final compost was ready to use within 12 days. During composting, temperature, CO_2 levels, and pressure rose due to microbial activity, resulting in satisfactory final compost that was acceptable for agricultural applications. Nitrates negatively correlated with CO_2, EC, and ammonium levels, while phosphate positively correlated with ammonium, EC, and CO_2 levels [108].

AC has already been tested under different aeration methods using cotton gin waste [109], straw and sheep manure [110], poultry litter [111], and sawdust [100], and literature reveals that various methods have been used for AC in the past. Windrow and forced-air composting are of more significant concern as both yield similar results. Energy and cost consumption in forced air composting is higher than in windrow composting. Perforated PVC pipes were laid under waste for air circulation through air pumps to aid in waste degradation. The temperature achieved by thermophilic and mesophilic organisms was also lower in forced air composting than in windrow composting. Higher temperatures are necessary for the elimination of harmful pathogens from waste. Some studies reveal that higher temperatures can be achieved in windrow composting but controlling this temperature is complicated by the frequent turnings. Manual turning involves intensive labor application as compared to mechanical turning. On the other hand, mechanical turning saves time and reduces cost as compared to manual turning. The degree of humification in mechanically or manually turned windrows (compost) is far greater than in forced aerated (air pumping) compost. Because of rapid evaporation, the moisture content requirement of forced aerated compost is much higher than turned composting. The amount of ideal OM and particle size of compost is higher in manually or mechanically turned end-stage compost than in forced aerated compost because no turning is carried out in aerated compost. According to studies on mechanically or manually turned AC compost, it is superior in all aspects compared to forced aerated composting.

Table 5. Chemical properties of different materials in various aeration methods.

Aeration Method	Materials	Temperature Achieved (°C)	pH	Total N	Total P	Total K	C/N Ratio	Reference
Turning	Pig waste Sawdust	67	6.7	18–27 g/kg	N/A	N/A	N/A	[100]
Air pump	Pig waste Saw dust	60	6.8	18–27 g/kg	N/A	N/A	N/A	
Air blower	Poultry manure Wood shaving Waste feed Feathers	63	7.0	16.31 g/kg	15.57 g/kg	19.78 g/kg	15	[101]
Mechanical turning	Olive oil husk Grape stalks	65	7.3	0.95–1.17%	N/A	N/A	46	[102]
Centrifugal ventilator	Olive oil husk Grape stalks	54	7.1	1.08–1.27%	N/A	N/A	46	
Forced aeration	Waste from olive mill Sheep litter Grape stalks	63	9.5	1.94%	0.9 g/kg 2.5 g/kg 1.3 g/kg	2.4 g/kg 2.7 g/kg 2.8 g/kg	15.5	[103]
Windrow turning	Waste from olive mill Sheep litter Grape stalks	68	9.2	1.89%	0.9 g/kg 2.5 g/kg 1.3 g/kg	2.4 g/kg 2.7 g/kg 2.8 g/kg	15	
Centrifugal ventilator	Poultry manure Bamboo biochar Wheat straw	55	9.0	57% at the final stage			10.3	[104]

Table 6. Physicochemical characteristics of different materials in vessels.

Vessel	Materials	Temperature Achieved (°C)	Total N	C/N Ratio (%)	pH	Moisture Content (%)	Reference
Bin-1	Swine waste Rice straw	60	19.30 g/kg	5.15 decrease	8.01	45 to 65	[61]
Bin-2	Swine waste Rice straw	60	18.62 g/kg	4.57 decrease	8.03	45 to 65	
Bin-1	Green waste	64	200 mg/kg	20 ± 1	8.5	81 ± 33	
Bin-2	Green waste Paper waste	70	120 mg/kg	25 ± 1	8.5	72 ± 2	[105]
Bin-3	Green waste Biosolids	72	700 mg/kg	27 ± 3	8.9	44 ± 11	
Sleeve-1	Green waste Sewage sludge	>45	44.9% loss	10.9 decrease	6.5 at sleeve opening	54.4	[106]
Sleeve-2	Green waste Sewage sludge	>45	42.9% loss	11.8 decrease	7.2 at sleeve opening	38	
Sleeve	Olive mill waste Green waste	55	1.05%	21.5	8.2 at sleeve opening	55	[107]
Vessel	Mixed food waste	53	250 mg/kg	11	6.94	72	[108]

3.3. Anaerobic Composting (AnC)

AnC degrades OM in the presence of micro-organisms without oxygen utilization [112–114]. AnC occurs in two steps. For instance, cow dung was fed daily in a digester for biogas generation to utilized in the first step. A byproduct of the digester was slurry, referred to as digested, from which all GHGs were eliminated. The product was utilized effectively for composting and reduced environmental pollution. The slurry was mixed with shredded browns enough moisture to carry out further decomposition. The period of AnC is comparatively more extended than AC, and the schematic diagram is shown in Figure 5.

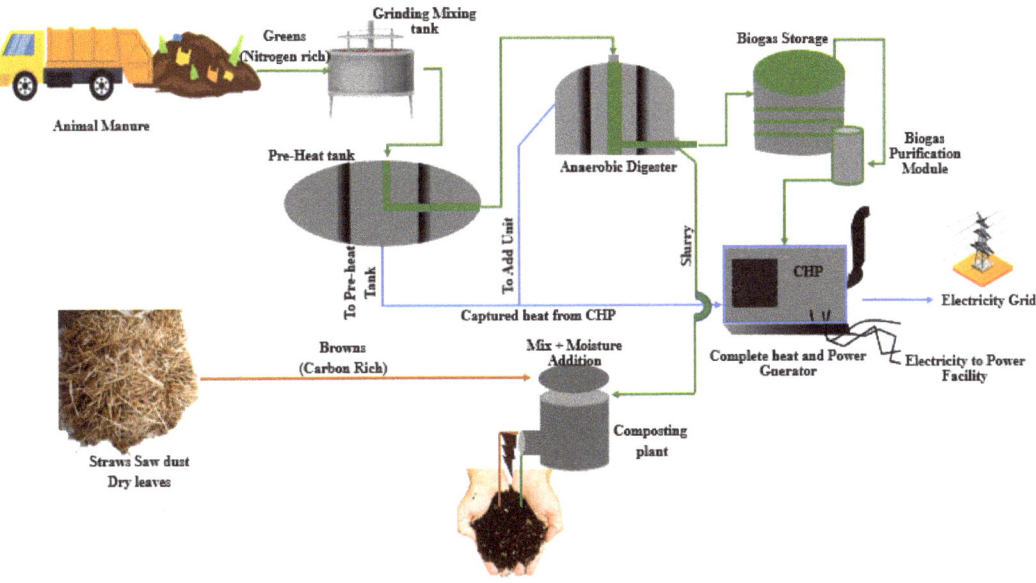

Figure 5. Schematic diagram of AC.

In the past, AnC has been used for the degradation of kitchen waste, fly ash and crop deposits [115], garden and animal waste [116], municipal waste [117], sawdust and pig manure [118], and sewage sludge [119]. When AnC was utilized for the above-mentioned feedstocks, it reduced GHG emissions since it occurs in an oxygen-free environment [120]. GHGs are regarded as harmful gases contributing to global warming, eutrophication, and acidification if high amount enter the atmosphere [121]. Methane produced by anaerobic digestion can be exploited as an energy source for either electricity production or combustion. Anaerobic digestion prevents environmental pollution as the generated methane is used and thus gets removed by burning. AnC requires a high amount of moisture and N-enriched material (animal manure, food waste, and sewage sludge) for the successful completion and generation of beneficial end-user products. Due to the high moisture at the end of AnC, compost tea is produced that can be used as liquid fertilizer that is enriched in nutrients (N, P, and K). In addition, the emission of volatile compounds (terpenes, ethers, and esters) during active composting periods in AnC is negligible [122]. The preparation of the final product from anaerobic compost requires a longer time. As a drawback, the final product of anaerobic digestion contains E. coli and Salmonella, which are hazardous to human health. Another negative impact of anaerobic digestion is the production of odors.

4. Emerging Composting

Two-stage composting is a technology that incorporates two diverse methods into a single composting method to improve the finished product quality, process speed, and environmental impact of traditional composting. It is a novel approach to bio-fertilizer use. Various two-stage techniques have been considered, including combining two composting technologies, e.g., VC and conventional composting. However, in this paper, AnC followed by AC is reviewed. AnC is termed primary composting (PC) in the two-stage composting process, and the AC process is termed secondary composting (SC).

AnC followed by AC is a comparatively innovative idea in two-stage composting. As an initial effort, [96] investigated the transformation of OM and the kinetics of sewage sludge composting in two stages using grass and rape straw. The whole procedure required 217 days, with 10 days in the bioreactor for OM oxidation and waste sanitation and 207 days in the windrow for compost maturation. The concept of two-stage composting is recent and research on its economic and environmental effects is minimal. Most of the studies are aimed at increasing process reliability and product consistency. Table 7 summarizes recent studies into two-stage composting and the results obtained using various additives. A steady higher temperature of about 70 °C was established in PC during two-stage composting [123]. Within 40 days of switching to SC, the temperature fell from 50 to 30 °C. There will be less N loss from SC at this temperature at the mesophilic stage.

Moreover, since a higher temperature was observed during PC, N loss and GHG emission in the bioreactor can be minimized or regulated within a minimal range. According to [124], more thermophilic phases were noted, with one occurring during the PC and two to four occurring during the SC. Most of the thermophilic steps that followed were over 55 °C. Bamboo vinegar was added to the compost throughout SC to decrease the hazard of N degradation.

The thermophilic temperature range was between 55 and 70 °C in all the studies mentioned in Table 7. The high temperatures completed the maturation of the pile, ensuring the compost's safe use as a bio-fertilizer. Similarly, various feedstocks, e.g., pig manure, poultry manure, and other supplemental waste substrates, can be used in co-composting operations [125,126]. When AC is replaced with two-stage composting, there is a reduction in processing time. The addition of a bulking agent, particle size reduction, and aeration rate change are all essential considerations in determining the process performance and the finished product's consistency [127]. On the other hand, AnC followed by AC will minimize the area, labor, and time required for AC, along with the capital cost and power depletion. Two-stage composting also reduces GHG pollution and waste conveyance costs if PC is controlled at waste occurrence locations to decrease the total of waste before transport to the location for SC. Two-stage composting may be a novel way to manage organic waste at home or market. The organic waste can be collected and composted in a digester near the city. The incompletely composted waste from several cities could be transported to AC sites for further treatment. AC alone has adverse effects on the environment and results in GHGs, including ammonia, methane, and nitrous oxide, which may cause ozone depletion and global warming. On the other hand, AnC eliminates harmful GHG emissions into the atmosphere and it is a great energy resource. This is consistent with the literature about two-stage composting (AnC followed by AC) because AnC cannot be utilized directly without further treatment. It is rich in ammonia content which could burn crops. AC overcomes this hazard and converts the byproducts into useful resources. Two-stage composting yields better results than aerobic, anaerobic, and VC in terms of humic substances, OM, energy generation (heat and electricity), environmental protection, and nutrient-enriched end products.

Table 7. Chemical properties of different materials.

Materials	Final N	Temperature (°C)	CH$_4$	C/N Ratio (%)	Final P	Final K	pH	EC	Reference
Rice straw	0.78	35.6	346 mL g VS^{-1}	29.6	N/A	N/A	N/A	N/A	[63]
Dairy manure Corn stover Tomato residue	31.6 kg	N/A	1186 gm	N/A	47.4 kg	279.1 kg	N/A	N/A	[121]
Kitchen waste Garden waste Paper waste	60 g/ton	N/A	100 m^3/ton	N/A	N/A	N/A	8.6	1.8	[122]
Banana Cow dung Poultry waste	2.09%	57	N/A	12.8 12.6 12.9	11.86%	0.39%	>9.0	0.59 0.57 0.65	[123]
Pig manure	1045 mg/kg	N/A	18.6 mL/day	8.7	N/A	N/A	8.5	N/A	[124]
Food waste Inoculum	N/A	37	2.27 m^{3}m^{-3}/day	8.7	N/A	N/A	7.1	N/A	[125]

5. Comparison

Bio-waste combinations, including kitchen, garden, and paper wastes, have been utilized for aerobic and two-stage composting (combined anaerobic/AC). AC was carried out under forced aerated conditions for twelve weeks and two-stage composting in which AnC (PC) continued without aeration for three weeks while AC (SC) continued for the last two weeks. The results showed that AC yielded 742 g/ton of explosive gases, while two-stage composting yielded 236 g/ton and 44 g/ton of volatile gases (esters, terpenes, ethers, compounds). Biogas produced during the anaerobic phase utilized in combustion resulted in 99% removal of combustible gases. Two-stage composting was an attractive method for reducing volatile gas emissions [122]. Another study of prepared of anaerobic compost of the paper industry and urban solid waste was assessed in total ammonia, germination indices, volatile organic acids, and total oxygen uptake. The results showed that the application of prepared anaerobic compost was less effective unless anaerobic composting was followed by AC, which yielded better results [96]. The efficiency of AC and AnC was evaluated using different combinations of banana peel [plain banana peel (B), inoculated banana peel mixed with cow dung (BC), and poultry litter (BP)]. It was suggested that the decomposition rate in AC is faster than AnC at this scale with increased N and K content as follows: BP > BC > B [124]. The effect of AnC prepared from pig waste was assessed in neutralizing high chlorine content in soil due to polychlorinated biphenyl. If the chlorine content was significantly higher than the limit, then di-chlorination would be inhibited. Soil-to-organic waste ratio was 2:3, the C to N ratio was 20. At a moisture content of 60%, di-chlorination was the highest at 1 mg/kg [123].

A study was conducted to investigate the adverse effects [acidification potential (AP), eutrophication potential (EP), and global warming potential (GWP)] of various organic waste treatments (dairy manure, corn stover, and tomato residue). All treatment techniques used anaerobic digestion followed by composting. The results showed that if AnC was used before composting, EP, GWP, and AP were reduced. If AC and composting were used alone, the harmful potential concentration increased in the ecosystem. If the farm was equipped to use anaerobic digestion, then followed with composting would be suitable for all life cycle impact categories [121]. The performance of MUSTAC (Multistep sequential batch two-phase AnC) was evaluated, and the processes involved including hydrolysis, acidification, post-treatment, and methane recovery. This process was utilized for treating inoculated food waste using AnC. MUSTAC and anaerobic digestion were assessed in terms of environmental constraints. MUSTAC yielded the best results in reducing volatile emissions with high methane conversion efficiency attained in a relatively short period (Table 8). The product obtained could be used for soil improvement. MUSTAC has proven to produce value-added products with high efficiency and reliability [124]. A study was performed to assess the suitability of increasing biogas from composting rice straw with the effects of primary temperature, C/N ratio, and substrate on the finished product. Concentrations of lignin, cellulose, and hemicellulose in the rice straw were sufficiently degraded. The initial concentration of the parameters mentioned above significantly affected bio gasification, as mentioned in Table 8. Methano-bacteria, clostridia, and beta proteo-bacteria were the microbial communities included in anaerobic compost, providing valuable information about microbial behavior and the independent effects of anaerobic digestion [63].

Table 8. Recent developments in two-stage composting.

Time (d)	Waste	Composting System	Amendment	Remark Outcome	Reference
PC:1 SC: 207	Dewatered sewage sludge	1—Aerated bioreactor (PC: 1 m^3) 2—Weekly turned Windrow (SC: 0.8 m^3)	Different proportions of rape straw and grass.	1—Feedstock composition affects process succession. 2—Rape straw raised temperature of compost and formation of humic acid.	[125]
PC: 6 SC: 24	Green waste	1—PC: non-covered digester, automated turning, and watering (daily) 2—SC: Windrow, turned and saturated every 30 d.	Brown sugar and calcium superphosphate in various proportions.	Proposed two-stage composting produce higher-quality compost in limited time. Adding 0.5% brown sugar and 6% calcium superphosphate to compost during SC increased consistency.	[126]
PC: 6 SC: 24	Green waste	1—PC: non-covered digester, automated turning, and (daily) 2—SC: windrow, turn, and water every 3 d.	Different proportions of rhamnolipid (RL) and initial compost particle size (IPS).	1—Addition of 0.15% RL and particle size of 15 mm IPS increased aeration and water permeability, resulting in higher micro-organism numbers and enzyme activities, thus speeding up degradation process. 2—Mature compost of greater efficiency accomplished in just 24 d.	[126]
PC: 10 SC: 170	Dewatered sewage sludge	1—Aerated bioreactor (PC: 1 m^3) 2—Weekly turned windrow (SC: 0.8 m^3)	Aeration rate in bioreactor (0.5 and 1.0 L/min kg dm) was changed.	1—A greater aeration rate in bioreactor resulted in OM losses. 2—Compost was safe to use as soil amendment because results exhibited low levels of heavy metals, low possible environmental risk, and suitable sanitary consistency.	[128]

6. Conclusions and Recommendations

This review examines the management of AWM through various composting processes—conventional and emerging composting—and composting stages, the composting of crop residue waste, MSW, and BMW, as well as the underlying mechanisms, and the factors influencing composting. In addition, it compares conventional composting [vermicomposting (VC), aerobic (AC), and anaerobic (AnC)] with new composting techniques (two-stage composting). AW must be treated quickly and effectively for the sustainable growth of agriculture and environmental habitats. There are numerous ways to make valuable products from this massive volume of waste, but some are more cost-effective and/or rational. Composting is the most cost-effective and environment-friendly AWM practice, preferable to landfilling, burning, and open-dumping of agricultural and farm wastes. Composting is crucial for recycling waste into resources, preserving environmental quality, and safeguarding public health. These methods include the recycling of AWM to increase soil fertility and the production of biofertilizers through different processes. The summary of composting phases and critical waste substrates demonstrates that composting is the most effective method for AWM. The literature reveals some conventional and emerging composting processes. In conventional composting, VC humifies in 3 to 4 months which cannot fulfill fertilizer demand. AnC also requires 2 to 3 months to prepare for daily biogas and slurry production. Rapid compost preparation with a fine and higher degree of humification can only be done through AC. In this study, several past studies examine recent developments in organic composting. Several recommendations were made to improve technical development. This critical review also highlights current advancements in composting for AWM, makes recommendations to aid its technological development and acknowledge its benefits, and will boost the scientific community's interest in composting processes.

BSF larvae are best tool for AWM, which can also decompose AW quickly. The increased lignocellulosic component of AWMs limits their decomposition. Comparatively, VC with BSF reduces GHG emissions 47-fold. Future research should focus on discovering the ideal settings for BSF larvae to evolve, flourish, and handle MSW and crop wastes in subtropical regions. BSF can be employed to decompose recalcitrant AWM substrates. Centipedes and pill bugs are sometimes utilized in composting. Composting with these insects would help these plants and environments survive. They could be composted instead of discarded. Insects should be examined to see whether they can aid in macro- or micronutrient enrichment of compost. Compost can release pathogen-killing enzymes. As composts include several nutrients, they avoid providing mono nutrients. Before planting, a soil analysis can reveal nutrient deficiencies. Mono fertilizers extracted from compost fertilizer would reduce nutrient waste.

Farmers can use inoculum that degrades complex biodegradables to speed up the composting process. More study is needed to identify an odor-trapping technique to overcome the compost processing-related air quality issue. Implementing composting and CO_2 capture should reduce GHG emissions. AnC or other composting processes could use an odor-trapping device. Researchers and businesses have long recognized the waste potential as a source of raw materials. The chemical components of waste are of particular concern. In the past decade, a shift has been made from composting organic fractions for crop production to anaerobic digestion, which can produce methane as an energy source. Due to government incentives, European waste firms have changed their investments to anaerobic digestion systems. These government incentives may encourage new composting innovations, e.g., incorporating bioenergy technologies (anaerobic digestion, biochar). Bioenergy byproducts may be composted to maximize their economic, agricultural, and environmental value. To make compost more acceptable, anti-nematodes, viricides, bactericides, and fungicides generated from plants may also be added. By avoiding pesticides, organic farming would continue to be promoted. Slow-decomposing materials can be composted separately from other materials so that the composting period of the latter is not prolonged. There is a need for additional research to discover whether substances that require longer decomposition times tend to mineralize over time. Slow mineralizing minerals that serve as a long-term source of nutrients could be advantageous to biennial and perennial plants; this theory's validity should be studied further. This research could reveal the nutritional benefits of leaves that decompose slowly and will help determine whether they should be composted. Composting rather than burning agricultural waste is garnering more attention in developing nations. Before spreading compost onto the soil, it should be frequently evaluated for maturity and pollutants to prevent introducing potential hazards to the soil and other living things. Finally, additional trials are required to determine how to accelerate the composting process. Even though the two-stage composting process developed in the past continues to be an emerging composting process, the best practices will aid in sustaining the composting process.

Author Contributions: Conceptualization, U.W.H. and S.H.; methodology, S.H., M.W., R.N. and S.A.; formal analysis, M.W.; investigation, A.N., U.W.H., P.T.H. and M.W.; resources, S.H., M.S. and U.W.H.; data curation, M.W., A.N. and S.A.; writing—original draft preparation, S.H., S.A., H.A.L. and M.W.; supervision, U.W.H., M.S. and S.H.; project administration, U.W.H., S.H. and M.S.; funding acquisition, S.H., R.N. and U.W.H.; writing—review and editing, S.A., H.A.L., R.N. and A.N. All authors have read and agreed to the published version of the manuscript.

Funding: This research received no external funding.

Data Availability Statement: Data used to support the study's findings can be obtained from the corresponding author upon request.

Acknowledgments: The authors would like to express their gratitude to The Joint Graduate School of Energy and Environment (JGSEE); King Mongkut's University of Technology Thonburi; the Center of Excellence on Energy Technology and Environment (CEE); the Ministry of Higher Education, Science (MHESI) Research and Innovation and Department of Mathematics; and Muhammad Nawaz Shareef of the University of Agriculture, Multan, Pakistan for their support and technical help. We are thankful to Muhammad Faheem from the Department of Environmental Science and Engineering, School of Environmental Studies, China University of Geosciences for English editing services.

Conflicts of Interest: The authors declare no conflict of interest.

References

1. Maqsood, A.; Abbas, J.; Rehman, G.; Mubeen, R. The paradigm shift for educational system continuance in the advent of COVID-19 pandemic: Mental health challenges and reflections. *Curr. Res. Behav. Sci.* **2021**, *2*, 100011. [CrossRef]
2. Hashim, S.; Waqas, M.; Rudra, R.P.; Khan, A.A.; Mirani, A.A.; Sultan, T.; Ehsan, F.; Abid, M.; Saifullah, M. On-Farm Composting of Agricultural Waste Materials for Sustainable Agriculture in Pakistan. *Scientifica* **2022**, *2022*, 5831832. [CrossRef] [PubMed]
3. Batool, S.A.; Chuadhry, M.N. The impact of municipal solid waste treatment methods on greenhouse gas emissions in Lahore, Pakistan. *Waste Manag.* **2009**, *29*, 63–69. [CrossRef] [PubMed]
4. Munir, A.; Nawaz, S.; Bajwa, M.A. Farm manure improved soil fertility in mungbean-wheat cropping system and rectified the deleterious effects of brackish water. *Pak. J. Agric. Sci.* **2012**, *49*, 511–519.
5. Cogger, C.; Hummel, R.; Hart, J.; Bary, A. Soil and Redosier Dogwood Response to Incorporated and Surface-applied Compost. *HortScience* **2008**, *43*, 2143–2150. [CrossRef]
6. Ventorino, V.; Pascale, A.; Fagnano, M.; Adamo, P.; Faraco, V.; Rocco, C.; Fiorentino, N.; Pepe, O. Soil tillage and compost amendment promote bioremediation and biofertility of polluted area. *J. Clean. Prod.* **2019**, *239*, 118087. [CrossRef]
7. Karak, T.; Bhattacharyya, P.; Paul, R.K.; Das, T.; Saha, S.K. Evaluation of Composts from Agricultural Wastes with Fish Pond Sediment as Bulking Agent to Improve Compost Quality. *CLEAN–Soil Air Water* **2013**, *41*, 711–723. [CrossRef]
8. Monte, M.; Fuente, E.; Blanco, A.; Negro, C. Waste management from pulp and paper production in the European Union. *Waste Manag.* **2009**, *29*, 293–308. [CrossRef]
9. Aruna, G.; Kavitha, B.; Subashini, N.; Indira, S. An observational study on practices of disposal of waste Garbages in Kamakshi Nagar at Nellore. *Int. J. Appl. Res.* **2018**, *4*, 392–394.
10. Külcü, R.; Yaldiz, O. The composting of agricultural wastes and the new parameter for the assessment of the process. *Ecol. Eng.* **2014**, *69*, 220–225. [CrossRef]
11. Nkwachukwu, O.I.; Chima, C.H.; Ikenna, A.O.; Albert, L. Focus on potential environmental issues on plastic world towards a sustainable plastic recycling in developing countries. *Int. J. Ind. Chem.* **2013**, *4*, 34. [CrossRef]
12. Azim, K.; Soudi, B.; Boukhari, S.; Perissol, C.; Roussos, S. Composting parameters and compost quality: A literature review. *Org. Agric.* **2018**, *8*, 141–158. [CrossRef]
13. Qasim, W.; Lee, M.H.; Moon, B.E.; Okyere, F.G.; Khan, F.; Nafees, M.; Kim, H.T. Composting of chicken manure with a mixture of sawdust and wood shavings under forced aeration in a closed reactor system. *Int. J. Recycl. Org. Waste Agric.* **2018**, *7*, 261–267. [CrossRef]
14. Food and Agriculture Organization (FAO). *Composting Process and Techniques*; Martínez, M.M., Pantoja, A., Eds.; Food and Agriculture Organization (FAO): Rome, Italy, 2021.
15. Mehta, C.M.; Sirari, K. Comparative study of aerobic and anaerobic composting for better understanding of organic waste management: A mini review. *Plant Arch.* **2018**, *18*, 44–48.
16. Misra, R.; Roy, R.; Hiraoka, H. *On-Farm Composting Methods*; UN-FAO: Rome, Italy, 2003.
17. Dróżdż, D.; Malińska, K.; Kacprzak, M.; Mrowiec, M.; Szczypiór, A.; Postawa, P.; Stachowiak, T. Potential of fish pond sediments composts as organic fertilizers. *Waste Biomass Valorization* **2020**, *11*, 5151–5163. [CrossRef]
18. Żukowska, G.; Mazurkiewicz, J.; Myszura, M.; Czekała, W. Heat energy and gas emissions during composting of sewage sludge. *Energies* **2019**, *12*, 4782. [CrossRef]
19. Avidov, R.; Saadi, I.; Krassnovsky, A.; Hanan, A.; Medina, S.; Raviv, M.; Chen, Y.; Laor, Y. Composting municipal biosolids in polyethylene sleeves with forced aeration: Process control, air emissions, sanitary and agronomic aspects. *Waste Manag.* **2017**, *67*, 32–42. [CrossRef]
20. Ren, X.; Zeng, G.; Tang, L.; Wang, J.; Wan, J.; Wang, J.; Deng, Y.; Liu, Y.; Peng, B. The potential impact on the biodegradation of organic pollutants from composting technology for soil remediation. *Waste Manag.* **2018**, *72*, 138–149. [CrossRef]
21. Martin, C.C.S.; Brathwaite, R.A. Compost and compost tea: Principles and prospects as substrates and soil-borne disease management strategies in soil-less vegetable production. *Biol. Agric. Hortic.* **2012**, *28*, 1–33. [CrossRef]

22. Kalamdhad, A.S.; Singh, Y.K.; Ali, M.; Khwairakpam, M.; Kazmi, A. Rotary drum composting of vegetable waste and tree leaves. *Bioresour. Technol.* **2009**, *100*, 6442–6450. [CrossRef]
23. Belyaeva, O.; Haynes, R. Chemical, microbial and physical properties of manufactured soils produced by co-composting municipal green waste with coal fly ash. *Bioresour. Technol.* **2009**, *100*, 5203–5209. [CrossRef] [PubMed]
24. Neklyudov, A.D.; Fedotov, G.N.; Ivankin, A.N. Aerobic processing of organic waste into composts. *Appl. Biochem. Microbiol.* **2006**, *42*, 341–353. [CrossRef]
25. Smith, J.L.; Collins, H.P.; Bailey, V.L. The effect of young biochar on soil respiration. *Soil Biol. Biochem.* **2010**, *42*, 2345–2347. [CrossRef]
26. Sayara, T.; Basheer-Salimia, R.; Hawamde, F.; Sánchez, A. Recycling of Organic Wastes through Composting: Process Performance and Compost Application in Agriculture. *Agronomy* **2020**, *10*, 1838. [CrossRef]
27. Hadar, Y.; Papadopoulou, K.K. Suppressive Composts: Microbial Ecology Links Between Abiotic Environments and Healthy Plants. *Annu. Rev. Phytopathol.* **2012**, *50*, 133–153. [CrossRef] [PubMed]
28. Sundberg, C.; Yu, D.; Franke-Whittle, I.; Kauppi, S.; Smårs, S.; Insam, H.; Romantschuk, M.; Jönsson, H. Effects of pH and microbial composition on odour in food waste composting. *Waste Manag.* **2013**, *33*, 204–211. [CrossRef] [PubMed]
29. Trupiano, D.; Cocozza, C.; Baronti, S.; Amendola, C.; Vaccari, F.P.; Lustrato, G.; Di Lonardo, S.; Fantasma, F.; Tognetti, R.; Scippa, G.S. The effects of biochar and its combination with compost on lettuce (*Lactuca sativa* L.) growth, soil properties, and soil microbial activity and abundance. *Int. J. Agron.* **2017**, *2017*, 3158207. [CrossRef]
30. Pane, C.; Palese, A.M.; Celano, G.; Zaccardelli, M. Effects of compost tea treatments on productivity of lettuce and kohlrabi systems under organic cropping management. *Ital. J. Agron.* **2014**, *9*, 153. [CrossRef]
31. Coelho, L.; Osório, J.; Beltrão, J.; Reis, M. Efeito da aplicação de compostos orgânicos no controlo de infestantes na cultura de Stevia rebaudiana e nas propriedades do um solo na região do Mediterrâneo. *Rev. Ciências Agrárias* **2019**, *42*, 111–120.
32. Pane, C.; Spaccini, R.; Piccolo, A.; Celano, G.; Zaccardelli, M. Disease suppressiveness of agricultural greenwaste composts as related to chemical and bio-based properties shaped by different on-farm composting methods. *Biol. Control* **2019**, *137*, 104026. [CrossRef]
33. Uyizeye, O.C.; Thiet, R.K.; Knorr, M.A. Effects of community-accessible biochar and compost on diesel-contaminated soil. *Bioremediat. J.* **2019**, *23*, 107–117. [CrossRef]
34. Pose-Juan, E.; Igual, J.M.; Sánchez-Martín, M.J.; Rodríguez-Cruz, M.S. Influence of Herbicide Triasulfuron on Soil Microbial Community in an Unamended Soil and a Soil Amended with Organic Residues. *Front. Microbiol.* **2017**, *8*, 378. [CrossRef]
35. He, Z.; Yang, X.; Kahn, B.A.; Stoffella, P.J.; Calvert, D.V. Plant nutrition benefits of phosphorus, potassium, calcium, magnesium, and micronutrients from compost utilization. *Compost. Util. Hortic. Crop. Syst.* **2001**, 307–320. [CrossRef]
36. Manirakiza, N.; Şeker, C. Effects of compost and biochar amendments on soil fertility and crop growth in a calcareous soil. *J. Plant Nutr.* **2020**, *43*, 3002–3019. [CrossRef]
37. Clay, D.E.; Alverson, R.; Johnson, J.M.; Karlen, D.L.; Clay, S.; Wang, M.Q.; Bruggeman, S.; Westhoff, S. Crop Residue Management Challenges: A Special Issue Overview. *Agron. J.* **2019**, *111*, 1–3. [CrossRef]
38. Sarkar, P.; Chourasia, R. Chourasia, Bioconversion of organic solid wastes into biofortified compost using a microbial consortium. *Int. J. Recycl. Org. Waste Agric.* **2017**, *6*, 321–334. [CrossRef]
39. Gajalakshmi, S.; Abbasi, S.A. Solid Waste Management by Composting: State of the Art. *Crit. Rev. Environ. Sci. Technol.* **2008**, *38*, 311–400. [CrossRef]
40. Huang, W.; Hall, S.J. Elevated moisture stimulates carbon loss from mineral soils by releasing protected organic matter. *Nat. Commun.* **2017**, *8*, 1–10. [CrossRef] [PubMed]
41. Ramachandra, T.; Bharath, H.; Kulkarni, G.; Han, S.S. Municipal solid waste: Generation, composition and GHG emissions in Bangalore, India. *Renew. Sustain. Energy Rev.* **2018**, *82*, 1122–1136. [CrossRef]
42. Ayilara, M.S.; Olanrewaju, O.S.; Babalola, O.O.; Odeyemi, O. Waste Management through Composting: Challenges and Potentials. *Sustainability* **2020**, *12*, 4456. [CrossRef]
43. Glawe, U.; Visvanathan, C.; Alamgir, M. Solid waste management in least developed Asian countries–a comparative analysis. In Proceedings of the International Conference on Integrated Solid Waste Management in Southeast Asian Cities, Siem Reap, Cambodia, 5–7 July 2005.
44. Sharholy, M.; Ahmad, K.; Mahmood, G.; Trivedi, R.C. Municipal solid waste management in Indian cities—A review. *Waste Manag.* **2008**, *28*, 459–467. [CrossRef]
45. Visvanathan, C.; Trankler, J. Municipal solid waste management in Asia: A comparative analysis. In Proceedings of the Workshop on Sustainable Landfill Management, Chennai, India, 3–5 December 2003.
46. Hussain, M.; Butt, A.R.; Uzma, F.; Ahmed, R.; Irshad, S.; Rehman, A.; Yousaf, B. A comprehensive review of climate change impacts, adaptation, and mitigation on environmental and natural calamities in Pakistan. *Environ. Monit. Assess.* **2020**, *192*, 1–20. [CrossRef] [PubMed]
47. Sivaramanan, S. Global Warming and Climate change, causes, impacts and mitigation. *Cent. Environ. Auth.* **2015**, *2*, 1–26. [CrossRef]
48. Gunaruwan, T.L.; Gunasekara, W.N. Management of Municipal Solid Waste in Sri Lanka: A Comparative Appraisal of the Economics of Composting. *NSBM J. Manag.* **2016**, *2*, 27. [CrossRef]
49. Steen, E.; Brooks, M. Medical Waste Management—A review. *J. Environ. Manag.* **2015**, *163*, 98–108.

50. Manzoor, J.; Sharma, M. Impact of Biomedical Waste on Environment and Human Health. *Environ. Claims J.* **2019**, *31*, 311–334. [CrossRef]
51. Patil, P.M.; Mahamuni, P.P.; Shadija, P.G.; Bohara, R.A. Conversion of organic biomedical waste into value added product using green approach. *Environ. Sci. Pollut. Res.* **2019**, *26*, 6696–6705. [CrossRef]
52. Dinesh, M.S.; Geetha, K.S.; Vaishmavi, V.; Kale, R.D.; Krishna-Murthy, V. Ecofriendly treatment of biomedical wastes using epigeic earthworms. *J. Indian Soc. Hosp. Waste Manag.* **2010**, *9*, 5–20.
53. Ayuso, M.; Hernández, T.; Garcia, C.; Pascual, J. Stimulation of barley growth and nutrient absorption by humic substances originating from various organic materials. *Bioresour. Technol.* **1996**, *57*, 251–257. [CrossRef]
54. Vuorinen, A.H.; Saharinen, M.H. Evolution of microbiological and chemical parameters during manure and straw co-composting in a drum composting system. *Agric. Ecosyst. Environ.* **1997**, *66*, 19–29. [CrossRef]
55. Guerra-Rodríguez, E.; Vázquez, M.; Díaz-Raviña, M. Co-composting of barley wastes and solid poultry manure. *Bioresour. Technol.* **2000**, *75*, 223–225. [CrossRef]
56. Keeling, A.; McCallum, K.; Beckwith, C. Mature green waste compost enhances growth and nitrogen uptake in wheat (*Triticum aestivum* L.) and oilseed rape (Brassica napus L.) through the action of water-extractable factors. *Bioresour. Technol.* **2003**, *90*, 127–132. [CrossRef] [PubMed]
57. Iranzo, M.; Cañizares, J.V.; Roca-Perez, L.; Sainz-Pardo, I.; Mormeneo, S.; Boluda, R. Characteristics of rice straw and sewage sludge as composting materials in Valencia (Spain). *Bioresour. Technol.* **2004**, *95*, 107–112. [CrossRef] [PubMed]
58. Dresbøll, D.B.; Thorup-Kristensen, K. Delayed nutrient application affects mineralisation rate during composting of plant residues. *Bioresour. Technol.* **2005**, *96*, 1093–1101. [CrossRef] [PubMed]
59. Gabrielle, B.; Da-Silveira, J.; Houot, S.; Michelin, J. Field-scale modelling of carbon and nitrogen dynamics in soils amended with urban waste composts. *Agric. Ecosyst. Environ.* **2005**, *110*, 289–299. [CrossRef]
60. Anto, H.; Trivedi, U.; Patel, K. Glucoamylase production by solid-state fermentation using rice flake manufacturing waste products as substrate. *Bioresour. Technol.* **2006**, *97*, 1161–1166. [CrossRef]
61. Zhu, N. Effect of low initial C/N ratio on aerobic composting of swine manure with rice straw. *Bioresour. Technol.* **2007**, *98*, 9–13. [CrossRef]
62. Jusoh, M.L.C.; Manaf, L.A.; Latiff, P.A. Composting of rice straw with effective microorganisms (EM) and its influence on compost quality. *Iran. J. Environ. Health Sci. Eng.* **2013**, *10*, 17. [CrossRef]
63. Yan, Z.; Song, Z.; Li, D.; Yuan, Y.; Liu, X.; Zheng, T. The effects of initial substrate concentration, C/N ratio, and temperature on solid-state anaerobic digestion from composting rice straw. *Bioresour. Technol.* **2015**, *177*, 266–273. [CrossRef]
64. Eklind, Y.; Beck-Friis, B.; Bengtsson, S.; Ejlertsson, J.; Kirchmann, H.; Mathisen, B.; Nordkvist, E.; Sonesson, U.; Svensson, B.H.; Torstensson, L. Chemical characterization of source-separated organic household wastes. *Swed. J. Agric. Res.* **1997**, *27*, 167–178.
65. Kimetu, J.M.; Lehmann, J.; Ngoze, S.O.; Mugendi, D.N.; Kinyangi, J.M.; Riha, S.; Verchot, L.; Recha, J.; Pell, A.N. Reversibility of soil productivity decline with organic matter of differing quality along a degradation gradient. *Ecosystems* **2008**, *11*, 726–739. [CrossRef]
66. Larney, F.J.; Olson, A.F.; Miller, J.J.; DeMaere, P.R.; Zvomuya, F.; McAllister, T.A. Physical and chemical changes during composting of wood chip-bedded and straw-bedded beef cattle feedlot manure. *J. Environ. Qual.* **2008**, *37*, 725–735. [CrossRef] [PubMed]
67. Yuan, J.-H.; Xu, R.-K.; Qian, W.; Wang, R.-H. Comparison of the ameliorating effects on an acidic ultisol between four crop straws and their biochars. *J. Soils Sediments* **2011**, *11*, 741–750. [CrossRef]
68. Zhao, X.-L.; Li, B.-Q.; NI, J.-P.; Xie, D.-T. Effect of four crop straws on transformation of organic matter during sewage sludge composting. *J. Integr. Agric.* **2016**, *15*, 232–240. [CrossRef]
69. Abbas, A.; Naveed, M.; Azeem, M.; Yaseen, M.; Ullah, R.; Alamri, S.; Ain Farooq, Q.U.; Siddiqui, M.H. Efficiency of wheat straw biochar in combination with compost and biogas slurry for enhancing nutritional status and productivity of soil and plant. *Plants* **2020**, *9*, 1516. [CrossRef] [PubMed]
70. Kapoor, J.; Sharma, S.; Rana, N. Vermicomposting for organic waste management. *Int. J. Recent Sci. Res.* **2015**, *6*, 7956–7960.
71. Hřebečková, T.; Wiesnerová, L.; Hanč, A. Changes of enzymatic activity during a large-scale vermicomposting process with continuous feeding. *J. Clean. Prod.* **2019**, *239*, 118127. [CrossRef]
72. Huang, K.; Xia, H.; Zhang, Y.; Li, J.; Cui, G.; Li, F.; Bai, W.; Jiang, Y.; Wu, N. Elimination of antibiotic resistance genes and human pathogenic bacteria by earthworms during vermicomposting of dewatered sludge by metagenomic analysis. *Bioresour. Technol.* **2020**, *297*, 122451. [CrossRef]
73. Sinha, R.K.; Bharambe, G.; Chaudhari, U. Sewage treatment by vermifiltration with synchronous treatment of sludge by earthworms: A low-cost sustainable technology over conventional systems with potential for decentralization. *Environmentalist* **2008**, *28*, 409–420. [CrossRef]
74. Gupta, R.; Garg, V. Stabilization of primary sewage sludge during vermicomposting. *J. Hazard. Mater.* **2008**, *153*, 1023–1030. [CrossRef]
75. Meena, K.; Renu, B. Vermitechnology for sewage sludge recycling. *J. Hazard. Mater.* **2009**, *161*, 948–954.
76. Loh, T.C.; Lee, Y.C.; Liang, J.B.; Tan, D. Vermicomposting of cattle and goat manures by Eisenia foetida and their growth and reproduction performance. *Bioresour. Technol.* **2005**, *96*, 111–114. [CrossRef] [PubMed]
77. Plaza, C.; Nogales, R.; Senesi, N.; Benitez, E.; Polo, A. Organic matter humification by vermicomposting of cattle manure alone and mixed with two-phase olive pomace. *Bioresour. Technol.* **2008**, *99*, 5085–5089. [CrossRef] [PubMed]

78. Ghosh, M.; Chattopadhyay, G.; Baral, K. Transformation of phosphorus during vermicomposting. *Bioresour. Technol.* **1999**, *69*, 149–154. [CrossRef]
79. Garg, V.; Kaushik, P. Vermistabilization of textile mill sludge spiked with poultry droppings by an epigeic earthworm Eisenia foetida. *Bioresour. Technol.* **2005**, *96*, 1063–1071. [CrossRef]
80. Pramanik, P.; Ghosh, G.; Ghosal, P.; Banik, P. Changes in organic–C, N, P and K and enzyme activities in vermicompost of biodegradable organic wastes under liming and microbial inoculants. *Bioresour. Technol.* **2007**, *98*, 2485–2494. [CrossRef]
81. Sen, B.; Chandra, T. Chemolytic and solid-state spectroscopic evaluation of organic matter transformation during vermicomposting of sugar industry wastes. *Bioresour. Technol.* **2007**, *98*, 1680–1683. [CrossRef]
82. Yadav, A.; Garg, V. Feasibility of nutrient recovery from industrial sludge by vermicomposting technology. *J. Hazard. Mater.* **2009**, *168*, 262–268. [CrossRef]
83. Subramanian, S.; Sivarajan, M.; Saravanapriya, S. Chemical changes during vermicomposting of sago industry solid wastes. *J. Hazard. Mater.* **2010**, *179*, 318–322. [CrossRef]
84. Yadav, K.D.; Tare, V.; Ahammed, M. Vermicomposting of source-separated human faeces for nutrient recycling. *Waste Manag.* **2010**, *30*, 50–56. [CrossRef]
85. Tripathi, G.; Bhardwaj, P. Comparative studies on biomass production, life cycles and composting efficiency of Eisenia fetida (Savigny) and Lampito mauritii (Kinberg). *Bioresour. Technol.* **2004**, *92*, 275–283. [CrossRef] [PubMed]
86. Yadav, A.; Garg, V. Recycling of organic wastes by employing Eisenia fetida. *Bioresour. Technol.* **2011**, *102*, 2874–2880. [CrossRef] [PubMed]
87. Das, D.; Bhattacharyya, P.; Ghosh, B.; Banik, P. Bioconversion and biodynamics of Eisenia foetida in different organic wastes through microbially enriched vermiconversion technologies. *Ecol. Eng.* **2016**, *86*, 154–161. [CrossRef]
88. Biruntha, M.; Karmegam, N.; Archana, J.; Karunai Selvi, B.; John Paul, J.A.; Balamuralikrishnan, B.; Chang, S.W.; Ravindran, B. Vermiconversion of biowastes with low-to-high C/N ratio into value added vermicompost. *Bioresour. Technol.* **2020**, *297*, 122398. [CrossRef]
89. Padmavathiamma, P.K.; Li, L.Y.; Kumari, U.R. An experimental study of vermi-biowaste composting for agricultural soil improvement. *Bioresour. Technol.* **2008**, *99*, 1672–1681. [CrossRef] [PubMed]
90. Busato, J.G.; Lima, L.S.; Aguiar, N.O.; Canellas, L.P.; Olivares, F.L. Changes in labile phosphorus forms during maturation of vermicompost enriched with phosphorus-solubilizing and diazotrophic bacteria. *Bioresour. Technol.* **2012**, *110*, 390–395. [CrossRef] [PubMed]
91. Sonowal, P.; Khwairakpam, M.; Kalamdhad, A.S. Vermicomposting of solid pulp and paper mill sludge (SPPMS) using Eudrilus Eugeniae Earthworm. *Int. J. Environ. Sci.* **2014**, *5*, 502.
92. Vig, A.P.; Singh, J.; Wani, S.H.; Dhaliwal, S.S. Vermicomposting of tannery sludge mixed with cattle dung into valuable manure using earthworm Eisenia fetida (Savigny). *Bioresour. Technol.* **2011**, *102*, 7941–7945. [CrossRef]
93. Manaf, L.A.; Jusoh, M.L.C.; Yusoff, M.K.; Ismail, T.H.T.; Harun, R.; Juahir, H.; Jusoff, K. Influences of Bedding Material in Vermicomposting Process. *Int. J. Biol.* **2009**, *1*, 81. [CrossRef]
94. Suthar, S.; Singh, S. Vermicomposting of domestic waste by using two epigeic earthworms (Perionyx excavatus and Perionyx sansibaricus). *Int. J. Environ. Sci. Technol.* **2008**, *5*, 99–106. [CrossRef]
95. Suthar, S. Vermicomposting potential of Perionyx sansibaricus (Perrier) in different waste materials. *Bioresour. Technol.* **2007**, *98*, 1231–1237. [CrossRef] [PubMed]
96. Kalemelawa, F.; Nishihara, E.; Endo, T.; Ahmad, Z.; Yeasmin, R.; Tenywa, M.M.; Yamamoto, S. An evaluation of aerobic and anaerobic composting of banana peels treated with different inoculums for soil nutrient replenishment. *Bioresour. Technol.* **2012**, *126*, 375–382. [CrossRef] [PubMed]
97. Yang, W.-Q.; Zhuo, Q.; Chen, Q.; Chen, Z. Effect of iron nanoparticles on passivation of cadmium in the pig manure aerobic composting process. *Sci. Total Environ.* **2019**, *690*, 900–910. [CrossRef] [PubMed]
98. He, X.; Han, L.; Fu, B.; Du, S.; Liu, Y.; Huang, G. Effect and microbial reaction mechanism of rice straw biochar on pore methane production during mainstream large-scale aerobic composting in China. *J. Clean. Prod.* **2019**, *215*, 1223–1232. [CrossRef]
99. Zhang, L.; Li, L.; Sha, G.; Liu, C.; Wang, Z.; Wang, L. Aerobic composting as an effective cow manure management strategy for reducing the dissemination of antibiotic resistance genes: An integrated meta-omics study. *J. Hazard. Mater.* **2020**, *386*, 121895. [CrossRef]
100. Tiquia, S.; Tam, N. Composting of spent pig litter in turned and forced-aerated piles. *Environ. Pollut.* **1998**, *99*, 329–337. [CrossRef]
101. Tiquia, S.M.; Tam, N.F. Characterization and composting of poultry litter in forced-aeration piles. *Process. Biochem.* **2002**, *37*, 869–880. [CrossRef]
102. Baeta-Hall, L.; Sàágua, M.C.; Bartolomeu, M.L.; Anselmo, A.M.; Rosa, M.F. Bio-degradation of olive oil husks in composting aerated piles. *Bioresour. Technol.* **2005**, *96*, 69–78. [CrossRef]
103. Cayuela, M.; Sánchez-Monedero, M.; Roig, A. Evaluation of two different aeration systems for composting two-phase olive mill wastes. *Process Biochem.* **2006**, *41*, 616–623. [CrossRef]
104. Liu, N.; Zhou, J.; Han, L.; Ma, S.; Sun, X.; Huang, G. Role and multi-scale characterization of bamboo biochar during poultry manure aerobic composting. *Bioresour. Technol.* **2017**, *241*, 190–199. [CrossRef]
105. Roberts, P.; Edwards-Jones, G.; Jones, D.L. In-vessel cocomposting of green waste with biosolids and paper waste. *Compos. Sci. Util.* **2007**, *15*, 272–282. [CrossRef]

106. Liang, C.; Das, K.C.; McClendon, R.W. The influence of temperature and moisture contents regimes on the aerobic microbial activity of a biosolids composting blend. *Bioresour. Technol.* **2003**, *86*, 131–137. [CrossRef] [PubMed]
107. Avidov, R.; Saadi, I.; Krasnovsky, A.; Medina, S.; Raviv, M.; Chen, Y.; Laor, Y. Using polyethylene sleeves with forced aeration for composting olive mill wastewater pre-absorbed by vegetative waste. *Waste Manag.* **2018**, *78*, 969–979. [CrossRef] [PubMed]
108. Makan, A.; Fadili, A.; Oubenali, M. Interaction of physicochemical parameters during pressurized in-vessel composting of food waste. *Bioresour. Technol. Rep.* **2020**, *10*, 100350. [CrossRef]
109. Díaz, M.; Madejón, E.; López, F.; López, R.; Cabrera, F. Composting of vinasse and cotton gin waste by using two different systems. *Resour. Conserv. Recycl.* **2002**, *34*, 235–248. [CrossRef]
110. Solano, M.; Iriarte, F.; Ciria, P.; Negro, M.J. SE—Structure and environment: Performance characteristics of three aeration systems in the composting of sheep manure and straw. *J. Agric. Eng. Res.* **2001**, *79*, 317–329. [CrossRef]
111. Brodie, H.L.; Carr, L.E.; Condon, P. A Comparison of static pile and turned windrow methods for poultry litter compost production. *Compos. Sci. Util.* **2000**, *8*, 178–189. [CrossRef]
112. Liu, H.; Huang, X.; Yu, X.; Pu, C.; Sun, Y.; Luo, W. Dissipation and persistence of sulfonamides, quinolones and tetracyclines in anaerobically digested biosolids and compost during short-term storage under natural conditions. *Sci. Total Environ.* **2019**, *684*, 58–66. [CrossRef]
113. Xu, Z.; Li, G.; Huda, N.; Zhang, B.; Wang, M.; Luo, W. Effects of moisture and carbon/nitrogen ratio on gaseous emissions and maturity during direct composting of cornstalks used for filtration of anaerobically digested manure centrate. *Bioresour. Technol.* **2020**, *298*, 122503. [CrossRef]
114. Meng, X.; Yan, J.; Zuo, B.; Wang, Y.; Yuan, X.; Cui, Z. Full-scale of composting process of biogas residues from corn stover anaerobic digestion: Physical-chemical, biology parameters and maturity indexes during whole process. *Bioresour. Technol.* **2020**, *302*, 122742. [CrossRef]
115. Mandpe, A.; Paliya, S.; Kumar, S.; Kumar, R. Fly ash as an additive for enhancing microbial and enzymatic activities in in-vessel composting of organic wastes. *Bioresour. Technol.* **2019**, *293*, 122047. [CrossRef] [PubMed]
116. Arrigoni, J.P.; Paladino, G.; Garibaldi, L.A.; Laos, F. Inside the small-scale composting of kitchen and garden wastes: Thermal performance and stratification effect in vertical compost bins. *Waste Manag.* **2018**, *76*, 284–293. [CrossRef] [PubMed]
117. Zhang, H.; McGill, E.; Gomez, C.O.; Carson, S.; Neufeld, K.; Hawthorne, I.; Smukler, S. Disintegration of compostable foodware and packaging and its effect on microbial activity and community composition in municipal composting. *Int. Biodeterior. Biodegrad.* **2017**, *125*, 157–165. [CrossRef]
118. Troy, S.M.; Nolan, T.; Leahy, J.J.; Lawlor, P.G.; Healy, M.G.; Kwapinski, W. Effect of sawdust addition and composting of feedstock on renewable energy and biochar production from pyrolysis of anaerobically digested pig manure. *Biomass-Bioenergy* **2013**, *49*, 1–9. [CrossRef]
119. Zorpas, A.A.; Loizidou, M. Sawdust and natural zeolite as a bulking agent for improving quality of a composting product from anaerobically stabilized sewage sludge. *Bioresour. Technol.* **2008**, *99*, 7545–7552. [CrossRef]
120. Obersky, L.; Rafiee, R.; Cabral, A.R.; Golding, S.D.; Clarke, W.P. Methodology to determine the extent of anaerobic digestion, composting and CH4 oxidation in a landfill environment. *Waste Manag.* **2018**, *76*, 364–373. [CrossRef]
121. Li, Y.; Manandhar, A.; Li, G.; Shah, A. Life cycle assessment of integrated solid state anaerobic digestion and composting for on-farm organic residues treatment. *Waste Manag.* **2018**, *76*, 294–305. [CrossRef]
122. Smet, E.; Van Langenhove, H.; De Bo, I. The emission of volatile compounds during the aerobic and the combined anaerobic/aerobic composting of biowaste. *Atmos. Environ.* **1999**, *33*, 1295–1303. [CrossRef]
123. Zhang, L.; Sun, X.; Tian, Y.; Gong, X. Effects of brown sugar and calcium superphosphate on the secondary fermentation of green waste. *Bioresour. Technol.* **2013**, *131*, 68–75. [CrossRef]
124. Shin, H.S.; Han, S.K.; Song, Y.C.; Lee, C.Y. Multi-step sequential batch two-phase anaerobic composting of food waste. *Environ. Technol.* **2001**, *22*, 271–279. [CrossRef]
125. Kulikowska, D.; Klimiuk, E. Organic matter transformations and kinetics during sewage sludge composting in a two-stage system. *Bioresour. Technol.* **2011**, *102*, 10951–10958. [CrossRef] [PubMed]
126. Zhang, L.; Sun, X. Effects of rhamnolipid and initial compost particle size on the two-stage composting of green waste. *Bioresour. Technol.* **2014**, *163*, 112–122. [CrossRef] [PubMed]
127. Kumar, S.; Negi, S.; Mandpe, A.; Singh, R.V.; Hussain, A. Rapid composting techniques in Indian context and utilization of black soldier fly for enhanced decomposition of biodegradable wastes—A comprehensive review. *J. Environ. Manag.* **2018**, *227*, 189–199. [CrossRef] [PubMed]
128. Kulikowska, D.; Gusiatin, Z.M. Sewage sludge composting in a two-stage system: Carbon and nitrogen transformations and potential ecological risk assessment. *Waste Manag.* **2015**, *38*, 312–320. [CrossRef]

Disclaimer/Publisher's Note: The statements, opinions and data contained in all publications are solely those of the individual author(s) and contributor(s) and not of MDPI and/or the editor(s). MDPI and/or the editor(s) disclaim responsibility for any injury to people or property resulting from any ideas, methods, instructions or products referred to in the content.

MDPI AG
Grosspeteranlage 5
4052 Basel
Switzerland
Tel.: +41 61 683 77 34

Processes Editorial Office
E-mail: processes@mdpi.com
www.mdpi.com/journal/processes

Disclaimer/Publisher's Note: The statements, opinions and data contained in all publications are solely those of the individual author(s) and contributor(s) and not of MDPI and/or the editor(s). MDPI and/or the editor(s) disclaim responsibility for any injury to people or property resulting from any ideas, methods, instructions or products referred to in the content.